International Legal Approaches to the Promotion of the Conservation and Sustainable Use of Forests

Europäische Hochschulschriften Recht

European University Studies in Law

Publications Universitaires Européennes de Droit

Band/ Volume **6750**

Yevhenii Yatsukhnenko

International Legal Approaches to the Promotion of the Conservation and Sustainable Use of Forests

PETER LANG

Lausanne - Berlin - Bruxelles - Chennai - New York - Oxford

Bibliographic Information published by the Deutsche Nationalbibliothek
The Deutsche Nationalbibliothek lists this publication in the Deutsche Nationalbibliografie; detailed bibliographic data is available online at
http://dnb.d-nb.de.

Library of Congress Cataloging-in-Publication Data
A CIP catalog record for this book has been applied for at the Library of Congress.

ISSN 0531-7312
ISBN 978-3-631-89810-9 (Print)
E-ISBN 978-3-631-90593-7 (E-PDF)
E-ISBN 978-3-631-90594-4 (E-PUB)
10.3726/b21048

Published by Peter Lang GmbH, Berlin, Deutschland

info@peterlang.com - www.peterlang.com

This publication has been peer reviewed.

Table of Contents

Foreword

Global forest ecosystems are fascinating from every point of view. They are real masterpieces of nature and impress the observer with their outstanding beauty. They constitute highly complex ecosystems, home to myriads of living creatures. They are characterized by highly complex physical and chemical processes, so that science has not yet been able to comprehend all their mysteries. The importance of forest ecosystems for the prosperity of all living beings on the Earth, including humans, cannot be underestimated. Being aware of all these issues inspired me to research into the role of international environmental law in the promotion of sustainable use and conservation of forests.

The current Ph. D. Thesis was completed at the Institute of Environmental and Technology Law (IUTR) in the City of Trier, Germany, in the years 2017–2021, whereby the main text was accomplished in 2020. Subsequently, the dissertation was updated in accordance to the newest developments in the sphere of international environmental law in April 2021. The current Ph. D. Thesis was accepted as dissertation by the Law Department of the University of Trier during the summer semester 2022.

I would like to express my particular gratitude to my Ph. D. supervisor Mr. Professor Dr. Alexander Proelß (University of Hamburg), who accompanied me on my way to the successful accomplishment of my doctoral research, and sustained me in every aspects of my research and of my stay in Germany. I am also really grateful to my second assessor Mr. Prof. Dr. Meinhard Schröder (University of Trier) for his advices.

I would like to express my gratitude as well to Mr. Senior-Professor Dr. habil. Dr. h.c. mult. Peter-Christian Müller-Graff (University of Heidelberg) for his advice and for his support at the beginning of my studies in Germany. I am thankful for the suggestions given to me by Mr. Professor Alessandro Fodella (University of Trento) and Mr. Professor Attila Massimiliano Tanzi (University of Bologna).

Furthermore, I am particularly grateful to my father Mr. Serhii Yatsukhnenko and my mother Mrs. Alla Yatsukhnenko, who always supported and sustained me during my studies and my research, and in my life in general.

I would also like to express my gratitude to the Konrad-Adenauer Foundation and Dorothee-Wilms Foundation that supported my research and my stay in Germany with financial funding and ideational support. In particular, I want to acknowledge the support of Mr. Dr. Simon Backovsky and Mrs. Dr. Christina Krause.

Finally, special thanks to my friends Daniela Intignano, Josh DeWitt, Vanessa Amadori, Chiara De Rosa, Royston Anthony Lin, Roberta Marenaci, and Aurora Paladini for their moral support, and, particularly, for their help with the proof-reading of my dissertation.

Trier, 15.05.2023
Yevhenii Yatsukhnenko

List of Abbreviations

ABS	Access and benefit-sharing
CBD	Convention on Biological Diversity
CDM	Clean Development Mechanism
CIFOR	Center for International Forestry Research
CITES	Convention on International Trade in Endangered Species of Wild Fauna and Flora
COP	Conference of the Parties
CMP	Conference of the Parties serving as the meeting of the Parties to the Kyoto Protocol
CSR	Corporate Social Responsibility
EASAC	European Academies Science Advisory Council
EUROFOR	European Rapid Operational Force
FAO	Food and Agriculture Organization of the United Nations
FERI	Forest Ecosystem Restoration Initiative
FREL	Forest Reference Emission Level
FRL	Forest Reference Level
FSC	Forest Stewardship Council
ICDP	Integrated Conservation and Development Projects
ILCs	Indigenous and local communities
ITTA	International Tropical Timber Agreement
ITTO	International Tropical Timber Organisation
IUCN	International Union for Conservation of Nature
JMA	Joint Mitigation Approaches
KP	Kyoto Protocol
MAT	Mutually agreed terms
MOP-COP	Meeting of the Conference of the Parties serving as the Meeting of the Parties to the Nagoya Protocol on ABS
OECD	Organisation for Economic Cooperation and Development
PES	Payments for Ecosystem Services
PIC	Prior informed consent

RECIEL	Review of European, Comparative and International Environmental Law
RED	Reducing emissions from deforestation
REDD+	Reducing emissions from deforestation and forest degradation, and the role of conservation, sustainable management of forests, and enhancement of forest carbon stocks in developing countries
UNCTAD	United Nations Conference on Trade and Development.
UNESCO	United Nations Educational, Scientific and Cultural Organization
UNFF	United Nations Forum on Forests
WWF	World Wild Fund

Introduction

A. Deforestation Crisis

Global forest ecosystems are the most important terrestrial biodiversity hotspots. They fulfil multiple socio-economic and environmental functions. Moreover, their conservation is particularly important from ecological, cultural, ethical and other points of view. Forests play a crucial role in natural environmental cycles and substantially influence the flows of energy between the atmosphere, soil, and ocean. Forests are sinks of carbon dioxide. They protect soils, maintain and promote freshwater resources, improve air quality, maintain and preserve soil fertility.[1] Furthermore, forests likely play a significant role in the formation and distribution of atmospheric precipitations.[2] Forests supply humanity with food, clean water, construction and raw materials. A lot of medicines and other biochemical substances are developed and manufactured based on genetic resources coming from forests. According to the report of the World Bank, 1,6 Billion of people depend on forests for food, employment, livelihood, and income generation.[3] 75% of the fresh water consumption comes from forests.[4]

Unsustainable forest use has already resulted in large-scale deforestation and forest degradation. Many unsustainably managed forests are currently unable to fulfill their multiple environmental functions as efficiently as natural forests do. Deforestation and forest degradation have adverse impacts on the human environment, on socio-economic development, and on livelihoods. For example, deforestation causes dangerous climate changes on the regional and global level, the erosion of soils and desertification, the loss of biodiversity, the elevated risk

1 Eikermann, Anja, Forests in International Law, 2015, pp. 27 ff.; Luyssaert, Sebastian/Schulze, E Detlef/Börner, Annett/Knohl, Alexander/Hessenmöller, Dominik/Law, Beverly E./Ciais, Philippe/Grace, John, Old-Growth Forests as Global Carbon Sinks, in: Nature, Volume 455, Number 7210, 2008, p. 213.

2 *Makarieva, Anastasia/Gorshkov, Viktor,* The Biotic Pump: Condensation, Atmospheric Dynamics and Climate, in: International Journal of Water, Volume 5, Number 4, 2010, pp. 365 ff.

3 *The World Bank,* Why Forests are Key to Climate, Water, Health, and Livelihoods, The World Bank, March 18, 2016. – Internet resource – http://www.worldbank.org/en/news/feature/2016/03/18/why-forests-are-key-to-climate-water-health-and-livelihoods [Last visit: 30.04.2023]

4 Ibid.

of floods, the aggravation of poverty, etc.[5] Deforestation can create serious obstacles for sustainable development. A prominent example of the adverse impacts that deforestation has on regional and local levels can be seen in the highly deforested Haiti, which is currently experiencing a disastrous socio-economic situation. In contrast, the nearby Dominican Republic has seen the promotion of forest protection contribute to its relative economic prosperity.[6] Deforested areas are exceptionally prone to natural disasters such as floods and landslides. For instance, deforestation was one of the main causes of catastrophic floods in Venezuela in 1999 and in Haiti in 2004.[7] These local adverse impacts of deforestation and forest degradation easily evolve into long-term instability factors in international relations. Those States that have suffered damage from deforestation, are likely to require additional financial and humanitarian assistance from the international community. Moreover, deforestation can generate direct adverse impacts on a global level, including global climate change and the loss of biodiversity. In some cases, natural catastrophes caused by deforestation create additional hazards for the international community. For example, extensive forest fires in Indonesia in 2015 provoked by unsustainable forest management resulted in dangerous air contamination in South-Eastern Asia.[8] Against this backdrop, the apparently intrastate problem of deforestation acquires a substantial international dimension. Consequently, forest ecosystems should be protected for the benefit of mankind as a whole.

B. Problem Identification

The short-term revenues arising from unsustainable forest management and forestland conversions can incite States and private actors to neglect environmental issues while exploiting forests. Local communities can be forced to manage forests unsustainably in order to support their livelihoods. Multiple environmental

5 *Eikermann* [n. 1], pp. 27 ff.; *Luyssaert/Schulze/Börner/Knohl/Hessenmöller/Law/Ciais/Grace* [n. 1], p. 213; *Makarieva/Gorshkov* [n. 2], pp. 365 ff.; *Butler, Rhett A.*, Consequences of Deforestation, in: Mongabay, 22 July 2012. – Internet resource – http://rainforests.mongabay.com/0901.html [Last visit: 30.04.2023]

6 *Diamond, Jared*, Collapse: How Societies Choose to Fail or Survive, 2005, pp. 110 ff.

7 Ibid.

8 *Balch, Oliver*, Indonesia's Forest Fires: Everything You Need to Know, in: The Guardian, 11 November 2015. – Internet resource – https://www.theguardian.com/sustainable-business/2015/nov/11/indonesia-forest-fires-explained-haze-palm-oil-timber-burning [Last visit: 30.04.2023]

functions of forests can be disregarded in this process. Therefore, in order to promote forest conservation, the principles and modalities, according to which the forest managing entities operate, should be modified in a way allowing proper evaluation of the multiple functions and environmental benefits of forests while promoting sustainable livelihoods and income diversification. A lack of resources and political will to introduce sustainable forest management and to ensure the protection of natural forests and vulnerable forest ecosystems is particularly severe in developing States. The promotion of sustainable forest management and forest protection may result in competitive disadvantages whenever the protection of globally relevant ecological functions of forests does not bring any revenues and benefits in monetary and political terms. It is also noteworthy that in many cases not a single State or a private company that is subject to its jurisdiction, but a multinational company is accountable for the exploitation of forest ecosystems and their degradation. The main commitment that such a company assumes is to increment revenues of its stakeholders.[9] Due to the lack of strong environmental legislation in headquarter States and in States where the main activities of the company unfold, multinational companies often commit themselves to ensure the socio-economic and environmental sustainability only voluntary.[10] Such commitments are usually not sufficient in order to halt and reverse global forest loss. Due to a lack of economic incentives, States, private actors and multinational companies exploiting forests refuse to promote the conservation of forests at the expense of their short-term economic interests.[11] Furthermore, national legislation and international law fail to deliver strong political signals indicating that single States and the international community are committed to forest conservation, and that the existing initiatives, mechanisms, and projects are designed to be applied on a long-term basis.

In addition to major multinational companies, small forest owners and users are also accountable for deforestation. They are forced to manage forests unsustainably and to convert forestlands by a lack of income diversification and by the desire for enhanced revenues, as well as by a lack of knowledge concerning environmental functions of forests.[12] Forestland conversion is often the only

9 *Humphreys, David*, Deforestation and the Crisis of Global Governance, 2006, pp. 214 ff.

10 *United Nations Conference on Trade and Development,* The Social Responsibility of Transnational Corporations, UNCTAD, United Nations, 1999, pp. 7 ff.

11 *Humphreys* [n. 9], pp. 214 ff.

12 *World Wild Fund*, Causes of Deforestation, 2016. – Internet resource – http://www.worldwildlife.org/threats/deforestation [Last visit: 30.04.2023]

source of income for farmers in developing States.[13] The financial and technical expertise and scientific knowledge needed for forest conservation is often absent.[14] Furthermore, weak legislation and a lack of forest monitoring systems in forest-rich developing States are among major drivers of deforestation. Deforestation and forest degradation are also linked to the lack of socio-economic development. According to some estimates, a large percentage of illegally harvested wood is used for fuel.[15] Therefore, the improvement of local livelihoods and the socio-economic development in forest-rich States should be seen as an indispensable part of forest conservation.

Due to a lack of financial and technological resources, those developing States that seek to promote forest protection largely rely on the assistance provided by developed ones. According to the existing approaches, the strengthening of forest conservation requires, *inter alia*, a monetary evaluation of multiple forest functions. In this way, forest owners and users can be motivated to safeguard forests, particularly to conserve natural forest ecosystems. This monetary evaluation of multiple functions of forests can be achieved through international cooperation, while a lot of ecological services are consumed on a global rather than on a regional level. Logically, international law as an instrument for the coordination of the peaceful coexistence and cooperation between States must deliver legal provisions for the development and implementation of international mechanisms that would deliver necessary financial and technological resources and promote a monetary evaluation of environmental forest functions. In other words, an international legal regime for forests is needed. This regime has already come into being.[16]

However, the existence of controversial interests in relation to forests impedes the development of the international legal regime for forests.[17] Developed States partially rely on the supply of timber and non-timber forest goods and agricultural produce cultivated on converted forestlands. On the other hand, they are also interested in the preservation of genetic resources of forests, while they also want to ensure for their private companies the free access thereto. Another major interest of developed States is the promotion of forests' role as carbon sinks and stores.[18] In Developed States, despite more sustainable forest management go

13 Ibid.

14 Ibid.

15 Ibid.

16 See Subsection D.IV of Chapter 2.

17 *Eikermann* [n. 1], pp. 9 ff.; *Humphreys* [n. 9], pp. 9 ff.

18 *Humphreys* [n. 9], pp. 9 ff.

into practice, which means sufficient number of trees is planted to replace the wood that is removed from forests, there are still rising conflicts between forest's roles to the environment and its industrial exploitation. This is because biodiversity issues and the natural resilience of forest ecosystems are widely neglected in the process.[19]

At the same time, developing States envisage to achieve socio-economic development. They seek to tackle their balance of payment deficits on the cost of deforestation and environmental degradation.[20] Forest-dependent people need forests in order to sustain their livelihoods. Among them, indigenous and local communities (ILCs) are especially dependent on forest ecosystems, which are an integral part of their everyday lives.[21] On the other hand, private companies seek to ensure highest possible revenues from forest exploitation.[22] It is worth noting that the present and future generations are particularly interested in the preservation of forests with consideration not only of their socio-economic and environmental functions, but also of their intrinsic value.

These controversial interests of actors engaged in forest utilization is one of the main reasons for the highly fragmented and ambivalent character of the international legal regime for forests. This regime consists of non-legally binding documents specific to forests, among which the Rio Forest principles[23] of 1992 and the Non-Legally Binding Instrument on All Types of Forests (NLBI) of 2007[24] are especially worth mentioning. Moreover, it includes the forest relevant provisions of international environmental treaties that regulate other environmental issues. No single legally binding treaty specific to forests has ever been adopted so far. The international legal regime for forests is characterized by a low level of effectiveness, while deforestation rates remain alarmingly high.

The international legal regime for combating climate change and the one for sustainable use and conservation of biodiversity deliver the basis for elaborating a more comprehensive approach to forests. The Convention on Biological

---

19 See Subsection A.I of Chapter 1.

20 *Humphreys* [n. 9], p. 10.

21 Ibid.

22 *Eikermann* [n. 1], pp. 9 ff.

23 Non-Legally Binding Authoritative Statement of Principles for a Global Consensus on the Management, Conservation and Sustainable Development of All Types of Forests, adopted on 14 June 1992, in: ILM, Volume 31, 1992, pp. 881 ff.

24 United Nations, Non-Legally Binding Instrument on All Types of Forests, GA/RES/62/98, 17 December 2007. – Internet resource – https://documents-dds-ny.un.org/doc/UNDOC/GEN/N07/469/65/PDF/N0746965.pdf?OpenElement [Last visit: 30.04.2023]

Diversity (CBD)[25] recognizes the sovereign rights of developing States over biodiversity and natural habitats, and aims to promote the sustainable use and conservation of biodiversity compounds. The United Nations Framework Convention on Climate Change (UNFCCC) refers to forests as sinks and reservoirs of greenhouse gases.[26] In 2007, the Conference of Parties of the UNFCCC (COP UNFCCC) announced the development of the REDD+-mechanism.[27] REDD+ stands for "reducing emissions from deforestation and forest degradation and the role of conservation, sustainable management of forests and enhancement of forest carbon stocks". REDD+ is designed to be applied in developing States.

It is noteworthy that the CBD provides the legal basis for the development and implementation of the access and benefit-sharing regime (ABS) in relation to genetic resources.[28] This regime presupposes that a biodiversity-rich State grants access to genetic resources under its jurisdiction to other States, which possess financial and technological resources needed for the development and manufacturing of biodiversity-based products.[29] In exchange, developing States receive a certain share of benefits arising from the commercialization and application of these products.[30] Moreover, users of genetic resources are supposed to provide the State of origin of such resources with financial and technological resources needed for industrial, scientific, and socio-economic development. The ABS regime can strengthen the conservation of forests, insofar as it promotes the application of sustainable forestry practices. Forest-rich States can be further motivated to conserve forest ecosystems under their jurisdiction as they will be able to receive benefits arising from the utilization of genetic resources of forests.

However, the international legal framework for the application of REDD+ and ABS is highly fragmented. The provisions of the CBD and the Nagoya Protocol dealing with ABS often contain only vague wording, such as "the Parties should…" or the "Parties are encouraged to…". States have systematically avoided

25 Convention on Biological Diversity, adopted on 5 June 1992 and entered into force on 29 December 1993, in: 1760 UNTS 79.

26 United Nations Framework Convention on Climate Change, adopted on 9 May 1992 and entered into force on 21 March 1994 in: 1771 UNTS 107, Art. 3.3.

27 *Parker, Charlie/Mitchell, Andrew/Trivedi, Mandar/Mardas, Niki*, The Little REDD+ Book, 2009, pp. 12 ff.

28 CBD, Articles 15–19.

29 CBD, Article 15.1–2; *Krohn, Susan Nicole*, Die Bewahrung tropischer Regenwälder durch völkerrechtliche Kooperationsmechanismen, 2002, pp. 108 ff.

30 CBD, Article 19.

strong legal commitments.[31] The controversial interests of different groups of forest users and owners hinder the development and implementation of relevant projects and mechanisms. REDD+ and ABS can be attractive both for forest-rich States, ILCs, and foreign investors, but only if they articulate and balance the conflicting interests of relevant stakeholders. Otherwise, developing States and forest dependent people can be discriminated against in the process of project implementation. They can be deprived of their traditional user and owner rights in relation to forest resources. In the worst case, ILCs can be evicted from their lands with no monetary compensation provided to them. Foreign investors are often in a better position to pursue their rights while negotiating the terms and conditions of access.[32]

Moreover, the implementation of REDD+ can in certain cases aggravate the problem of deforestation, especially in relation to the ongoing conversion of natural forests. The international legal framework for REDD+ does not contain a legally binding definition of terms "forest" and "deforestation". Therefore, a monoculture tree plantation established after a natural forest was cleared can potentially be eligible for REDD+ finance.

Consequently, REDD+ and ABS do not automatically ensure the conservation of forest ecosystems. Despite their potential to contribute to forest protection, their development is a new challenge for international environmental law. The pertinent legal provisions must ensure that projects effectively promote environmental conservation and socio-economic development in the course of their implementation.

Both REDD+ and ABS can be criticized for a lack of holistic approach to forests. Forest ecosystems should not be regarded solely as carbon sinks or pools of genetic resources. The consideration of multiple benefits of forests can be promoted through the establishment of links and synergies between REDD+ and ABS. However, even in this case, relevant activities would not go beyond the scope of the UNFCCC and the CBD. Consequently, the development of new mechanisms specific to forests, whose main scope would consist in the promotion of forest conservation, can be proposed. It is necessary to ensure that

31 *Fodella, Allessandro*, Recent Developments on Access and Benefit Sharing Relating to Genetic Resources (ABS) in International Law, in: *Casonato/Busatta/Penasa/Piciocchi/Tomasi* (a cura di), Il biodiritto e i suoi confini: Definizioni, dialoghi, interazioni, Università degli Studi di Trento, 2014, pp. 94 ff.

32 *Oldham, Paul/Hall, Stephen/Forero, Oscar*, Biological Diversity in the Patent System, in: PLoS ONE, Volume 8, Issue 11, 2013, e78737.

international legal mechanisms dealing with forests do not favor the conversion of natural forests and the establishment of monoculture plantations. A lack of political will and the controversial interests of forest owners and users currently obstacle the development of these mechanisms. All negotiations held so far on a potential forest convention have merely resulted in the adoption of non-legally binding documents. Even where resources needed for forest conservation are available, developing States willingly commit themselves to promote forest conservation only after developed States undergo legally binding commitments to provide significant financial and technological resources. These resources must serve to compensate the loss of revenues that results from the cessation of unsustainable forestry practices. However, developed States are ready to provide financial assistance only on a voluntary basis and/or through market-based mechanisms.

Against this backdrop, the further development of the international legal regime for forests is more likely to take place in the spheres where international consensus already exists, such as REDD+ and ABS. Consequently, the promotion of synergies between ABS and REDD+ and the development of joint projects can be desirable. It should be mentioned in this regard that the possibility of integrating REDD+ into the Biotrade Initiative is already discussed.[33] The Biotrade Initiative was established by the UNCTAD. It is aimed to promote the implementation of the CBD.[34] Thus, Joint REDD+ Biotrade projects can contribute to the promotion of links between two regimes.

Against this background, the present research is inspired by the understanding that (1) alarmingly high deforestation rates promote destruction of global forest ecosystems and compromises sustainable development in many areas, (2) there is no international legal solution available that would effectively tackle the problem of deforestation, and (3) any imaginable solution should consider in a holistic manner multiple benefits and functions of forest ecosystems, as well as forest-related socio-economic issues.

---

33 *United Nations Conference on Trade and Development*, Biodiversity and Climate Change: Integrating REDD+ into BioTrade Strategies, UNCTAD, United Nations, 2014. – Internet resource – http://unctad.org/en/PublicationsLibrary/ditcted2014d6_en.pdf [Last visit: 30.04.2023]

34 *United Nations Conference on Trade and Development*, Training Manual on Developing Joint Biotrade and REDD+ Projects, UNCTAD, United Nations, 2014. – Internet resource – http://unctad.org/en/PublicationsLibrary/ditcted2015d1_en.pdf [Last visit: 30.04.2023]

C. Scope and Course of the Research and Main Research Questions

Considering the previously identified problem of forest loss, the scope of this research is to analyze the newest international approaches to forest protection. REDD+ and ABS are in the foreground of this dissertation. It will be examined, whether and to what extent these international legal mechanisms promote forest conservation and sustainable forest use, whether they are able to promote socio-economic development, and to address the controversial interests of forest users and owners. For this purpose, gaps and deficiencies of the existing international legal regime for forests will be analyzed. As this analysis includes the existing mechanisms that have failed to combat deforestation effectively, Chapter 2 that is dedicated to this analysis is predominantly descriptive. The vague and superficial character of the existing international legal regime for forests is characterized by a low level of efficiency due to the fact that the drivers of deforestation are not effectively tackled by the relevant provisions. Therefore, the drivers of deforestation are also analyzed in this research. For this purpose, Chapter 1 provides a brief insight into the current conditions of global forest ecosystems and the main hazards to their existence and well-being. Ecological and socio-economical functions of forests are briefly analyzed, while effective forest conservation mechanisms should be based on the proper understanding of multiple forest functions.

Subsequently, the REDD+-mechanism, the ABS regime and their potential to promote forest conservation are examined. Considering that both Mechanisms are derived from two different international legal regimes relevant to forests, at first, they will be analyzed independently from each another.

While analyzing ABS, it is necessary to consider the general approach of the CBD to forest protection and sustainable forest use, because the conservationist provisions of this treaty can contribute to the promotion of forest conservation. The protection of both biodiversity and natural habitats is crucial for the conservation of forests. Against this backdrop, it will be analyzed, to what extent the CBD considers the natural links between different compounds of forest ecosystems, and ensures the compliance with its provisions.

As the next step, the particularities of the development and implementation of ABS will be analyzed. ABS must be attractive for the owners and users of genetic resources of forests. The users of genetic resources are expecting legal certainty and transparency.[35] Consequently, it is necessary that the legal framework for

35 *von Saint André, Susanne*, Genetische Ressourcen und traditionelles Wissen: Zugang,

ABS provides the minimum standards for the development and implementation of national legal provisions for ABS in forest-rich States. The providers of genetic resources will need assistance while implementing ABS.[36]

Moreover, investors expect that their intellectual property rights in relation to ABS-projects will be protected by patents.[37] At the same time, there is a number of ethical, political and socio-economic issues in relation to the patentability of biodiversity-based products. The interests and rights of providers of genetic resources will also be considered.

Another particular problem exists in relation to the traditional forest-related knowledge (TFRK) of forest-dependent people.[38] TFRK is often needed for the development of biodiversity-based products from genetic resources.[39] However, the original owners of TFRK are often not rewarded for their contribution to the product development.[40] Therefore, the ability of legal provisions guiding the implementation of ABS to pursue the rights of all relevant stakeholders will be assessed.[41] For this purpose, such documents as the CBD, the Nagoya Protocol,[42] and Bonn Guidelines[43] will be analyzed.

Subsequently, the REDD+-mechanism will be addressed in greater detail. It will be analyzed to which extent the REDD+-mechanism provides for the incorporation of forest-related activities into international climate change mitigation

---

Teilhabe und Rechtsdurchsetzung, 2013, p. 33; *Fodella* [n. 31], pp. 101 ff.

36 *Fodella* [n. 31], pp. 101 ff.

37 *Lawson, Charles*, The Role of Patents in Biodiversity Conservation, in: Nature Biotechnology, Volume 27, 2009, pp. 994 f.

38 *von Saint André* [n. 35], p. 23.

39 *Koutouki, Konstantia/Rogalla von Bieberstein, Katharina,* The Nagoya Protocol: Sustainable Access and Benefits-Sharing for Indigenous and Local Communities, in: Vermont Journal of Environmental Law, Volume 13, 2012, p. 519.

40 Ibid.

41 *Jospeh, Reji K.*, International Regime on Access and Benefit Sharing: Where Are We Now?, in: Asian Biotechnology and Development Review, Volume 12, Number 3, 2010, pp. 79 f.; *von Saint André* [n. 35], pp. 28 f.

42 Nagoya Protocol on Access to Genetic Resources and the Fair and Equitable Sharing of Benefits Arising from their Utilization to the Convention on Biological Diversity, adopted on 29 October 2010 and entered into force on 12 October 2014, in: 3009 UNTS No. 30619.

43 Secretariat of the Convention on Biological Diversity, Bonn Guidelines on Access to Genetic Resources and Fair and Equitable Sharing of the Benefits Arising out of their Utilization, CBD, 2002. – Internet resource – https://www.cbd.int/doc/publications/cbd-bonn-gdls-en.pdf [Last visit: 30.04.2023]

efforts and how forest conservation can be promoted thereby. For this purpose, the legal framework for REDD+ will be examined. In this context, it will be analyzed whether and to which extent the REDD+-mechanism is able to articulate and balance out conflicting interests of different groups of forest owners and users. Public and private entities investing in REDD+-projects plead for further legal certainty and transparency, whereas forest-rich developing States and forest-dependent people put a special emphasis on the land tenure and development issues.[44] The implementation of REDD+ is subject to the provision of financial and technological support to developing States. The technological support is particularly necessary in order to assure a transparent and accurate estimation of net forest emissions and carbon stocks changes in the course of the implementation of REDD+.[45] For this reason, the question will be answered whether the existing legal framework for REDD+ adequately promotes fundraising and technology transfer.

The other side of the coin is that the implementation of REDD+ can create new hazards for forest ecosystems. Certified emissions reductions units can arise not only from sustainable forest management, forest conservation, and near-natural afforestation, but also from monoculture tree plantations. These biodiversity-poor plantations cannot fulfill the full range of ecological functions of forests. Due to their decreased vitality, resilience, and regeneration ability, their carbon stocks are highly unsustainable. On the other hand, these plantations can deliver timber and non-timber forest products, and can reduce the anthropogenic pressure on natural forests. However, in order to safeguard multiple functions and benefits of forests, it is necessary to conserve natural forests, and to promote near-natural afforestation. Consequently, it will be analyzed, whether and to which extent the REDD+-related provisions of international documents promote the conservation of natural forests and of forest biodiversity. The rights of ILCs can be violated by REDD+ projects. The eviction of forest-dependent people has already taken place in certain cases.[46] Against this background, it

44 *Streck, Charlotte/Conway, Darragh*, Forestry and Agriculture under the UNFCCC: A Jigsaw Waiting to be Assembled, in: *Gray, Kevin R./Tarasofsky, Richard/Carlarne, Cinnamon P.* (eds.), The Oxford Handbook of International Climate Change Law, 2016, pp. 575 ff.

45 *Sandker, Marleke*, Emerging Approaches to Forest Reference Emission Levels and/or Forest Reference Levels for REDD, FAO, 2014, pp. 4 f.

46 *Ahmed, Nafeez*, World Bank and UN Carbon Offset Scheme 'Complicit' in Genocidal Land Grabs – NGOs, 3 July 2014. – Internet resource – https://www.theguardian.com/

will be analyzed whether and to which extent REDD+ ensures forest conservation, promotes non-carbon benefits of forests, contributes to sustainable development, and protects the rights of ILCs. While analyzing the legal framework for REDD+, the question is answered, whether the lack of stronger commitments to implement REDD+ can be remedied through the explicit reference to this mechanism in Article 5 of the Paris Agreement,[47] which entered into force on 4 November 2016.

Chapter 5 of this research is dedicated to the analysis of options for the further development of the international legal regime for forests through synergies between REDD+ and biodiversity-related mechanisms, such as ABS. The potential outcome of this synergetic implementation of different mechanisms for forest conservation and for the promotion of rights of forest-dependent people will be examined. Considering that these synergies are unlikely to consider the full range of multiple benefits of forests, other approaches to the development of forest-related international law are analyzed in Chapter 6. These approaches include the adoption of a legally binding convention on forest conservation,[48] the adoption of an overarching convention for cooperation between different forest-related international treaties, the promotion of regional initiatives, and other options.[49] The practical applicability and legal enforceability of different approaches will be examined.

---

environment/earth-insight/2014/jul/03/world-bank-un-redd-genocide-land-carbon-grab-sengwer-kenya [Last visit: 30.04.2023]

47 The Paris Agreement, adopted on 12 December 2015 and entered into force on 4 November 2016, in: UNTS No. 54113.

48 *Schulte zu Sodingen, Beate*, Der völkerrechtliche Schutz der Wälder, 2002, pp. 306 f.

49 *Eikermann, Anja*, Forests in International Law, 2015, pp. 183 ff.

Chapter One Protection of Forest Ecosystems as a Challenge for International Law

A. Forests as a Part of Biosphere and Their Role in the Human Environment

I. Current Conditions and Classification of Forest Ecosystems

In 2015 the total area covered by forests was estimated by FAO to be 3,999 billion hectares or 30,6 % of the global land area.[50] This number is lower than the values for 1990, when the forest area cover was estimated to be 4,128 billion hectares, which corresponds to the 31,6% of the global land area.[51] Forests are distributed unequally between States with approximately 53% of them situated in the Russian Federation, Brazil, Canada, the United States of America and China.[52] On the other hand, 64 States with a combined population of more than 2 billion people have a forest area that does not exceed 10% of their territory.[53]

The world's forest ecosystems include different types of forests. Each forest type fits particularly the climate conditions of a specific geographic region. Although there are many different types of forest ecosystems, they can be divided in three major groups corresponding to the major geographical zones, including tropical, temperate and boreal forests.

Boreal forests dominate the natural landscape in higher latitudes of the northern Hemisphere with exceptionally low winter temperatures. The northern border of the boreal zone corresponds with the end of the general forest vegetation distribution that is replaced by the arctic vegetation zones.[54] Boreal forests are the most extended in the world covering a territory of approximately 25 million square kilometers, representing 11% of the terrestrial surface.[55] Most

---

50 FAO, Global forest assessment 2015, 2016, p. 16.

51 Ibid.

52 Ibid.

53 Ibid.

54 *Thomas, Sean C./MacLellan, James*, Boreal and Temperate Forests, in: *Owes, John N./ Lund, Gyde H.* (eds.), Encyclopedia of Life Support Systems. Forests and Forest Plants, Volume 1, 2009, p. 152.

55 Ibid, p. 154.

of them grow on the territory of Russia, China, Mongolia, the Baltic Republics, Fennoscandia, and Canada. However, typical boreal forests also grow in more southern latitudes, involving the territories of the United States, Japan, Korea, Poland, Ukraine, Romania, Hungary, Slovakia, Czech Republic, Austria, Switzerland, Germany, as well as Chile and Argentina. Aside from the northern coniferous tree species, broad-leaved deciduous tree species also play an important role in boreal forests.[56]

Temperate forests are common in the temperate climate zone. Mild temperatures and a prominent seasonality characterize the temperate climate.[57] The structure and other typical characteristics of temperate forests correspond to this particular seasonal climate regime. The most representative tree species of temperate forests include winter deciduous trees and evergreen coniferous species, which are physiologically inactive during the unfavorable winter season. The evergreen temperate forests are common in the regions with a strong mild maritime climate.[58]

Tropical forests are found in tropical regions,[59] which extend between the Tropic of Cancer and the Tropic of Capricorn (23°27⊠ latitude north and south). The other criteria for distinguishing a tropical region refers to the 20 °C isotherm, which indicates the area between approximately 26° latitude south and 30° latitude north.[60] However, the major weakness of these approaches is that they do not consider local geographical and climatic variations within tropical regions. For example, they are not applicable to mountainous regions.[61] For this reason, another criterion is currently applied, according to which tropical regions are those where the daily thermal oscillation is higher than the annual thermal oscillation.[62] Tropical regions can be divided into seasonal and humid tropics.[63] The typical landscape type of seasonal tropics is the savanna. Scarce forest ecosystems of seasonal tropics are exposed to the so-called savannification. The most biodiversity-rich forest ecosystems are located in humid tropics.[64]

---

56 Ibid.

57 Ibid.

58 Ibid.

59 *Seling, Irene/Spathelf, Peter*, Tropical and Subtropical Forests, in: *Owes, John N./Lund, Gyde H.* (eds.), Encyclopedia of Life Support Systems. Forests and Forest Plants, Volume 1, 2009, p. 177.

60 Ibid.

61 Ibid.

62 Ibid.

63 Ibid, p. 178.

64 Ibid.

A relatively closed canopy of trees preventing the sunlight to achieve the ground characterizes these forests.[65] Nearly half of the world's tropical forests are concentrated in Southern and Central America. Other important tropical forest regions are Southwestern Asia and Central Africa.[66]

According to another classification, forests are classified based on the level of disturbances caused by human activities. From this point of view, forests are usually divided into three major groups:

Natural forests cover approximately 36% of the total forest area.[67] These forests are characterized by a natural biodiversity level in addition to their outstanding vitality and regeneration ability. Moreover, they fit into the local climate conditions, and can better withstand extreme weather fluctuations and pest infestations.[68] Natural forests are predominantly represented by the boreal and tropical forest types.[69] In comparison to those forests that were damaged by unsustainable forest management or by natural disasters, healthy natural forest ecosystems are able to fulfill more ecological functions both qualitatively and quantitatively.[70] They can also deliver precious timber and non-timber products. However, natural forests should be managed in a way that minimizes adverse environmental impacts on them. Taking into account the relatively small share of natural forests in the global forest landscape, further industrial exploitation of these forests should be avoided. At any time, the benefits of ecological functions fulfilled by natural forests highly outweigh those temporary advantages that are associated with the logging activities in these forests.

The total area of near natural or naturally regenerating forests is estimated to be 57% of the global forest area.[71] The capacity to fulfill multiple ecological functions characterizes also this type of forests, though it is significantly lower than

---

65 Ibid.

66 Ibid.

67 *FAO 2016* [n. 1], p. 5.

68 *Lehmann, Susanne*, Schutz der Wälder – Nationale Verantwortung tragen und global handeln, 2007, p. 5; *Gamfeldt, Lars/Snäll, Tord/Bagchi, Robert/Jonsson, Micael/Gustafsson, Lena/Kjellander, Petter/Ruiz-Jaen, María C./Fröberg, Mats/Stendahl, Johan/Philipson, Christopher D./Mikusiński, Grzegorz/Andersson, Erik/Westerlund, Bertil/Andrén, Henrik/Moberg, Fredrik/Moen, Jon/Bengtsson, Jan*, Higher Levels of Multiple Ecosystem Services Are Found in Forests With More Tree Species, in: Nature communications, Volume 4, 2013, p. 1345.

69 *FAO 2016* [n. 50], p. 28.

70 *Lehmann* [n. 68], p. 5.

71 *FAO 2016* [n. 50], p. 27.

the same ability of natural forests. This holds true for their biodiversity level, vitality and regeneration ability.[72] However, when naturally regenerated forests are left to develop naturally and undisturbed, they tend to acquire the basic characteristics that are common for natural forest ecosystems.[73] Noteworthy, some forests classified as natural were severely damaged by natural disasters (such as the Ice Age, hurricanes, natural forest fires, volcanic eruptions, etc.) in prehistorical times. Subsequently, they have successfully regenerated naturally. However, in certain cases, especially after a large-scale deforestation, forests are unable to regenerate for a long period of time. As a result, they can be replaced by shrubs and arid grasslands. Consequently, it is particularly important to preserve and maintain natural forest ecosystems, and to promote their regeneration.

Planted forests cover the remaining 7% of the global forest area.[74] They are usually characterized by a minimal level of biodiversity, reduced vitality and low regeneration capacity. They are also particularly prone to forest fires, pest infestations and other natural hazards.[75] However, they often fulfill a number of ecological functions, including the protection of soil and fresh water resources. Planted forests deliver timber and non-wood forest products.[76] In the cases where natural tree compositions are chosen for afforestation, a planted forest can gradually evolve in a near-natural one.

It is noteworthy that the share of planted forests remains low on the global scale. On the other hand, a general increase of the share of planted forests has taken place in the last decades. For example, in 1990 the share of planted forests in the global forest cover was 4%, in year 2005 it achieved 6%. In 2015, the share of planted forests had already presented 7% of the global forest cover.[77] This data demonstrates a worrying development, as the current level of scientific knowledge about forest ecosystems does not permit to artificially restore natural forests. This means that planted forests are not able to fulfill multiple ecological functions, although they still can contribute to the reduction of the anthropogenic pressure on natural forests.[78] In several cases, the attempts to plant

72 *Lehmann* [n. 68], p. 5.

73 See *Taki, Hisatomo/Yamaura, Yuichi/Okabe, Kimiko/Maeto, Kaoru*, Plantation vs. Natural Forest: Matrix Quality Determines Pollinator Abundance in Crop Fields, in: Scientific Reports, Volume 1, 2011, pp. 132 ff.

74 *FAO 2016* [n. 50], p. 28.

75 *Lehmann* [n. 68], p. 5.

76 *Lehmann* [n. 68], p. 5.

77 *FAO 2016* [n. 50], p. 28.

78 *Lehmann* [n. 68], p. 5.

artificial forests were unsuccessful because the local climate conditions were not sufficiently considered and the use of invasive tree species combined with the prevalence of monoculture plantations resulted in a lacking vitality of planted forests.[79] On the contrary, the experience with the near-natural afforestation in Niger justifies that this strategy can be far more cost-effective and ecologically friendly. It is important to note that Niger has simultaneously achieved the restoration of near-natural forest cover and the improvement of the life conditions of local farmers without big financial expenditures.[80]

II. Forest Ecosystems as an Integral Part of the Environment

Global forests are critical natural ecosystems that consist of numerous biological species. These species interact in diverse and multiple ways. Circulation of energy and substances constantly occur between living and nonliving components of forest ecosystems, including solar energy, atmosphere, soil, etc. In addition, with the development of culturally shaped forms of the human society, forest ecosystems interact strongly with the human civilization, which has significant impact on them.

1. Forest Ecosystems and Biodiversity

Forest ecosystems outstand for their exceptionally high level of biodiversity. Tropical rainforests can compete in this aspect with maritime wildlife hotspots, such as coral reefs.[81] Forests host more than 80% of terrestrial biodiversity.[82]

---

79 *Hua, Fangyuan/Wang, Xiaoyang/Fisher, Brendan/Wang, Lin/Zhu, Jianguo/Tang, Ya/Yu, Douglas W./Wilcove, David S.*, Opportunities for Biodiversity Gains under the World's Largest Reforestation Programme, in: Nature Communications, Volume 7, Issue 1, 2016, Art. 12717. – Internet resource – https://www.nature.com/articles/ncomms12717#citeas [Last visit: 30.04.2023]; *Xu, Jianchu*, China's New Forests Aren't as Green as They Seem, in: Nature, Volume 477, 2011, p. 371.

80 *Polgreen, Lydia*, In Niger, Trees and Crops Turn Back the Desert, in: New York Times, 11 February 2007. – Internet resource – http://www.nytimes.com/2007/02/11/world/africa/11niger.html [Last visit: 30.04.2023]

81 CBD COP Decision II/2 "Forests and biological diversity", Decision adopted by the Conference of the Parties to the Convention on Biological Diversity at its second session, held at Jakarta from 6 to 17 November 1995, United Nations, 1995. – Internet resource – https://www.cbd.int/decisions/cop/?m=cop-02 [Last visit: 30.04.2023]

82 *Moreales-Hidalgo, David/Oswalt, Sonja N./Somanathan, E.*, Status and Trends in Global Primary Forest, Protected Areas, and Areas Designated for Conservation of Biodiversity from the Global Forest Resources Assessment 2015, in: Forest Ecology and Management, Volume 352, 2015, pp. 68, 73 ff.

Despite the fact that boreal forests are characterized by a comparatively lower biodiversity level, they offer refuge to a bulk of endemic species. These high levels of biodiversity enable forest ecosystems to fulfill their multiple environmental functions.[83]

2. *Influence of Forest Ecosystems on the Regional and Global Climate*

Forests largely influence the chemical composition of the atmosphere. They act as sinks and reservoirs of greenhouse gases, including the CO_2. Forests produce oxygen and consume CO_2, while storing it in the forest biomass and soil. As a result, forests effect the regional and global temperature regime by influencing the CO_2 levels in the atmosphere. In 2015, carbon stocks in the above- and below-ground biomass were estimated to be 296 Gigatons.[84] According to the FAO estimates, forest carbon stocks are continuing to decrease at the rate of 0,3 Gt per year,[85] which is lower than the value for the period 1990–2010, when the average annual loss of forest carbon stocks was estimated to be 0,5 Gt per year.[86] The main reason for this change is the reduction of deforestation rates in Asia and in Latin America, where only Brazil managed to achieve a considerable reduction of forest carbon loss rates from 240 million tons CO_2 per year to 80 million tons per year.[87] However, the recent increase in deforestation rates in this State in 2018–2020 reveals an alarming upward trend.

At any time, carbon dioxide emissions from forests continue to play a major role. Estimates reveal that between 11% and 20% of anthropogenic greenhouse gas emissions are caused by deforestation and forest degradation.[88] However, these estimations do not take into account the whole extent of the impacts of deforestation on the global climate. One more comprehensive analyze suggests that if deforestation was totally stopped, the natural forest regeneration was promoted, and standing mature natural forests were left intact, tropical forests alone

83 Ibid.

84 *FAO 2016* [n. 50], p. 30.

85 Ibid.

86 Ibid.

87 Ibid.

88 *Goodman, Rosa C./Herold, Martin*, Why Maintaining Tropical Forests Is Essential and Urgent for a Stable Climate, CGD Working Paper 385, 2014, pp. 1 ff. – Internet resource – https://www.cgdev.org/sites/default/files/CGD-Climate-Forest-Paper-Series-11-Goodman-Herold-Maintaining-Tropical-Forests.pdf [Last visit: 30.04.2023]

could have captured one third of anthropogenic greenhouse gas emissions.[89] Another recent research indicates that reforestation represents the most efficient and cost-effective way to mitigate climate change caused by the CO_2 emissions.[90] Indeed, it is argued that a large-scale forest regrowth and near-natural reforestation is the only approach that permits to keep the global warming below 1,5°C degree mark.[91] At the same time, other researches have shown that technological solutions are far less effective than the conservation and restoration of forests.[92] Moreover, they can have immense adverse impacts on the environment and on wildlife conservation.[93]

It should be noted that the misunderstanding in relation to the forest's role in the carbon cycle continues to prevail. Firstly, some researchers and environmental activists affirm that newly planted forests can sink the carbon dioxide

89 *Seymour, Frances/Busch, Jonah*, Why Forests? Why Now?: The Science, Economics, and Politics of Tropical Forests and Climate Change, Center for Global Development, 2016, pp. 27 ff., pp. 59 ff., pp. 121 ff.; see *Bush, Jonah/Engelmann, Jens*, Tropical Forests Offer up to 24–30 Percent of Potential Climate Mitigation, Center for Global Development, 4 November 2014 – Internet resource – https://www.cgdev.org/blog/tropical-forests-offer-24%E2%80%9330-percent-potential-climate-mitigation [Last visit: 30.04.2023]; *Seymour, Frances*, What If We Had a Safe and Natural Way to Capture and Store Carbon?, Center for Global Development, 4 November 2014. – Internet resource – https://www.cgdev.org/blog/what-if-we-had-safe-and-natural-way-capture-and-store-carbon [Last visit: 30.04.2023]

90 *Bastin, Jean-Francois/Finegold, Yelena/Garcia, Claude/Mollicone, Daniloe/Rezende, Marcelo/Routh, Devin/Zohner, Constantin M./Crowther, Thomas W.*, The Global Tree Restoration Potential, in: Science, Volume 365, Issue 6448, 2019, p. 78.

91 Ibid.

92 *EASAC*, Negative Emission Technologies: What Role in Meeting Paris Agreement Targets?, EASAC Policy report 35, February 2018, pp. 5 ff.; *Carrington, Damian*, "Silver bullet" to suck CO2 from air and halt climate change ruled out, in: Guardian, 1 February 2018. – Internet resource – https://www.theguardian.com/environment/2018/feb/01/silver-bullet-to-suck-co2-from-air-and-halt-climate-change-ruled-out [Last visit: 30.04.2023]; *Fuss, Sabine/Lamb, William F./Callaghan, Max W./|Hilaire, Jerome/Creutzig, Felix/Amann, Thorben/Beringer, Tim/Carcia, Wagner de Oliveira/Hartmann, Jens/Khanna, Tarun/Luderer, Gunnar/Nemet, Gregory F./Rogelj, Joeri/Smith, Pete/Vicente, Jose Luis Vicente/Wilcox, Jennifer/Dominguez, Maria del Mar Zamora/Minx, Jan C.*, Negative Emissions – Part 2: Costs, Potentials and Side Effects, in: Environmental Research Letters, Volume 13, Number 6, 2018, pp. 1 ff.

93 *Fuss/Lamb/Callaghan/Hilaire/Creutzig/Amann/Beringer/et. al.* [n. 92], pp. 2 f.

more effectively than old-growing natural forests.[94] Following this logic, certain forest owners and users may argue that some near-natural and natural forests can be replaced by planted forests and monoculture plantations. However, they do not take into account the fact that the clearing of natural forests always results in CO_2 emissions from biomass and forest soil. The absorption of CO_2 by naturally regenerating and artificially planted forests cannot compensate the emissions caused by deforestation.[95] The problem intensifies when monoculture plantations are planted instead of near-natural a/reforestation. These plantations are often cleared before them reaching the stage of maturity. Moreover, even if young growing forests apparently sink more CO_2 than natural forests, the sustainability of forest stocks in these forests is usually lower.[96] Planted forests are far more prone to forest fires, pest infestations and other natural hazards, which usually generate significant CO_2 emissions. As a result, a planted forest can become a net source of greenhouse gas emissions.

Additionally, there is growing evidence that forest ecosystems are able to considerably influence the water cycle in the atmosphere, as well as the formation and distribution of atmospheric precipitations, which is likely to have repercussions on the global and local temperatures. Through the complex processes of evaporation and transpiration, forests enrich the atmosphere above them with water vapor.[97] The concentration of water vapor in air masses above forests largely exceeds its concentration above the ocean. This circumstance likely generates the movement of air masses that are saturated with air moisture into the forested regions, even from the ocean.[98] The impact of forests on the movement of moist air masses can be the subject of further research, which goes beyond the scope of this dissertation. However, the available data suggests that the

94 *Luyssaert, Sebastiaan/Schulze, E.-Detlef/Börner, Annett/Knohl, Alexander/Hessenmöller, Dominik/Law, Beverly E./Ciais, Philippe/Grace, John*, Old-Growth Forests as Global Carbon Sinks, in: Nature, Volume 455, 2008, p. 213.

95 Ibid.

96 Ibid.

97 *Makarieva, Anastassia M./Gorshkov, Victor G./ Sheil, Douglas/Nobre, Antonio Donato/ Bunyard, Peter/Li, B. Larry*, Why Does Air Passage over Forest Yield More Rain? Examining the Coupling between Rainfall, Pressure, and Atmospheric Moisture Content, in: Journal of Hydrometeorology, Volume 15, Issue 1, 2014, pp. 422 f.; *Makarieva, Anastasia/Gorshkov, Viktor*, The Biotic Pump: Condensation, Atmospheric Dynamics and Climate, in: International Journal of Water, Volume 5, Number 4, 2010, pp. 366 f.; *Nobre, Antonio Donato*, The Future Climate of Amazonia, 2014, p. 11.

98 *Nobre* [n. 97], p. 11.

evapotranspiration may have large repercussions on the forests' role in climate change mitigation and adaptation.[99] The reason for it is that water vapor is an important greenhouse gas in the Earth's atmosphere.[100] Additionally, water vapor can absorb thermal energy and transport it to other geographical regions, contributing to its distribution across the globe.[101] The ability of forests to influence the water vapor concentration in the atmosphere depends on different factors, including the prevailing tree species in a given ecosystem, medium tree height and the closeness of the forest canopy.[102] The optimal combination of these factors is usually achieved only in natural forest ecosystems. That is why the ability of naturally regenerating and planted forests to influence the concentration of water vapor in the atmosphere is usually lower than the same ability of natural forests.[103]

In addition to this, forests influence the global and regional temperature by changing the albedo of the Earth's surface. The albedo characterizes the amount of solar energy that is reflected back to space without contributing to the heat accumulation on the Earth[104]. This ability can have different effects in boreal, temperate and tropical regions respectively. In northern latitudes, the larger albedo of the Earth that is caused by deforestation causes local temperature decrease and promotes the spread of everlasting ice cover. The same process in tropical latitudes can promote local and regional temperature increase.[105] In both cases, the change of the surface albedo that is caused by deforestation will have unpredictable impacts on the global and regional climates. The cooling effect in the north can apparently be desirable in the light of climate change mitigation/adaptation. However, the associated changes in the global atmosphere

---

99 *Sheil, Douglas*, Forests, Atmospheric Water and an Uncertain Future: The New Biology of the Global Water Cycle, in: Forest Ecosystems, Volume 5, 2018. – Internet resource – https://forestecosyst.springeropen.com/track/pdf/10.1186/s40663-018-0138-y [Last visit: 30.04.2023]

100 *Schneider, Tapio/O'Gorman, Paul A./Levine, Xavier J.*, Water Vapor and the Dynamics of Climate Changes, in: Review of Geophysics, Volume 48, Issue 3, 2010, pp. 1. ff. – Internet resource – https://agupubs.onlinelibrary.wiley.com/doi/epdf/10.1029/2009RG000302 [Last visit: 30.04.2023]

101 Ibid, pp. 2 f.

102 Ibid.

103 Ibid.

104 *Lee, Xuhui*, Observed Increase in Local Cooling Effect of Deforestation at Higher Latitudes, in: Nature, Volume 479, Number 7373, 2011, p. 384.

105 Ibid, p. 385.

circulation patterns are difficult to predict.[106] It should also be taken into account that throughout history global climate was exposed not only to warming, but also to drastic cooling events (ice ages). The devastating effects of such events are difficult to predict. For this reason, the cooling effect generated by deforestation in the north should draw more attention. It should not be understood only as a positive phenomenon. Further research is needed in order to examine the climate effects of albedo changes due to deforestation in the upper latitudes. However, this research goes beyond the scope of this dissertation.

At any time, it should be admitted that those land areas that are afforested in a near-natural manner and sustainably managed, may also have positive impacts on climate change mitigation by absorbing CO_2 and influencing the concentration of water vapor and the atmospheric movement of moist air masses. However, they cannot replace natural forests.

3. *Influence of Forests on Global Precipitation Patterns*

As it was mentioned above, another important function that forests fulfill is their ability to influence global precipitation patterns. Although scientists previously denied any relationship between forest ecosystems and atmospheric precipitations,[107] the newest scientific research points to a clear correlation between them.[108] In this regard, the ability of forests to saturate atmospheric air with tiny electricized mineral particles that are produced by plants, fungi, and microorganisms that inhabit forest ecosystems, is especially important. Those electricized mineral particles serve as cores for the formation of raindrops and snowflakes. Researches justify that these particles can significantly influence the global and regional precipitation patterns.[109] Additionally, the aforementioned

106 Ibid.

107 *Schulte zu Sodingen, Beate*, Der völkerrechtliche Schutz der Wälder, 2002, pp. 52 f.

108 *Makarieva/Gorshkov/Sheil/Nobre/Bunyard/Li* [n. 97], pp. 442 ff.; *Wright, Jonathon S./Fu, Rong/Worden, John R./Chakraborty, Sudip/Clinton, Nicholas E./Risie, Camille/Sun, Ying/Yin, Lei*, Rainforest-initiated Wet Season Onset over the Southern Amazon, in: Proceedings of the National Academy of Sciences of the United States of America, Volume 114, Number 32, 2017, pp. 8481 ff.

109 *Pöhlker, Christopher/Wiedemann, Kenia T./Sinha, Bärbel/Manabu, Shiraiwa/Gunthe, Sachin S./Smith, Mackenzie/Su, Hang/Artaxo, Paulo/Chen, Qi/Cheng, Yafang/Elbert, Wolfgang/Gilles, Mary K./Kilcoyne, Arthur L. D./Moffet, Ryan/Weigand, Markus/Martin, Scot/Pöschl, Markus/Andreae, Meinrat O.*, Biogenic Potassium Salt Particles as Seeds for Secondary Organic Aerosol in the Amazon, in: Science, Volume 337, Issue 6098, 2012, p. 1075.

forest´s capacity to notably influence the humidity of upper atmosphere levels through the complex process of evapotranspiration plays an important role in the formation of rainclouds.[110] Evidence suggests that the change in the concentration of water vapor can have a significant influence on the velocity and direction of the movement of air masses.[111] This means that forests can largely determine global and regional precipitation patterns, as they largely influence the movement of moist air masses into those regions where dry climate conditions would otherwise prevail.[112]

The analysis of the forests' role in global precipitation patterns goes beyond the scope of this dissertation. Despite this, the available data suggests that deforestation provokes dangerous and abrupt changes in precipitation patterns even thousands of kilometers away.[113] For example, tropical deforestation likely causes a reduction of rainfall in the USA and in other areas across the globe.[114]

4. *Socio-Economic Functions of Forest Ecosystems*

One more side to the environmental role of forests is that they also provide humanity with many precious goods, deliver multiple benefits and contribute to the eradication of poverty and unemployment. According to the World Bank reports, 75% of the accessible fresh water for the industrial, agricultural, domestic and ecological needs come from forests.[115] About one-third of the world's

---

110 *Wright/Fu/Worden/Chakraborty/Clinton/Risie/Sun/Yin* [n. 108], pp. 8481 ff.; *Ellison, David/Morris, Cindy E./Locatelli, Bruno/Sheil, Douglas/Cohen, Jane/Murdiyarso, Daniel et. al.* Trees, forests and water: Cool insights for a hot world, in: Global Environmental Change, Volume 43, 2017, pp. 51 ff.; *Makarieva/Gorshkov/Sheil/Nobre/Bunyard/Li* [n. 97], pp. 442 ff.

111 See Subsection A.II.3 of this Chapter.

112 *Makarieva/Gorshkov/Sheil/Nobre/Bunyard/Li* [n. 97], pp. 442 ff.; *Wright/Fu/Worden/Chakraborty/Clinton/Risie/Sun/Yin* [n. 108], pp. 8481 ff.; *Ellison/Morris/Locatelli/Sheil/Cohen/Murdiyarso et. al.* [n. 110], pp. 53 ff.

113 *Spracklen, Dominik V./Arnold, S. R./Taylor, C. M.*, Observations of increased tropical rainfall preceded by air passage over forests, in: Nature, Volume 489, Number 7415, 2012, pp. 282 ff.

114 *Bettwy, Mike*, Tropical Deforestation Affects Rainfall in the U.S. and Around the Globe, NASA, 13 September 2005. – Internet resource – https://www.nasa.gov/centers/goddard/news/topstory/2005/deforest_rainfall.html [Last visit: 30.04.2023]

115 *The World Bank*, Why Forests are Key to Climate, Water, Health, and Livelihoods, the World Bank, 18 March 2016. – Internet resource – http://www.worldbank.org/en/news/feature/2016/03/18/why-forests-are-key-to-climate-water-health-and-livelihoods [Last visit: 30.04.2023]

big cities receive fresh water from forest ecosystems. Numerous industrial production systems, human communities and natural ecosystems depend on forests in this regard.[116] Multiple forest functions, including the influence on global and regional precipitation patterns, the interception of atmospheric moisture, the protection of fertile soils, and groundwater recharge determine the forest's ability to maintain and promote freshwater resources. These forest functions are of particular importance in relation to the current scarcity of freshwater resources, by which 80% of the world's population is facing water insecurity.[117]

Aside from freshwater, forests also provide humanity with food, medicine, valuable timber and non-timber products, and other goods and services. The enclosure of forests' resources promotes employment. 1,3 billion people depend on forests directly or indirectly for food, energy, employment and income.[118] An estimated number of 300–350 million people live directly in dense forest areas, and depend almost entirely on forests for subsistence.[119] Rural households in dense, forested regions receive, on average, 22% of their income from forests, a number that is higher than the contribution of wage labor, livestock or self-owned businesses. Approximately half of the income from forests results from the direct harvesting of food, medicine, fuel, fodder, construction materials and medicines. In several forested regions, payments for ecological services of forests is another significant source of income for locals. Moreover, at least hundreds of millions of people outside of forests also depend on forest ecosystems for medicines, food, construction materials, etc.[120]

B. Drivers of Deforestation and Its Consequences

I. Anthropogenic Threats to Forest Ecosystems

Forest ecosystems are prone to natural hazards, such as pest infestations, hurricanes, forest fires, volcanic eruptions and others. Natural adaptation mechanisms determine the forests' resilience and enhance their regeneration capacity.

116 Ibid.

117 Ibid.

118 *The World Bank*, Enhance Livelihoods of Forest Communities, The World Bank, 16 March 2016. – Internet resource – http://www.worldbank.org/en/news/feature/2016/03/18/why-forests-are-key-to-climate-water-health-and-livelihoods [Last visit: 30.04.2023]

119 Ibid.

120 Ibid.

Plant and animal species are able to substitute one another to fulfil certain naturally determined roles in the process of forest regeneration.[121] Moreover, forest ecosystems can respond to the unfavorable changes of environmental conditions by modifying their vital activity, for example, by changing the leaf area index, by opening and closing the plant stomata, etc.[122] The ability to withstand unfavorable environmental conditions is much higher for a forest ecosystem than for single trees outside of forests. The highest resilience level is achieved through the interaction between plant and animal species that constitute forest ecosystems.[123]

However, human activities create new hazards to forest ecosystems. As a result, forest ecosystems are exposed to such amounts of stress that exceed their natural resilience and regeneration capacity. The biggest threat, in this regard, is unsustainable industrial logging and forestland conversion.[124] It causes a drastic decline in the forest's capacity to fulfill ecological functions and it might ultimately lead to the complete destruction of forest ecosystems. With the reduction of industrial and agricultural activities, forest ecosystems can regenerate even on degraded soils in deforested areas, but this is a very time-consuming and complicated process.[125] On the other hand, the large-scale deforestation can supposedly promote, in several cases, such extensive regional climate changes that the natural afforestation becomes very complicated and herbaceous and shrub vegetation replaces the forest.[126]

---

121 *Chazdon, Robin L.*, Beyond Deforestation: Restoring Forests and Ecosystem Services on Degraded Lands, in: Science, Volume 320, 2008, p. 1458.

122 Ibid.

123 Ibid.

124 Ibid, p. 1459.

125 *Chazdon, Robin L./Guariguata, Manuel R.*, Natural Regeneration as a Tool for Large-scale Forest Restoration in the Tropics: Prospects and Challenges, in: Biotropica, Volume 48, Issue 6, 2016, pp. 716 ff.; *Fröhlich, M./Quednau, H. D.*, Statistical Analysis of the Distribution Pattern of Natural Regeneration in Forests, in: Forest Ecology and Management, Volume 73, 1995, pp. 46 ff.; *Harmer, Ralph/Morgan, Geoff*, Development of Quercus Robur Advance Regeneration Following Canopy Reduction in an Oak Woodland, in: Forestry: An International Journal of Forest Research, Volume 80, Issue 2, 2007, pp. 137 ff.; *FAO*, Restoring Forest Lasndscapes through Assisted Natural Regeneration (ANR) – A Practical Manual, 2019, pp. 4 ff.; *Chazdon, Robin L.,/Bodin, Blaise/Guariguata, Manuel/Lamb, David/Walder, Bethanie/Chokkalingam, Unna/Shono, Kenichi*, Partnering with Nature: The Case for Natural Regeneration in Forest and Landscape Restoration, FERI Policy Brief, 2017, pp. 5 ff.

126 *Nobre, Antonio Donato*, The Future Climate of Amazonia, 2014, pp. 14 ff.

Industrial forest use and, to some extent, deforestation itself are inevitable and needed to satisfy a high demand for timber and non-timber products, as well as to deliver land areas for agricultural or industrial use and for infrastructure development.[127] The problem lies in the fact that the adverse impacts on the environment are not often considered in the process of forest management. Moreover, deforestation does not often bring positive socio-economic effects due to unsustainable management of converted forestlands.[128] In case converted forestlands are deemed for agricultural use, unsustainable cultivation and harvesting practices result in declining soil fertility. If the converted forestlands are thought to serve for infrastructure development and urban construction, the failure to consider the fragility of soils in deforested areas often leads to the failure of construction projects.[129] In many cases, the development projects simply do not exist or do not consider all opportunities for economic development of deforested lands. Developing States often spend their revenues from the sale of logged timber to finance their short-term State expenditures instead of investing in the long-term socio-economic development.[130]

The unsustainable use of forests and converted forestlands promotes further deforestation and forest degradation, which could be avoided if the already exploited forest resources and forestland areas were managed sustainably, and the associated revenues were invested in the long-term development strategies.

II. Consequences of Deforestation and Forest Degradation

The unsustainable use of forest resources, which causes extensive damage to forest ecosystems, together with large-scale deforestation result in long-term adverse environmental impacts. Combined with the misuse of revenues from forest utilization and the unsustainable management of converted forestlands, such long-term adverse environmental impacts lead to the emergence of an ecological, economic, social, and sometimes even political crisis.

Deforestation and forest degradation results in dangerous changes of the global and regional climate and in the increase of frequency of natural atmospheric hazards.[131] Moreover, it leads to the depletion of freshwater resources,

127 *Eikermann, Anja*, Forests in International Law, 2015, pp. 27 ff.

128 Ibid.

129 Ibid.

130 Ibid.

131 *Cramer, Wolfgang/Bondeau, Alberte/Schaphoff, Sibyll/Lucht, Wolfgang/Smith, Benjamin/Sitch, Stephen*, Tropical forests and the global carbon cycle: impacts of atmospheric carbon dioxide, climate change and rate of deforestation, in: Philosophical

the soil erosion, the destruction of natural habitats, and the drastic reduction of biodiversity.[132] All these factors lead to the loss of development opportunities, to increased poverty, and political instability. Research demonstrates the interdependence between deforestation and the frequency and scale of disastrous flooding events.[133]

As a remarkable example of adverse impacts of deforestation, the miserable living conditions of the local population in Haiti can be referred to. Here, the degradation of local forests has been encouraged by decades of shortsighted policies, grounded in a low productivity subsistence farming. The demand for more and more land where to cultivate cacao, coconut and other tropical products, indeed, is ever growing but this agricultural produce is destined to foreign markets, thus providing low income for the local residents.[134] Throughout history, the local economy depended largely on revenues from the export of cacao, coffee, mango, sugarcane and other tropical crops.[135] The production of these crops stimulates forestland conversion: local farmers are constrained to cut down forests to clear new land areas for agricultural purposes and for the production of charcoal, which serves as the main source of energy for most of Haitian households.[136] All these factors resulted in the degradation and loss of forests in many populated Haitian regions. Although research estimates the total tree cover of Haiti to be about 30% of the State territory, the quality of forests in the populated areas is often far too low to fulfil their multiple ecological functions.[137] As

---

Transactions of the Royal Society Biological Sciences, Volume 359, Issue 1443, 2004, pp. 331 ff.; *Sheil, Douglas/Bargués-Tobella, Aida/Ilstedt, Ulrik/Ibisch, Pierre L./Makarieva, Anastassia/McAlpine, Clive/Morris, Cindy E./Murdiyarso, Daniel/Nobre, Antonio D./Poveda, Germán/Spracklen, Dominick V./ Sullivan, Caroline A./Tuinenburg, Obbe A./van der Ent, Ruud J.*, Forest restoration: Transformative trees, in: Science, Volume 366, Issue 6463, 2019, pp. 316–317.

132 *Sheil/Bargués-Tobella/Ilstedt et. al.* [n. 131], pp. 331 ff.

133 *Bradshaw, Corey J. A./Sodhi, Navjot S./Peh, Kevin S./Brook, Barry W.*, Global evidence that deforestation amplifies flood risk and severity in the developing world, in: Global Change Biology, Number 13, 2007, pp. 2380 ff.

134 *Ellis, Allexandra*, The Anthropocene: Deforestation and its Consequences in Haiti, in: Foreign affairs Review, 10 April 2013. – Internet resource – http://foreignaffairsreview.co.uk/2013/04/haiti/ [Last visit: 30.04.2023]

135 Ibid.

136 Ibid.

137 *Churches, Christopher E./Wampler, Peter J./Sun, Wanxiao/Smith, Andrew J.*, Evaluation of Forest Cover Estimates for Haiti Using Supervised Classification of Landsat Data,

a result, many agricultural areas in Haiti are exposed to a large-scale erosion and soil degradation. The deforested areas of Haiti are especially prone to floods and landslides that have already induced enormous damage and losses to the local population.[138]

On the other hand, the nearby Dominican Republic used to promote industrial development that allowed to reduce human pressure on forests. Large-scale introduction of natural gas consumption for energy purposes led to an almost complete abandonment of charcoal. The protection of forests has without any doubt contributed to the relative economic prosperousness of the Dominican Republic. Indeed, the available data suggests that deforestation and forest degradation contributed to the worsening of poverty in Haiti and increased the proneness to natural disasters. Large-scale deforestation, alongside with disastrous torrential rains, have also provoked extensive and destructive floods and landslides in many other places around the world, for example, in Venezuela in 1999.[139]

It is already recognized that the deforestation and forest degradation contribute largely to desertification and soil erosion, which both result in the loss of fertile soils. Current estimates show that 12 million hectares of available agricultural land is lost every year.[140]

It is important to note that deforestation and forest degradation cause a number of globally relevant adverse impacts, such as global climate change and biodiversity loss.

III. Different Ethical Approaches to the Protection of Forest Ecosystems

Numerous and precious functions fulfilled by forest ecosystems and adverse impacts of deforestation signalize that there is an urgent need to conserve forests. At this time, controversies still exist on the issue of which ethical approach should guide the development and implementation of forest conservation

---

in: International Journal of Applied Earth Observation and Geoinformation, Volume 30, 2014, pp. 203 ff.

138 *Ellis* [n. 134].

139 *Rudenstine, Sasha/Galea, Sandro*, The Causes and Behavioural Consequences of Disasters: Models Informed by the Global Experience 1950–2005, 2012, pp. 33 f.

140 *Kamal, Baher*, Arable Land Lost at Unprecedented Rate: 33,000 Hectares… a Day!, in: Inter Press Service, 2016. – Internet source – http://www.globalissues.org/news/2016/08/16/22410 [Last visit: 30.04.2023]

strategies. Different ethical views are to be addressed at this stage of the research because they are useful for understanding the ratio behind the protection of forest ecosystems.

According to one approach, the protection of nature aims to preserve precious natural functions and benefits provided by ecosystems and consumed by humans. From this point of view, only humanity has an intrinsic value, whereas nature serves to support the human existence.[141] For this reason, this approach is called "anthropocentric". It presupposes that species, habitats and natural processes, whose conservation does not bring benefits to humanity, can be underestimated.[142] The anthropocentric approach permits to include the needs of forest dependent people into forest protection strategies. Additionally, this approach takes due regard to the fact that nowadays many ecosystems are influenced by human activities. Moreover, the anthropocentric rhetoric helps to address the problem of deforestation by focusing on the relevance of forest conservation for the benefits of the whole humanity. However, this approach can hardly justify the need to completely abandon industrial activities in natural forests in order to preserve natural habitats and forest biodiversity.

On the other hand, supporters of a so-called *ecocentric approach* aim to focus on the protection of natural ecosystems and biodiversity. According to this approach, all nature has its own intrinsic value, independently of its potential or actual usefulness to humanity.[143] Thus, humanity has the responsibility to protect nature and to preserve ecosystems and living species. Supporters of the ecocentric approach argue that the ecological footprint of humans surpasses that of all other species: humans are the unique living beings on the planet that considerably influence its wildlife and the only ones that have the cognitive ability to perceive the planet as an integral organism.[144] However, this approach does not consider the fact that every species relies on others and on the environment to survive. The aforementioned exceptional cognitive abilities of humans also presuppose the particular art of relationships between people and nature.[145] Most human activities modify the environment and, in many cases, result in adverse

141 *Kortetmäki, Teea,* Anthropocentrism versus Ecocentrism Revisited: Theoretical Confusions and Practical Conclusions, in: SATS: Northern European Journal of Philosophy, Volume 14, Issue 1, 2013, pp. 22 f.

142 Ibid.

143 Ibid, pp. 23 f.

144 Ibid, p. 29.

145 Ibid, p. 31.

environmental impacts. This means that if the ecocentric approach were strictly applied, every human activity would be prohibited. The ecocentric approach highlights that forest ecosystems possess an intrinsic value. The notion of the intrinsic value of nature is crucial for understanding that those species and ecosystems that are not currently useful to humanity must also be protected. However, the pure ecocentric approach can be difficult to implement, because the right to sustainably use natural resources cannot be denied to people. Instead, a sparing and gentle use of natural resources should be promoted.

Consequently, both approaches are important because they provide a different insight in the ethical principles behind the protection of nature.[146] It can be argued, however, that the differentiation between the anthropocentric and ecocentric approaches can be regarded as a philosophical issue deprived of any practical meaning. Actually, humanity and nature are deeply interrelated. Consequently, neither of these two approaches can be implemented without considering the other one. Indeed, both nature and human beings possess an intrinsic value. This is why only those approaches to nature conservation that promote the conservation of ecosystems and the improvement of livelihoods can be realistic. In order to ensure the balance between multiple goals of environmental governance, environmental and socio-economic issues should be integrated in the decision-making process. With the intrinsic value of nature in mind, large areas of untouched natural ecosystems should be preserved, whereas sustainable agricultural and forestry practices should be introduced on anthropized areas. Arguably, the conflict between the intrinsic value of nature and the intrinsic value of humanity is likely to arise only in a short-term perspective, as humans are likely to benefit from the conservation of nature on a long-term scale. From this point of view, the intrinsic value of nature and the role of natural resources in the existence of human civilization should be both taken into account when developing – and implementing – nature conservation strategies and policies.

IV. Drivers of Unsustainable Forest and Forest Land Exploitation and Ways to Combat Them

The briefly sketched ethical approaches to the relationship between humanity and nature deliver insights into the causes of unsustainable forest and forestland exploitation that lead to deforestation and forest degradation.

146 Ibid, p. 30.

Humans have exploited forest resources forever, as timber was used as fuel and construction material for ages. Due to the ever-growing need for new farmlands, human pressure on forests was high in certain regions since prehistorical times.[147] This resulted in the loss of forest cover in ancient civilization hotspots. During the medieval times, many communities in Europe were struggling with the scarceness of wood resources and rulers were forced to implement corresponding authoritative measures.[148] At the time, deforestation was not a serious threat on a global scale since it did not lead to the measurable reduction of forest cover across the globe. Nevertheless, major deforestation-induced disasters have already occurred in previous centuries. It is argued that the desertification of Australia and the Sahara that both had global environmental impacts were caused by human activities, such as unsustainable agricultural practices.[149]

However, the industrial revolution has led to an increased human pressure on forest ecosystems. In the last few centuries, major deforestation has occurred around the world. Recent decades have witnessed considerable forest loss in developing States. A bulk of natural forests was replaced by planted forests and monoculture plantations with reduced biodiversity. With new revelations about the ecological role of forests in mind, it could be expected that States, societies and private actors would apply sustainable forest use practices. What actually happened is that the methods and practices of forest management are only slowly improving. Despite considerable progress achieved in several parts of the

147 *Hughes, Donald/Thirgood, J. V.*, Deforestation, Erosion, and Forest Management in Ancient Greece and Rome, in: Journal of Forest History, Volume 26, Number 2, 1982, pp. 60 ff.; *Kaplan, Jed O./Krumhardt, Kristen M./Zimmermann, Niklaus*, The prehistoric and preindustrial deforestation of Europe, in: Quarterny Science Reviews, Volume 28, 2009, pp. 3016 ff.; *Neil, Roberts/Woodbridge, Jessie/Davis, Basil A. S./Kaplan, Jed O./Marquer, Laurent/Mazier, Florence/Nielese, Birgitte Anne/Trondman, Anna-Kari/Michelle, Leydet/Sugita, S./Fyfe, R. M./Gaillard, M.-J.*, Europe's lost forests: A pollen-based synthesis for the last 11,000 years, in: Scientific Reports, Volume 8, Issue 1, 2018. – Internet resource – https://www.nature.com/articles/s41598-017-18646-7.pdf [Last visit: 30.04.2023]

148 *Radkau, Joachim*, Wood and Forestry in German History: In Quest of an Environmental Approach, in: Environment and History, Volume 2, Number 1, Lammi Symposium special issue, February 1996, p. 67.

149 *Wright, David K.*, Humans as Agents in the Termination of the African Humid Period, in: Frontiers in Earth Science, Volume 5, 2017. – Internet resource – https://www.frontiersin.org/articles/10.3389/feart.2017.00004/full [Last visit: 30.04.2023]

planet (e. g., in Costa Rica[150]), unsustainable forest management is still prevalent in many regions, such as South-East Asia. In many parts of the world, a practice of establishing protected forest areas has only brought limited progress because, in many cases, the need for economic development outweighed environmental issues when selecting those forest ecosystems that should be protected.[151] For this reason, it is important to examine which factors obstacle the extension of protected forest areas and the introduction of sustainable forest management on a wider scale.

1. Underlying Causes of Deforestation and Forest Degradation

First of all, one major obstacle to the promotion of gentle forest use is the low level of environmental consciousness, resulting in a lack of regard to environmental issues in the decision-making process and the pursuit of short-term economic goals at the cost of long-term environmental degradation. In many cases, governments, forest owners and users, and other stakeholders do not have incentives to promote the protection of forests as deforestation, and forest degradation, are highly profitable in the short-term perspective for a limited number of engaged actors.[152] Despite this, adverse effects of deforestation and forest degradation can be sharply felt on both global and regional levels. The short-term economic benefits associated with unsustainable logging, and forest clearance, outweigh the adverse environmental impacts from the subjective point of view of a limited circle of forest owners and users. It should not be forgotten that, in certain cases, local inhabitants are constrained to manage forests unsustainably as the only way to sustain their livelihoods. Moreover, international trade in unsustainably harvested timber, and crops, grown on converted forest-lands is an important source of cash inflows for some developing States. Consequently, socio-economic factors and the underestimation of environmental value of forests should be regarded as major driving forces of deforestation and

---

150 *Watts, Jonathan*, Costa Rica recognised for biodiversity protection, in: The Guardian, 25 October 2010. – Internet resource – https://www.theguardian.com/environment/2010/oct/25/costa-rica-biodiversity [Last visit: 30.04.2023]

151 *Muradian, Roldan/Pellegrini, Lorenzo/Aguilar, Bernardo/Agarwal, Bina*, Payments for Ecosystem Services and the Fatal Attraction of Win-Win Solutions, in: Conservation Letters, Volume 6, Issue 4, 2013, pp. 274 ff.

152 *Humphreys, David*, Deforestation and the Crisis of Global Governance, 2006, p. 3; *Farley, Joshua/Costanza, Robert*, Payments for ecosystem services: From local to global, in: Ecological Economics, Volume 69, 2010, p. 2064.

forest degradation. Additionally, poverty and underdevelopment are other major driving forces of deforestation and forest degradation.

Economic science tends to describe the conflict between the monetary revenues derived from deforestation, and the necessity to preserve ecological functions of forests, by applying denominations of public goods, and externalities, to the ecosystem services of forests. Based on economic theory, goods can be classified according to the criteria of excludability and rivalry.[153] Excludability means that some customers can be excluded from the use of a good. At the same time, rival goods are those, whose consumption by one person, or entity, limits the ability of others to consume it.[154] According to these criteria, goods can be divided into four categories. The first category includes public goods, which are non-excludable and non-rival, for example, public streetlights.[155] On the other hand, goods, which are non-excludable, but rival, are common goods.[156] Goods, that are non-rival, but excludable, are "club goods". Excludable and rival goods are private ones.[157] The ecological services provided by forests is a typical public good. It is impossible to exclude someone from the consumption of benefits delivered by forest's ability to influence the composition of the atmosphere, the formation and distribution of precipitations, to protect soils from erosion, etc. At the same time, the consumption of forest ecosystem services by one consumer cannot limit their availability for others, unless the forest ecosystem is deforested or degraded.

From this point of view, the promotion of ecosystem services of forests through gentle forest use, as well as their depletion through deforestation and forest degradation, can be regarded as externalities. The concept of externalities refers to non-intentional side effects that arise from the production or consumption behavior of economic agents.[158] Externalities can be either negative, for example, adverse environmental impacts, or positive, such as the promotion of ecological functions of forests.[159]

Thus, ecological functions of forests are often undermined in the decision-making process. Markets fail to evaluate in monetary terms the environmental

153 *Humphreys*, Deforestation [n. 152], p. 3 f.; *Farley/Costanza* [n. 152], pp. 2064 f.

154 *Farley/Costanza* [n. 152], pp. 2064 f.

155 Ibid, p. 2065.

156 Ibid.

157 Ibid.

158 Ibid.

159 Ibid.

benefits delivered by forests. In many cases, forest owners and users are not rewarded for the promotion of these ecological functions of forests. They seldomly bear costs associated with the depletion of forest resources.[160]

2. *Ways to Address the Underlying Causes of Deforestation and Forest Degradation*

Different strategies are proposed to address the failure of markets in evaluating ecological functions of forest ecosystems. The first strategy consists of the application of external regulations of forest-related markets, such as trade restrictions, market bans and the limitation of those industrial activities that cause a damage or threat to forest ecosystems. While the administrative regulation of markets does not deliver any economic benefits, this strategy permits to consider both the benefits of the protection of forests for the public and the intrinsic value of forests. However, the implementation of these external mechanisms of forest industry and market regulation has been a challenging process, because it requires the introduction of strict compliance and monitoring measures.[161] Taking into account the globally relevant adverse impacts of deforestation and forest degradation, the bans and restrictions have to be applied to the international trade in forest and agricultural products as well.[162] However, the development, and implementation, of these restrictive measures is particularly challenging.[163] The reason for this is the particular nature of international legal mechanisms. Firstly, the consensus of States is needed to adopt any new treaty or regulation. The consensus on the introduction of international market restrictions and bans has been difficult to achieve.[164] States oppose themselves to any international regulations that would restrict their trade in forest goods and agricultural produce if the negotiated treaty does not deliver any benefits for them. Such benefits can be either monetary or non-monetary, such as the improvement of their political status. However, possible benefits rarely outweigh the losses associated with trade bans and restrictions.

One of the shortcomings of market bans, and restrictions, is that they do not promote the socio-economic development in forest-rich States. They just put

---

160 Ibid.

161 *Gale, Fred P.*, The tropical timber trade regime, 1998, pp. 120 ff.; *Schulte zu Sodingen* [n. 107], pp. 300 f.

162 *Schulte zu Sodingen* [n. 107], pp. 300 f.

163 Ibid.

164 Ibid.

restrictions on industrial and commercial activities without proposing any alternative development strategies.

Environmental issues have to be promoted on the demand part of the market as well. To incentivize and help diffusing knowledge about those products whose cultivation, harvesting or processing is associated with the risk of deforestation or forest degradation, voluntary environmental certifications (ecolabels) have been created.[165] Such environmental certifications can help customers to consider environmental issues in their own market behavior. Ecolabelling strategies can be reinforced through voluntary environmental commitments from manufacturers who can adopt a code of conduct known as the Corporate Social Responsibility (CSR).[166] Such strategies have been recognized as a source of added value for the enterprises, thus creating a competitive advantage for them as well. However, considering that investors are solely obliged to increase the corporate revenue for their shareholders, the environmental issues can be underestimated.[167] Voluntary organizations, which promote and fulfill the ecolabelling of forest products, such as the Forest Stewardship Council, have been criticized for the lack of transparency, for disregarding environmental issues and for the misuse of ecolabelling for the promotion of certain manufacturers.[168] Finally, despite the CSR and ecolabelling have been applied for decades,[169] deforestation rates remain unacceptably high.

To address the failure of markets in evaluating multiple forest functions, a viable strategy can be the development of market-inherent solutions. These mechanisms can create economic incentives for forest-rich States, forest owners and users to promote the protection of forests. They can also oblige them to bear

165 *Savilaasko, Sini/Cerutti, Paolo Omar/Zumaeta, Javier G. Montoya/Ruslandi/Mendoula, Edouard E./Tsanga, Raphael,* Timber Certification as a Catalyst for Change in Forest Governance in Cameroon, Indonesia, and Peru, in: International Journal of Biodiversity Science, Ecosystem Services & Management, Volume 13, Issue 1, 2017, pp. 117 ff.

166 *Campbell, John L.*, Why would corporations behave in socially responsible ways? An institutional theory of corporate social responsibility, in: The Academy of Management Review, Volume 32, Number 3, 2007, pp. 946 f.

167 *Humphreys, David*, From Corporate Social Responsibility to the Democratic Regulation of Transnational Corporations, in: The international journal of environmental, cultural, economic and social sustainability, Volume 5, Number 4, 2009, pp. 207 f.

168 FSC Watch, The 10 worst things about the Forest Stewardship Council – Internet resource – https://fsc-watch.com/2014/06/01/the-10-worst-things-about-the-forest-stewardship-council/ [Last visit: 30.04.2023]

169 Forest Stewardship Council was founded in 1993.

the costs of the adverse impacts of deforestation.[170] This process is also called the "internalization of externalities".[171] It can be achieved, among other things, through environmental taxation, and imposing financial penalties on companies, whose activities cause deforestation and forest degradation. While representing an external regulation, environmental taxation can be part of the general market-based approach to the protection of forests.[172] Taxation can tackle the market's failure to take into consideration environmental issues by incorporating adverse environmental impacts into the price. An environmental tax can incentivize companies, and consumers, to up-scale their efforts dedicated to the reduction of negative environmental impacts.[173] However, taxation is only applicable to major companies engaged in forest exploitation. It can only worsen the situation for farmers that are forced to clear forests because of the poor economic conditions in their States. Moreover, the introduction of environmental taxation is subject to the national legislation in forest-rich States.[174] However, the respective governments will likely oppose the adoption of such regulations, unless they will be further incentivized to promote the protection of forests. Under current conditions, they would not be ready to introduce environmental taxation because it would undermine their development opportunities.[175]

Consequently, these market-inherent mechanisms are desirable because they would integrate both development issues and environmental goals and create economic incentives for the protection of forests. There are different theoretical approaches regarding how this can be achieved. After the limits of nature conservation through the designation of protected areas were discovered, conservationists have proposed the concept of Integrated Conservation and Development Projects (ICDP).[176] The ICDP is perceived as an *indirect* approach to the promotion of the environmental conservation in developing States because

---

170 *Farley, Joshua/Costanza, Robert*, Payments for ecosystem services: From local to global, in: Ecological Economics 69, 2010, p. 2064.

171 Ibid.

172 *Braathen, Nil Axel/Greene, James*, Environmental Taxation: A Guide for Policy Makers, OECD, September 2011. – Internet resource – https://www.oecd.org/env/tools-evaluation/48164926.pdf [Last visit: 30.04.2023]

173 Ibid.

174 *Schulte zu Sodingen* [n. 107], pp. 300 ff.

175 Ibid.

176 *Muradian, Roldan/Pellegrini, Lorenzo/Aguilar, Bernardo/Agarwal, Bina*, Payments for Ecosystem Services and the Fatal Attraction of Win-Win Solutions, in: Conservation Letters, Volume 6, Issue 4, 2013, p. 274.

the creation of incentives for the conservation of forests is not directly related to the objectives of environmental policies. The ICDP strategy includes three main aspects. The first aspect consists of compensating local communities, and other actors, whose rights to use forest resources are restricted by forest conservation measures.[177] The second aspect foresees the creation of alternatives for economic development that would permit to avoid deforestation and forest degradation, including the agricultural intensification and the promotion of alternative livelihoods.[178] The third aspect consists of enhancing the monetary value of ecological functions of forests, as it is perceived by the economic actors, by developing previously unexplored markets and ecotourism.[179] ICDP was largely promoted after the Rio summit in 1992. However, on the eve of the Earth summit in Johannesburg in 2012, a number of scholars, and conservationists, argued that this strategy had only limited success for a number of reasons.[180] The locals have often perceived the opportunities created by ICDP as an additional source of income rather than a substitute for unsustainable forms of forest exploitation.[181] In many cases, the legislative, economical, technical resources as well as the technical infrastructure that are needed for the successful implementation of projects are not available in developing States.[182] In several cases, the needs, and aspirations, of local communities were not considered. The projects imposed severe restrictions on the rights of forest-dependent people to use forest resources, but the locals did not receive an adequate compensation.[183]

Consequently, environmentalists, and policy-makers, have developed new approaches to the promotion of the protection of forests in developing States, which foresee the application of direct, result-based, payment schemes for nature conservation.[184] The consumers of ecosystem services are supposed to provide

---

177 *CIFOR*, Integrated conservation and development: an overview, 2007. – Internet resource – http://www.cifor.org/conservation/publications/pdf_files/cambodia/Integrated%20conservation%20and%20development.pdf [Last visit: 30.04.2023].

178 Ibid.

179 Ibid.

180 *Muradian/Pellegrini/Aguilar/Agarwal* [n. 176], p. 274.

181 *Ferraro, Paul/Kiss, Agnes*, Direct payments to conserve biodiversity, in: Science, Volume 298, 2002, p. 1718.

182 Ibid.

183 *Wells, Michael/Guggenheim, Scott/Khan, Asmeen/Wardojo, Wahjudi/Jepson, Paul*, Investing in Biodiversity: A Review of Indonesia's Integrated Conservation and Development Projects, East Asia Region, World Bank, Washington D. C., 1998, p. 1.

184 *Ferraro/Kiss* [n. 181], p. 1718.

the necessary financial resources. Those strategies predominantly take the form of Payments for Ecosystem Services (PES). They are based on attempts to express the value of ecological functions of forest ecosystems in economic terms. PES can be defined as a voluntary, conditional agreement, between at least one provider, and one consumer, over a well-defined ecosystem service.[185] In relation to forests, PES presupposes that forest owners, and users, will receive payments for their efforts dedicated to the preservation of forests and the promotion of afforestation. The payments should come from communities, municipalities and States that benefit from the ecosystem services of forests. PES can help addressing the failure of markets in considering the ecological functions of forests. In addition, they promote the poverty alleviation in developing States.

In contradiction to the common assumption of PES as a "market-based" approach, pure market-based projects are rarely implemented.[186] "Market" can be defined as "an actual or nominal place where forces of demand and supply operate, and where buyers and sellers interact (directly or through intermediaries) to trade goods, services, or contracts or instruments, for money or barter".[187] The market is characterized by high conditionality, high level of commoditization and the voluntary behavior of market actors.[188] Ecosystem services usually do not fulfill the criteria of commoditization because most of them, by nature, are either public goods or common pool resources.[189] In addition, the level of conditionality and voluntariness is relatively low because potential "buyers" cannot cease consuming ecosystem services and the "sellers" cannot interrupt their supply unless their activity destroys, or damages, a forest ecosystem.[190]

Consequently, PES projects rarely operate through markets with competitive forces. These projects are based on bilateral agreements, which do not leave a lot of place for actors to compete with each other over demand, supply and price.[191]

---

185 *Wunder, Sven,* The efficiency of payments for environmental services in tropical conservation, in: Conservation Biology, Volume 21, Issue 1, 2007, pp. 48.

186 *Muradian, Ronald/Rival, Laura,* Between markets and hierarchies: the challenge of governing ecosystem services, in: Ecosystem Services, Volume 1 2012, p. 94; *Muradian/Pellegrini/Aguilar/Agarwal* [n. 176], p. 276; *Wunder, Sven*, When payments for the ecosystem services will work for conservation, in: Conservation Letters, Volume 6, Issue 4, 2013, p. 231.

187 See definition of the "market" at http://www.businessdictionary.com/definition/market.html [Last visit: 30.04.2023]

188 *Muradian/Rival* [n. 186], p. 95; *Muradian/Pellegrini/Aguilar/Agarwal* [n. 176], p. 276.

189 *Muradian/Pellegrini/Aguilar/Agarwal* [n. 176], p. 276.

190 Ibid.

191 *Wunder* [n. 185], pp. 48 ff.

In reality, competition usually occurs between the PES project and a technical solution, which could partially substitute ecosystem services provided by forests. For example, it can be evaluated, to which extent the ability of forests to maintain and enhance freshwater resources may be substituted by the construction of artificial water filtration systems. In this case, PES would fail to consider multiple functions of forest ecosystems and, furthermore, when artificial water filtration proves to be more cost-effective, forest owners and users will lose the incentives to promote forest conservation.

One should acknowledge that, in certain cases, the PES strategies can be applied as pure market mechanisms. For example, pure market solutions are applicable to those carbon emissions trading schemes that represent a uniform, globally valued, service with a high level of market integration.[192] In any case, PES can be regarded as a market-friendly solution as it permits to apply an economic criterion for tradable goods in relation to the environmental functions of forests, which previously were perceived as mere public goods.

Another common misunderstanding is that an independent monetary evaluation of ecological functions of forests always precludes the implementation of PES. In reality, the price for ecosystem services is often determined bilaterally by a "seller" and a "buyer". The monetary evaluation can be based either on monetary, or non-monetary, criteria, such as the improvement of local livelihoods through the consumption of ecosystem services.[193] Consequently, the PES approach often presupposes the subjective monetary evaluation of ecosystem services, which results in the lack of consideration of multiple ecosystem services of forests and of their intrinsic value. If the "buyer" is ready to pay only a smaller price and the "seller" of ecosystem services is ready to promote the protection of forests only upon condition that a higher monetary compensation be provided to him, a PES-agreement will hardly come into being. The lack of attention to multiple functions of forests and the lacking will to provide a fair compensation constitutes one major weakness of PES. Hereafter, other critical aspects will be analyzed.

3. *Negative Aspects of the PES Approach*

Despite all the benefits that PES can deliver for the protection of forests, this approach can be criticized from different points of view.

192 Ibid.

193 Ibid.

One problem is whether representatives of local communities that also benefit from ecological services of forests can be considered the only beneficiaries of PES.[194] They are likely to be both "sellers", and "consumers", but not "buyers" of ecological services. That way, local communities and owners of small and middle-sized farms and forestry industries can be incentivized to promote the conservation of forests. In such a case, large companies that also bear a large burden of responsibility for deforestation and forest degradation, may receive only a small share of payments.[195] Against this backdrop, those PES schemes that create incentives for forest conservation for local communities, local governments, and large companies should be developed. However, a stronger involvement of large companies in the PES schemes can lead to the undermining of the interests of ILCs, as businesses have enhanced opportunities to uphold their interests while implementing relevant projects.

It is argued that PES schemes are less susceptible to politicization, as the agreements between "sellers" and "buyers" usually include dispute resolution clauses.[196] However, what actually happens is that both the competent legislative and executive authorities and those groups that are particularly empowered to uphold their interests can play a major role in the development and implementation of PES. For this reason, the interplay of political forces can influence the modalities of PES, shaping the design of payment schemes and designating the main beneficiaries of payments.[197] For example, national PES schemes in Costa Rica and Mexico were often aimed at strengthening the historical stewardship over forests instead of promoting environmental additionality.[198] The projects were often limited to areas with low deforestation risk and with low opportunity costs. In the case of Costa Rica, large landowners were often disproportionately represented among the PES participants.[199] It is argued that the development

---

194 See *Jack, B. Kelsey/Kousky, Carolyn/Sims, Katharine R. E.*, Designing payments for ecosystem services: Lessons from previous experience with incentive-based mechanisms, in: Proceedings of the National Academy of Sciences of the United States of America, Volume 105, Number 28, 2008, pp. 9465 ff.

195 Ibid, p. 9468.

196 *Muradian/Pellegrini/Aguilar/Agarwal* [n. 176], p. 276.

197 Ibid.

198 *Sierra, Rodrigo/Russman, Eric*, On the Efficiency of Environmental Service Payments: A Forest Conservation Assessment in the Osa Peninsula, in: Ecological Economics, Volume 59, Issue 1, 2006, pp. 132 ff.

199 *Sanchez-Azofeifa, Arturo G./Pfaff, Alexander/Robalino, Juan Andreas/Boomhower, Judson P.*, Costa Rica's Payment for Environmental Services Program: Intention, Implementation, and Impact., in: Conservation Biology, Volume 21, Issue 5, 2007, pp. 1166,

of PES schemes was guided by the intention to gain the public support for the new initiative.[200] Similarly, the national PES schemes in Mexico envisage small farmers and indigenous communities as the main beneficiaries of payments, thus favoring historical stewardship over forest resources.[201] Consequently, the cost-effectiveness of PES policies and the environmental additionality of the relevant projects depend on the project design and on the promotion of interests of all relevant groups. If these requirements are not met, the PES approach will not represent the most cost-effective policy tool for achieving environmental goals.[202]

On the other hand, there is a risk of violation and abuse of human rights while implementing PES. In order to avoid it, it is necessary to be mindful of the land tenure issues and to the user rights of local communities. Otherwise, the PES approach can result in the aggravation of poverty instead of promoting local livelihoods.

Another problem with PES policies is that the projects in question can under certain circumstances discourage forest owners and users to pursue environmental protection instead of motivating them. The project outcome depends on the negotiated price for ecosystem services, the beneficiaries of payments as determined by the treaty, the local understanding of fairness and equity, and on political, socio-economical, and cultural factors.[203] There is a risk that PES will undermine the intrinsic motivation of forest owners and users to promote the protection of forests. It is possible because in certain cases the conservation of forests can be understood as a socially relevant mission, while the commitment

---

1168 ff.; *Porras, Inna*, Fair and Green? Social Impacts of Payments for Environmental Services in Costa Rica, 2010, pp. 11 ff.

200 *Muradian/Pellegrini/Aguilar/Agarwal* [n. 176], p. 276; *Sanchez-Azofeifa/Pfaff/Robalino/Boomhower* [n. 199], p. 1168.

201 *Muñoz-Piña, Carlos/Guevara, Alejandro/Torres, Juan Manuel/Bran, Josephina*, Paying for the Hydrological Services of Mexico's Forests: Analysis, Negotiations and Results, in: Ecological Economics, Volume 65, Issue 4, 2008, pp. 725 ff.; *McAfee, Kathleen/Shapiro-Garza, Elizabeth*, Payments for Ecosystem Services in Mexico: Nature, Neoliberalism, Social Movements and the State, in: Annals of the Association of American Geographers, Volume 100, 2010, pp. 580.

202 *Sanchez-Azofeifa/Pfaff/Robalino/Boomhower* [n. 199], pp. 1168 ff.

203 *Gneezy, Uri/Rustichini, Aldo*, Pay Enough or Don't Pay at All, in: The Quarterly Journal of Economic, Volume 115, Number 3, 2000, pp. 793 ff.; *Fehr, Ernst/Falk, Armin*, Psychological Foundations of Incentives, in: European Economic Review, Volume 46, Issue 4–5, 2002, pp. 689 ff.

to serving the community inspires an environmentally friendly behavior. Such situation is possible in the areas where strong cultural traditions for social cooperation and nature protection exist. The evidence suggests that the application of PES in those areas can contribute to the erosion of intrinsic motivation and of other traditional institutions.[204] This can lead to a negative outcome for the promotion of forest conservation. Therefore, the relevant issues should be taken into account while developing and implementing PES-schemes.

The concept of PES foresees that the price for environmental functions of forests should exceed the potential revenue from environmentally harmful activities in order to ensure the successful implementation of projects. However, this can be difficult to achieve, especially from the point of view of growing prices for natural commodities.[205] However, the low price for ecosystem services can undermine the effectiveness of projects.

It is important to note that PES policies often embrace only one of the multiple ecological functions of forests without considering the complexity of forest ecosystems and the interdependence of their components. This can result in the promotion of one particular characteristic of forest ecosystems on the cost of others. For example, as it was already shown above, while promoting forests' ability to capture carbon dioxide, forest owners and users tend to underestimate the importance of forest biodiversity and of the near-natural and natural forests.[206]

The problem is also that the monetary evaluation of ecosystem services is a complicated process, which hardly renders possible an adequate estimation of the price of services provided by forests. It is hardly possible to express the ecological role of forests in monetary terms. In many cases, those entities and actors that apply PES deal with the evaluation of things, whose value cannot be expressed in monetary terms. This is why the PES approach fails to consider the intrinsic value of nature.

Consequently, it is recommended that alternative approaches to the protection of forests be developed. These approaches should be free from the above-mentioned deficiencies of PES policies. First of all, the mechanisms for the protection of forests should always take into account the rights and interests of

204 *Narloch, Ulf/Pascual, Unai/Drucker, Adam G.*, Collective Action Dynamics Under External Rewards: Experimental Insights from Andean Farming Communities, in: World Development, Volume 40, Issue 10, 2012, pp. 2098 ff.

205 *Muradian/Pellegrini/Aguilar/Agarwal* [n. 176], pp. 274 ff.

206 Ibid.

forest-dependent people. Secondly, they should be based on understanding the complexity and integrity of forest ecosystems, while maintaining and promoting the outstanding role of natural forests. These mechanisms should be aimed at ensuring that forest ecosystems will be protected in their integrity, and the anthropogenic pressure on forests ought to be gradually reduced. Considering that poverty and underdevelopment are among the underlying causes of deforestation and forest degradation, any effective approach to the protection of forests should foresee the compensation of forest owners and users for their efforts dedicated to the protection of forests. The firsthand owners and users of forest resources, particularly forest-dependent people, should benefit directly from financial transfers. Poverty should be alleviated, and local livelihoods should be promoted. In an ideal case, the amount of payments should be determined by considering the intrinsic value of nature and the development needs of local communities. The transfer of payments should be subject to the submission of a detailed analysis containing properly monitored, evaluated and reported achievements in the promotion of forest conservation and re/afforestation. The multiple benefits of forests should serve as an incentive for major stakeholders to join relevant projects. Decision-makers should also promote the integration of monetary and non-monetary incentives for the protection of forests in the relevant policies that are aimed at promoting the conservation of forests.

However, as long as no alternative approaches are developed, the implementation of existing PES schemes can be the most realistic approach to the promotion of the protection of forests. Thus, it is important to analyze to which extent PES-based approaches can promote the protection of forests and facilitate the transition to sustainable use of forest resources. It should be examined to which extent the developers of these mechanisms succeed in remedying the detected deficiencies of PES schemes by promoting the interests of forest-dependent people, the protection of natural ecosystems, and the intrinsic value of forests and their biodiversity.

C. Forest Protection as an International Legal Issue

It appears at first glance that the protection of forests that are subject to the national jurisdiction of a State, on whose sovereign territory they are situated, should be regarded as an exclusive internal affair of a State concerned. However, a profound analysis demonstrates that the problem of deforestation and forest degradation has evolved into a major international legal issue.

The adverse environmental impacts of deforestation are immediately visible worldwide. As stated above, deforestation can cause considerable and abrupt climate change thousands of kilometers away from the area where it takes place. At the same time, local adverse impacts of deforestation also have indirect adverse impacts on a global scale. The growing frequency and destruction force of natural disasters, soil erosion, floods, and droughts lead to the further impoverishment of local communities, to the loss of development chances, and to the deterioration of socio-economic situation in affected areas.[207] As a result, States affected by large-scale deforestation risk to turn into a source of long-term instability in international relations requiring enhanced development assistance and humanitarian aid. It is important to notice that the adverse impacts of deforestation can cause large-scale emigration from affected areas.

It should also be mentioned that the driving forces of deforestation have acquired an extensive international dimension. The demand for forest timber and non-timber products on international markets drives large-scale forest logging in tropics, even though a bulk of timber consumed in developed States is actually harvested in their domestic forests.[208] Forests in developed States are supposed to be managed sustainably, but what actually happens is that also these forests suffer from the major loss of vitality and regeneration capacity.[209] The problem is further aggravated by the fact that commercial timber species are often particularly important for the sustainable functioning of forest ecosystems.[210] From this point of view, it is difficult to determine the actual contribution of the international timber trade to the global forest loss.

The growing demand on the international markets for oil palm, cacao, soybeans, coconuts and other tropical crops cultivated on converted forestlands represent a much greater problem contributing to the worsening of deforestation crisis.[211] Underdevelopment issues play a major role in this process because the

---

207 *Barraclough, Solon L./Ghimire, Kléber*, The Social Dynamics of Deforestation in the Developing Countries: Principle Issues and Research Priorities, UNRISD, United Nations, 1990, p. 9.

208 See *Laurance, William F.*, Reflections on the Tropical Deforestation Crisis, in: Biological Conservation, Volume 91, Issues 2–3, 1999, pp. 110 ff.

209 *Schulte zu Sodingen* [n. 107], pp. 92 f.

210 *Humphreys*, Deforestation [n. 152], pp. 93 f.

211 *Schulte zu Sodingen* [n. 107], pp. 93 f.; *Maclean, Ruth*, Chocolate Industry Drives Rainforest Disaster in Ivory Coast, in: Guardian, 13 September 2017. – Internet resource – https://www.theguardian.com/environment/2017/sep/13/chocolate-industry-drives-rainforest-disaster-in-ivory-coast [Last visit: 30.04.2023]; *Johnson, Alaine*, Why Is Palm Oil Such a Problem Now?, in: Sentient Media, 30 April 2020. – Internet resource – https://sentientmedia.org/why-is-palm-oil-such-a-problem-now/ [Last visit: 30.04.2023]; *Claire, Asher*, Brazil Soy Trade Linked to Widespread Deforestation,

lack of modern sustainable agricultural and forestry practices combined with the absence of industrial and technological capacities for the production of technological goods leads to increased deforestation and forest land conversion in forest-rich States. The export of timber and tropical crops cultivated on converted forestlands is the main source of currency for these States.[212] Outdated agricultural technologies and practices cause further forestland conversions. Economic globalization also plays a significant role, as international trade and market liberalization promotes investments in mining industries and agricultural monoculture plantations in forest-rich States.[213] The lack of environmental safeguards and standards combined with the lack of incentives for the conservation of forests in forest-rich States permits investors to undermine the ecological functions of forests.[214] Consequently, there is a need for international mechanisms of the protection of forests that would promote the socio-economic development in forest-rich States and encourage them to change the modalities and practices of forest management. The whole international community, including developed States, can enjoy multiple benefits from maintaining and promoting multiple functions of forests in developing States. The same applies to the improvement of the conditions of boreal and temperate forests in the sovereign territory of developed States. Developed States possess financial and technological resources that are needed for promoting the conservation of global forest ecosystems. For instance, they can contribute to the promotion of the protection of forests through the provision of financial resources to those States and communities that promote the conservation of natural forests, the regeneration of degraded forests, and the afforestation of converted forestlands. The financial and technological support can be provided either as development assistance or as a part of PES schemes. In order to ensure the effectiveness, consistency, and transparency of relevant financial mechanisms, an international legal framework is required in order to regulate their implementation. International law as the legal order governing relations between States should provide such legal framework.

Notwithstanding this, international legal policies can create new threats for forests and cause further damage to them. For example, certain climate change

---

Carbon Emissions, in: Mongabay, 3 April 2019. – Internet resource – https://news.mongabay.com/2019/04/brazil-soy-trade-linked-to-widespread-deforestation-carbon-emissions/ [Last visit: 30.04.2023]

212 *Eikermann* [n. 127], pp. 20. ff.

213 *Laurance* [n. 208] p. 113.

214 Ibid.

mitigation policies are likely to aggravate the problem of deforestation.[215] The promotion of biomass conversion of coal plants aimed at replacing of fossil fuels with wooden pallets is likely to cause considerable harm to forest ecosystems. This issue can be mainly regulated by international conventions and agreements. Therefore, the promotion of synergies between the conservation and sustainable utilization of forests and climate change mitigation and adaptation is another aspect of international forest-related policies.[216]

The promotion of the conservation of global forest ecosystems is likely to require such changes of the modalities of functioning of the global economy, which will ensure the wider integration of environmental issues in the decision-making process. It is important, among other things, to promote the monetary evaluation of ecosystem services of forests. Under current conditions, States and multinational companies promoting the conservation and sustainable use of forests while refusing on short-term profits associated with deforestation can lose their comparative advantages on the global market.[217] This means that, even if a forest-rich State has enough resources to promote sustainable forest use, it will not have enough economic incentives to do so. Consequently, mechanisms should be developed that foresee the monetary and non-monetary benefits for governments, forest owners and users to protect forests. Regardless of how the global economy and development assistance mechanisms will be modified, it is important to integrate environmental issues in the international legal framework guiding the international aspects of forest use.

---

215 *United States Security and Exchange Commission*, Climate Consequences of Current Carbon Accounting Practices for Bioenergy, Case Study on Wood-Pellet Manufacturer Enviva, 2019. – Internet resource – https://www.sec.gov/rules/petitions/2019/ptn4-741-exa.pdf [Last visit: 30.04.2023]; *Sandbag*, Playing with Fire: an assessment of company plans to burn biomass in EU coal power stations, 2019. – Internet resource – https://ember-climate.org/wp-content/uploads/2019/12/2019-SB-Biomass-report-1.7b_DIGI.pdf [Last visit: 30.04.2023]; *Ambrose, Jillian*, Converting coal plants to biomass could fuel climate crisis, scientists warn, in: The Guardian, 2019. – Internet resource – https://www.theguardian.com/environment/2019/dec/16/converting-coal-plants-to-biomass-could-fuel-climate-crisis-scientists-warn?CMP=share_btn_fb&fbclid=IwAR2o7PAn_db_uObZkicQqKleJ-XEpdYmrxysRAmDMlm1UK-WBrF_jkpN51s [Last visit: 30.04.2023]; *Haberl, Helmut/Sprinz, Detlef/Bonazountas, Marc/Cocco, Pierluigi/Desaubies, Yves/Henze, Mogens et al.*, Correcting a fundamental error in greenhouse gas accounting related to bioenergy, in: Energy Policy, Volume 45, 2012, pp. 18 ff.

216 See Subsection C.II of Chapter 4.

217 *Humphreys*, Deforestation [n. 152], pp. 120 f.

To sum up, the globally relevant adverse impacts of deforestation and forest degradation, the international dimension of driving forces of deforestation, the scarcity of financial and technological resources in forest-rich States, and the lacking willingness to promote the forest conservation and gentle use of forest resources have contributed to the establishment of deforestation crisis as an international legal issue.

D. International Controversies over the Protection of Forests

A number of issues impede the development of the international legal regime for forests. One of the major controversies is predetermined by the deferring perception of the problem by developed and developing States. Developed States often oppose themselves to any legally binding commitments, which would oblige them to restrict the exploitation of forests on their own territory. They claim that forests under their jurisdiction are already sustainably managed with due regard to the environmental issues, and that no further legislative measures are necessary.[218] However, this statement is erroneous, as forests in developed States also show signs of extensive degradation. Natural temperate forests have become extremely rare, and managed forests are not able to fulfill all the multiple forest functions.[219] Moreover, developed States expect that developing States will undergo legally binding commitments to conserve natural forests and to reduce drastically the industrial exploitation of forests on their territories. From their side, developing States argue that such one-side restrictions constitute an attempt to introduce disguised international trade restrictions in order to promote the national forestry industries of developed States. Developing States are reluctant to any attempts to extend the concept of common heritage of mankind[220] on the forest ecosystems under their jurisdiction and have constantly reasserted their sovereign rights to exploit their forestry resources for the purposes of socio-economic development.[221] Against this backdrop, they expect that developed States will impose the same restrictions on their domestic forest industries and undergo reciprocal legally binding obligations to provide extensive financial

218 *Schulte zu Sodingen* [n. 107], pp. 181 f.

219 *Halkka, Antti/Lappalainen, Iiris,* Insight into Europe's Forest Protection, WWF report, 2001, pp. 10 f.

220 See Section E of Chapter 6.

221 *Krohn, Susan Nicole,* Die Bewahrung tropischer Regenwälder durch völkerrechtliche Kooperationsmechanismen, 2001, pp. 92 ff.

assistance to developing States in order to compensate the inevitable revenue losses due to forest use restrictions.[222] For their part, the developed States oppose themselves to such legally binding obligations to provide financial assistance from their public funds and are ready to undergo uniquely voluntarily political commitments to support developing States, as far as possible and as appropriate.[223] In many cases, financial support provided to forest-rich States is only sufficient to cover "agreed full incremental costs", which means the costs directly related to the implementation of forest conservation projects.[224] This means that the provided financial resources will not cover the costs related to the lost development opportunities.[225] Supposedly, the interests of developing and developed States can be reconciled through mutually beneficial mechanisms, such as PES. Depending on the price for ecological services as determined by a bilateral agreement, these mechanisms can deliver financial and technological resources for the conservation of forests and provide further development assistance for the forest-rich States. Currently, forests' ability to sink and store carbon dioxide and to provide genetic resources for the development of biodiversity-based products belong to those ecological services of forests that attract the attention of the international community. Developed States are interested in the preservation and enhancement of the global forests' carbon stocks. Additionally, the pharmaceutical cosmetics and food industries, and other branches often largely depend on the access to biodiversity resources originating from the tropical forests of developing States. As it was shown above, recent scientific research delivered evidence that forests play a major role in global and regional precipitation patterns. The stable supply of abandoned fresh water resources is an existential issue for humanity. However, the relevant role of forests has not yet fully entered the agenda of the international forest-related negotiations.

---

222 *Birdsall, Nancy/Kuczynski, Pedro Pablo*, Climate Talks: Rich Countries Should Pay to Keep Tropical Forests Standing, in: The Guardian, 2 December 2015. – Internet resource – https://www.theguardian.com/global-development/2015/dec/02/climate-talks-rich-countries-should-pay-to-keep-tropical-forests-standing [Last visit: 30.04.2023]

223 *Humphreys* [n. 152], pp. 150 f.

224 UNFCCC, Article 4.3.

225 *Verheyen, Roda Leiden*, Climate Change Damage and International Law, 2005, pp. 169 ff., *Tladi, Dire*, Sustainable Development in International Law: An Analysis of Key Enviro-Economic Instruments, 2007, pp. 227 ff.; *Zahar, Alexander*, Climate Change Finance and International Law, 2016, pp. 27 f.

Consequently, the ecosystem services of forests related to their ability to sink and store carbon dioxide and to their role as natural habitats are now shifting into the focus of international forest-related talks. However, attempts to tackle the global deforestation crisis with the help of PES schemes that consider only these two forest functions often create threats for local forest-dependent communities and for forest biodiversity, as they tend to undermine other ecosystem services of forests. Another problem is that the price set for stored carbon units and for the access to genetic resources can be too low in order to incite forest owners and users to promote the conservation of forests.

E. Conclusions

Despite decades of on-going deforestation and forest degradation, forest ecosystems still occupy a major part of the land surface. However, the global forest area continues to alarmingly decline even though deforestation rates have reduced in the decade 2010–2020. Additionally, the newest data suggests that these rates are rising again. The destruction of natural forests is the most worrying issue. Forest ecosystems fulfill multiple and diverse functions, all of which are crucial for the preservation of a safe and healthy environment, for the promotion of sustainable development, and for the well-being of humanity. Some of the forests' functions are already widely recognized, including their ability to sink and store CO_2, and the role of forests as biodiversity pools. Other functions, such as the protection of fertile soils and the role of forests in the formation and distribution of atmospheric precipitations are already discussed in scientific papers. However, they have not yet found sufficient recognition on the agenda of international forest-related negotiations.

Forests can be protected either from the point of view of their intrinsic value or for the purpose of the promotion of their multiple functions for the benefit of humanity. The most effective approaches to forest conservation should combine both ecocentric and anthropocentric dimensions.

The problem of deforestation has acquired a significant international dimension. The underlying causes of deforestation are either directly or indirectly related to the international trade, as well as to the underdevelopment and poverty issues. Consequently, potential remedies to the problem of deforestation ought to be delivered through international cooperation. The cooperation between States is necessary in order to promote the conservation of forests and to promote sustainable development. International legal regulations are necessary in order to promote the conservation of forests in a cooperative spirit, and to ensure the coordinated implementation of forest conservation policies in developing and developed States. Against this backdrop, there is a need for the international legal regime for forests.

However, the development of this legal regime is hindered by the opposing and conflicting perception of deforestation by different governments. For this reason, the problem of deforestation deeply incorporates both inter- and intra-generational equity dimensions. The next chapters will deal with the progress achieved in the development of the international legal regime for forests and the attempts to overcome the existing obstacles on the way to a holistic protection of forests.

Chapter Two International Legal Regime for Forests

A. Introduction

Forest-related issues first attracted general public attention in the early 60s. Since then, the issues related to the conservation and sustainable use of forests has entered the agenda of international negotiations. However, the reaching of a consensus on forest issues has been particularly difficult, as the trade in forest goods and agricultural produce continues to play a significant role in national economies of forest-rich States. Many decades passed before the first non-legally binding forest-specific document, the Rio Forest Principles, was adopted. Current Chapter has at its focus the development of the international legal regime for forests. Firstly, the developments in the period between the Stockholm Conference in 1972 and the Rio Summit in 1992 will be examined. Then, the forest-relevant outcome of the Rio Summit will be briefly considered. Finally, the post-Rio developments will be analyzed.

B. Stockholm Conference and Forests

The first international conference on environmental issues, that took place in Stockholm in 1972, had a major impact on the formation and further development of international environmental law. The Stockholm conference outlined the important environmental characteristics of forests, which determine their value as the largest, highly diverse, and complex terrestrial ecosystems.[226] It also highlighted the need for sound land and forest use policies, outlined the necessity for the continuous monitoring of global forest conditions, and promoted the introduction of forest management planning.[227] The main documents adopted in Stockholm have proven to be very important for the further development

---

226 *UNEP*, Global Environmental Outlook 3, Past, present and future perspectives, 2002, p. 90.

227 United Nations A/CONF.48/14/Rev.1, Report of the United Nations Conference on the human environment. Chapter 2, Action Plan for the human environment, Stockholm, 15–16 June 1972, Recommendations 24–28. – Internet resource – http://www.un-documents.net/aconf48-14r1.pdf [Last visit: 30.04.2023]

of the international legal regime for forests, even though they do not explicitly mention forests.

The Stockholm Declaration, the most important document to come out of the conference, contains general principles of international environmental law, which are relevant for the protection of forests and the exploitation of forest resources. In its Preamble, the Stockholm Declaration underlines the crucial importance both of the natural and artificial aspects of the environment for the well-being of people and the realization of basic human rights, including our right to life, liberty, and security.[228] Consequently, Principle 1 of the Declaration stipulates the human right to live in an "environment of a quality that permits a life of dignity and well-being".[229] Considering the numerous ecological and socio-economical functions of forests, it can be affirmed that this environment can be achieved only if the conservation of forests and the sustainable use of their resources is assured. In addition, Principle 1 also affirms that people bear a "solemn responsibility" to conserve and improve the environment for the benefit of present and future generations.[230] It is submitted that this responsibility cannot be fulfilled unless forest ecosystems are conserved and sustainably managed. Principle 2 affirms the need to preserve the natural resources of the Earth. It lists such components of natural environment as soil, water, air, flora, and fauna, and underlines the necessity to safeguard the representative samples of natural ecosystems.[231] It can be seen as problematic that the principle in question differentiates between the conservation of components of natural environment and the preservation of natural ecosystems. In this way, it ignores the fact that natural resources should be preserved *together* with natural ecosystems. The preservation of only "representative samples" of natural ecosystems cannot be sufficient. For this reason, it would be reasonable to put the emphasis on the protection of natural resources in sustainably managed or intact natural ecosystems. Such an approach to the preservation of components of natural environments can have numerous benefits. Firstly, natural resources can be preserved better as a part of natural ecosystems. Secondly, it is only as a part of natural ecosystems that flora and fauna can fulfill all their natural ecological functions to their full extent.

These circumstances are considered in Principle 4, which affirms that wildlife and natural habitats should be preserved and sustainably managed, and that

228 Ibid.
229 Ibid.
230 Ibid.
231 Ibid.

relevant issues should be considered within the context of socio-economic development. However, it should be noted that the wording of Principle 4 is weaker than the wording of Principle 2. Principle 4 only underlines that wildlife protection issues "*must receive importance*" in development activities; whereas Principle 2 stipulates that the components of natural environment "*must be safeguarded*" for the benefit of present and future generations".

The Stockholm Conference prioritized to a certain extent the necessity of protecting natural resources, including flora and fauna, over the need to protect natural ecosystems. This outcome was symbolic for the future development of international environmental law until the Rio Conference in 1992. In fact, the international environmental treaties that were adopted in the aftermath of the Stockholm Conference dealt predominantly with the conservation of certain components of natural environment, such as migratory species or representative samples of natural habitats, whereas the large-scale protection of natural ecosystems played only a secondary role. The adoption of the Convention on Biological Diversity[232] (CBD) in 1992 somehow changed the situation, as this convention tended to promote a comprehensive approach to ecosystems. Another important feature of the international environmental treaties that were adopted in the post-Stockholm period is their focus on the protectionist approach to nature conservation. According to this approach, legal bans and restrictions on the utilization of certain natural commodities should be introduced in order to ensure their preservation.[233] At the same time, the conservation of forests and other natural commodities has to be insured in order to make possible their utilization in the future.[234] However, this approach does not give the necessary consideration to the creation of incentives for the protection of forests.

Considering the potential conflict of interests that could arise from the contradictions between the necessity to protect the environment and the need for socio-economic development, especially in developing States, Principle 21 confirms the right of States to exploit their natural resources according to their national development needs. This right has, however, been relativized through the

232 Convention on Biological Diversity, adopted on 5 June 1992 and entered into force on 29 December 1993, in: 1760 UNTS 79.

233 *Schulte zu Sodingen, Beate*, Der Völkerrechtliche Schutz der Wälder, 2002, pp. 140 ff.; *Kiss, Alexandre/Shelton, Dinah*, International Environmental Law, Third Edition, 2004, pp. 48 ff.

234 *Eikermann, Anja*, Forests in International Law – Is There Really a Need for an International Forest Convention?, 2015, p. 33.

obligation to avoid environmental harm to the territories of other States and to the territories beyond national jurisdiction. However, this provision is difficult to apply to the protection of forests, because, unlike cases concerning the contamination of transboundary water resources and the atmosphere, it is difficult to prove that the damage to the environment of another State was caused by the deforestation on the territory of the State where it took place. It can be argued that deforestation and forest degradation result in the enhancement of natural disasters all over the globe. However, it is not yet possible to irrefutably link any single event to deforestation and forest degradation. Moreover, it is difficult to prove that the loss of biodiversity and the climate change caused by deforestation and forest degradation, both of which have adverse impacts on a global level, are cases of transboundary environmental harm according to Principle 21. Current international law does not provide sufficient instruments for collectively or unilaterally raising claims against any State that damages environmental goods of global importance, should no environmental harm be directly inflicted upon the territory of any specific State.[235] Indeed, the ICJ's judgment on the "Whaling in the Antarctic" has demonstrated that this kind of claims can become possible in the future.[236] However, in this particular case Japan was accused of breaching specific obligations under the International Convention for the Regulation of Whaling, which factually forbids commercial whaling. In the case of forests, raising of a similar claim would be further complicated by the fact that no forest-specific duties of prudence, diligence, and care ("duty to protect") are established under any international environmental agreement.[237] These duties are currently recognized under customary international law.[238] However, the right to raise a claim in case no damage was inflicted on the territory of an applicant State is explicitly recognized only for the violations of *erga omnes partes* obligations.[239]

---

235 United Nations A/RES/56/83, Resolution adopted by the General Assembly on 12 December 2001, Responsibility of States for internationally wrongful acts, United Nations, 2001, Annex, Articles 43 ff. – Internet resource – https://undocs.org/en/A/RES/56/83 [Last visit: 30.04.2023]; See *Scovazzi, Tulio,* State Responsibility for Environmental Harm, in: Yearbook of International Environmental Law, Volume 12, Issue 1, 2001, pp. 62 ff.

236 Whaling in the Antarctic (Australia v. Japan: New Zealand intervening), Judgement of 31.03.2014, ICJ Report 2014, pp. 226–300.

237 See Subsection C.II of Chapter 6.

238 Legality of the Threat or Use of Nuclear Weapons, Advisory Opinion, ICJ Reports 1996, pp. 241–242, Paragraph 29.

239 Questions relating to the Obligation to Prosecute or Extradite (Belgium v. Senegal), Judgment of 20 July 2012, ICJ Reports 2012, pp. 449 f., Paragraphs 68–70.

Moreover, it can be argued that according to Principle 21 national sovereignty over natural resources is asserted stronger than a State's obligation concerning the prevention of environmental harm.[240] Indeed, taking into account that the casual link between the deforestation and the damage to the territory of another States cannot be easily verified, it is submitted that the rights of States in relation to their forest resources are not efficiently restricted by the above-mentioned obligation.[241] Particularly in the case of forests, States are able to prioritize utilization issues over conservation.[242]

Principle 24 affirms that all States have to cooperate in order to protect and improve the environment. As it has already been stated in the previous chapter, deforestation and other environmental problems have a global dimension and cannot be resolved by one single State. Many forest-rich States do not have the necessary financial and technological resources needed in order to both conserve the forest ecosystems within their territories and to simultaneously promote socio-economic development. Consequently, there is a need for international cooperation between States in order for environmental protection, including the conservation and sustainable utilization of forests, to be promoted. Therefore, one of important achievements of the Stockholm Conference is that it has promoted international cooperation on environmental issues, by assuming that the global character of environmental crises requires extensive cooperation between States and international organizations.

It has been argued that the Stockholm Conference did not contribute directly to the creation and development of the international legal regime for forests.[243] However, as well as the adoption of general environmental principles, the Stockholm Conference has also made a number of recommendations that have had a direct relation to forests. Inter alia, the Conference recommended that States promote research activities in order to develop forest planning and management with due regard to the environmental role of forests.[244] States were also advised to modernize their forest management plans, policies and procedures in order

---

240 *Brunnée, Jutta*, A Conceptual Framework for an International Forests Convention: Customary Law and Emerging Principles, in: *Canadian Council on International Law* (ed.), Global Forests and International Law, 1996, pp. 41 ff.; *Humphreys, David*, The Elusive Quest for a Global Forests Convention in: RECIEL, Volume 14, Issue 1, 2005, p. 2.

241 *Brunnée* [n. 240], pp. 43 ff.; *Humphreys* [n. 240], pp. 2 f.; *Eikermann* [n. 234], p. 33.

242 *Brunnée* [n. 240], pp. 43 ff.; *Humphreys* [n. 240], pp. 2 f.

243 *Eikermann* [n. 234], p. 32.

244 *UNEP* [n. 226], p. 90.

to consider the multiple functions of forests and to reflect upon the benefits that forests provide.[245] Additionally, the Stockholm Conference advised the UN Secretary-General to take steps in order to promote cooperation between different United Nations bodies on the issues related to the diffusion of necessary innovative knowledge on the environmental aspects of forests and forest management.[246] The Secretary-General was also advised to take steps in order to promote the continuing surveillance, with the participation of Member States, of the changes of the world's forest cover.[247] The Conference also advised the FAO to coordinate international activities related to research and exchange information on threats to forest ecosystems, including information on forest fires, diseases, and pests[248].

In the aftermath of the Stockholm Conference in 1972, the international community adopted a number of international conventions, inspired by the Stockholm Declaration and Action Plan. Some of these conventions, which are relevant for forest protection, are further analyzed in the following.

I. Convention on the Conservation of Migratory Species of Wild Animals

The Convention on the Conservation of Migratory Species of Wild Animals[249] (CMS) was adopted in Bonn on 23 June 1979 and entered into force on 1 November 1983. The Convention focuses on the protection of migratory species, some of which contribute to the well-being of forest ecosystems, at least during their seasonal migrations.[250] The Convention makes a substantial contribution to the protection of certain species. For example, certain species were already protected by strict and enforced laws in their breeding sites and wintering grounds but were still over-hunted on their migration routes.[251] The CMS does not oblige the Parties to it to establish protected areas. However, it promotes the

245 United Nations A/CONF.48/14/Rev.1 [n. 227], recommendation 24.

246 Ibid.

247 Ibid.

248 Ibid.

249 Convention on the Conservation of Migratory Species of Wild Animals, adopted on 23 June 1979 and entered into force on 1 November 1983, in: 1651 UNTS 333.

250 *Schulte zu Sodingen* [n. 233], pp. 150 f.

251 *Lyster, Simon*, The Convention on the Conservation of Migratory Species of Wild Animals (The "Bonn Convention"), in: Natural Resources Journal, Volume 29, Issue 4, 1989, p. 979.

conservation and restoration of natural habitats, including forests, which serve as a temporary refuge for a number of endangered migratory species that are listed in Appendix 1 to the Convention.[252] Species, which are not endangered, but are still exposed or likely to be exposed to unfavorable survival conditions, are included in Appendix II. Survival conditions are defined as being *unfavorable* when they do not meet the requirements set out in Article 1(c) of the CMS. For example, this is the case when population fluctuations indicate that there is a threat that a migratory species cannot maintain itself as a stable component of its ecosystem; when a population is reducing or is likely to reduce, and when forests or other natural habitats that are necessary for the maintenance of a specific species' population are absent.[253] In addition, survival conditions are unfavorable when the size of a population and the distribution of a species are below the historical level or the level at which a sustainable ecosystem can exist.[254] The conservation of a migratory species, as is included in Appendix II, should be regulated by a special agreement between the Range States, which are also listed in Appendix II. The agreement should identify the main threats to the existence of a migratory species and establish measures and procedures for their protection.[255] Article 5 of the CMS stipulates that these measures should include, to an appropriate extent, the conservation and restoration of natural habitats, which serve as a temporary refuge for migratory species, as well as the protection of the habitats from disturbances. As it was mentioned above, forest ecosystems can play an important role as natural habitats of migratory species. In this regard, the CMS particularly emphasizes the necessity to control the introduction of new, exotic species into ecosystems.[256] The legally binding provisions on this issue are not common in international environmental law. As shown below, the pertinent provisions of forest-specific documents are not legally-binding. The provision in question deserves particular attention, because the introduction of new species may represent a significant threat to forest ecosystems. Some species have actually evolved to be endemic only for certain geographic and climatic regions. Their introduction to other regions can have unpredictable impacts on the local environment. Thus, this provision should be incorporated into other

252 Article 3.4(a).
253 Article 1(c).
254 Article 1.
255 Ibid.
256 Article 5.

environmental agreements as well. To sum up, by protecting migratory species, the CMS also indirectly promotes the conservation of forest ecosystems.

As of 1 June 2019, 128 States have ratified the CMS.[257] For the moment, a number of Agreements and Memorandums of Understanding on the protection of certain endangered species have been signed. Some non-signatories of the Convention have joined these CMS-instruments.[258]

II. Convention on Wetlands of International Importance Especially as Waterfowl Habitat (Ramsar Convention)

The Convention on Wetlands of International Importance Especially as Waterfowl Habitat[259] (Ramsar Convention) is dedicated to the protection of internationally important wetlands, especially as a habitat for aquatic birds. What was highly praised at the time of its adoption was the fact that the Ramsar Convention was the first international environmental treaty that didn't only deal with the protection of a sole species or a group of species, but also promoted the conservation of one specific type of ecosystem – wetlands.[260] The adoption of the Ramsar Convention was inspired by the fact that due to economic, political, and geographical reasons single States are often unable to protect the wetlands in their territories alone.[261] It is worth noting that the Convention deals only with the conservation of wetlands, without mentioning development issues any single time. The Ramsar Convention also protects a number of wetland forests, including mangrove forests.[262] Additionally, activities within the scope of the Convention include the promotion of river basin management, considering the specific role of forests in the formation and maintenance of rivers and wetlands.[263]

---

257 Source: Official Web-site of the Convention on the Conservation of Migratory Species of Wild Animals – Internet resource – http://www.cms.int [Last visit: 30.04.2023]

258 Ibid.

259 Convention on Wetlands of International Importance Especially as Waterfowl Habitat, adopted on 2 February 1971 and entered into force on 21 December 1975, in: 996 UNTS 245.

260 *Schulte zu Sodingen* [n. 233], p. 142; *Sands, Philippe*, Principles of international environmental law, 2012, p. 404.

261 *Navid, Daniel*, The International Law of Migratory Species: The Ramsar Convention, in: Natural Resources Journal, Volume 29, Issue 4, 1989, p. 1003.

262 *Schulte zu Sodingen* [n. 233], p. 142.

263 *Ramsar Convention Secretariat*, River basin management, Handbook 9, 4th edition, Gland, Switzerland, 2011, p. 27.

For this purpose, the Convention established a specific international protection mechanism, according to which each Contracting Party should designate suitable wetlands in its territory to be included in the List of Wetlands of International Importance.[264] The List is maintained by the Conference Bureau, which was established in accordance with Article 8 of the Convention. Currently, the International Waterfowl and Wetlands Research Bureau (IWRB) and IUCN jointly fulfill the relevant duties.[265] Article 2.1 provides guidance on how protected wetland areas should be designated and determined. The boundaries of each area of wetland should be accurately identified and delimited on a map; they can include costal and riparian zones neighboring the wetlands, and islands and other terrestrial formations of marine water deeper than 6 meters at low tide lying within the wetlands, especially when they serve as a habitat for water birds.[266] Moreover, wetlands should be designated to the list according to their international significance in terms of their ecology, zoology, botany, hydrology, and limnology, with special focus being given to their role as seasonal or permanent habitats for waterfowl.[267]

Currently, there are over 2,200 designated Ramsar sites with a total area of 2,5 Millions in the territories of 170 Ramsar Convention Parties.[268] The Convention reaffirms the exclusive sovereign rights of parties with regard to wetlands within their territories that are not prejudiced by their inclusion to the abovementioned List of Wetlands of International Importance.[269] Parties should develop and implement specific plans in order to promote the conservation of the wetlands included in the list in question, as well as the environmentally friendly use of all of the wetlands within their sovereign territory.[270] In case of urgent national interest, parties are allowed to eliminate or reduce the boundaries of a wetland included in the list, but any adverse impacts on the environment as a result thereof should be compensated by establishing new watershed protected

---

264 Article 2.1.

265 Resolution 3.1 of the third Conference of contracting Parties of the Convention on Wetlands, 1987. – Internet resource – https://www.ramsar.org/sites/default/files/documents/library/key_res_3.01e.pdf [Last visit: 30.04.2023]

266 Article 2.1.

267 Article 2.2.

268 Ramsar Convention, The List of Wetlands of International Importance, Ramsar Convention, June 4 2020. – Internet resource – https://www.ramsar.org/document/the-list-of-wetlands-of-international-importance-the-ramsar-list [Last visit: 30.04.2023]

269 Article 2.3.

270 Article 3.1.

areas and natural reserves for waterfowl.[271] Newly created protected watershed zones should be of the same size as those that have been removed.[272] However, the application of this provision can be particularly challenging in the case of mangrove forests, as they need a lot of time to regrow.

The Ramsar Convention has promoted the protection of some types of forest ecosystems, particularly those of mangrove forests. Under the influence of the Convention, the "Mangrove action Plan" was established as an international Network for the protection of mangrove forests.[273] One of the main weaknesses of the Ramsar Convention is the lack of a financial mechanism, which would promote its implementation. For this reason, the Secretariat was forced to deal with a lack of funds at the early stages of the Convention's implementation.[274]

Thus, the Ramsar Convention provides a comprehensive scheme for the protection of the wetlands of international importance, including mangrove forests. However, it can be seen as a shortcoming that it is the States themselves that are primarily responsible for the designation of protected wetland areas, and that the international bodies of the Convention do not have enough authority to influence this process. In addition, the positive effects of the Ramsar Convention on the protection of forests are limited by the scope of the Convention.

III. Convention for the Protection of the World Cultural and Natural Heritage

The Convention concerning the Protection of the World Cultural and Natural Heritage (WHC) was adopted on 23 November 1972 and entered into force on 17 December 1975.[275] The Convention has contributed to the protection of global forest ecosystems by promoting the conservation of certain representative samples of forest ecosystems,[276] as they are covered by its definition of natural

271 Article 4.2.

272 Ibid.

273 For more information about the "Mangrove Action Plan", visit the Official Internet Site of the Network for the protection of mangrove forests. – Internet resource – http://mangroveactionproject.org/ [Last visit: 30.04.2023].

274 *Lausche, Barbara J.*, Weaving a Web of Environmental Law: Contributions of the IUCN Environmental Law Programme, 2008.

275 Convention for the Protection of the World Cultural and Natural Heritage, adopted on 23 November 1972 and entered into force on 17 December 1975, in: 1037 UNTS 151.

276 *Schulte zu Sodingen* [n. 233], p. 144; *Krohn, Susan Nicole*, Die Bewahrung tropischer Regenwälder durch völkerrechtliche Kooperationsmechanismen, 2001, p. 132.

heritage sites.[277] For the purpose of the Convention, the term *natural heritage* encompasses natural formations of outstanding aesthetic and scientific value, as well as habitats of endangered or particularly valuable animal and plant species.[278] Moreover, it also includes natural sites and areas of "outstanding universal value" from a scientific, conservationist, or natural beauty point-of-view.[279] It can be argued that this definition of the term *natural heritage* leaves some room for interpretation. For example, the notions of "aesthetic value" or "outstanding universal value" require further clarification. Therefore, if interpreted widely, the notion of *natural heritage*, as it is defined by the Convention, can be extended to every forest ecosystem around the globe. However, the Convention does not intend to establish a robust and comprehensive protection regime for all the natural ecosystems on the planet, but to promote the protection of specifically designated natural formations that fulfill specific criteria established by its committee. It is up to States to identify and delineate the natural heritage sites within their sovereign territory.[280] Thereby, they can apply non-legally binding Operational Guidelines.[281] This procedure of designating natural heritage sites is similar to the process of identifying wetlands of international importance under the Ramsar Convention.[282] Every State bears the responsibility to conserve all the national heritage sites and areas within its national jurisdiction and to pass them on to future generations.[283] For this purpose, a State has to use the utmost of its own national resources. However, when appropriate, it can also use resources delivered through international cooperation, including financial, scientific, and technical ones.

After any natural site or area is recognized as being one of natural heritage, States can recommend including it to the World Heritage List.[284] For this purpose,

277 Article 2.

278 Ibid.

279 Ibid; *Eikermann* [n. 234], p. 82.

280 Article 3.

281 UNESCO WHC. 12/01, Operational Guidelines for the Implementation of the World Heritage Convention, UNESCO, World Heritage Centre, 2012. – Internet resource – https://whc.unesco.org/archive/opguide12-en.pdf [Last visit: 30.04.2023]; *Goodwin, Edward J.*, The World Heritage Convention, the Environment, and Compliance, in: Colorado journal of international environmental law and policy, Volume 20, Issue 2, 2009, pp. 158 ff.

282 *Schulte zu Sodingen* [n. 233], p. 144.

283 Article 4.

284 *Schulte zu Sodingen* [n. 233], p. 144.

alongside with the development and implementation of national measures aimed at the promotion, conservation and protection of designated places,[285] each State submits an inventory of sites and areas of cultural and natural heritage to the Intergovernmental Committee for the Protection of the Cultural and Natural Heritage of Outstanding Universal Value (hereinafter – the World Heritage Committee) for their inclusion in the World Heritage List.[286] Based on the inventories submitted, the World Heritage Committee has to establish and maintain the "World Heritage List". The list should include sites and areas that fit the criteria of being of "outstanding universal value", as defined by the Committee itself.[287] Accordingly, in order to be included in the "World Heritage List", natural heritage sites must be home to an outstanding natural phenomena or areas of exceptional aesthetic importance.[288] Otherwise, they must represent a major stage in the paleontological or geological history of the Earth.[289] The sites that contain significant natural habitats for the conservation of biodiversity can also be included in the List.[290] It can be argued that all these criteria can be applied to the major forest ecosystems around the world, particularly to the remaining natural forests. However, the Committee is able to include in the list only those natural sites submitted by parties. For their part, parties are not obliged to recognize as being of a natural heritage all the natural sites that satisfy the relevant definition and criteria of the Convention.

The World Heritage Committee represents part of the international compliance and enforcement system designated to assisting parties' efforts dedicated to the conservation of their cultural and natural heritage.[291] Another important element of this system includes the Fund for the Protection of the World Cultural and Natural Heritage of Outstanding Universal Value (hereinafter – the World Heritage Fund), established in accordance with Article 15 of the Convention. Financial resources for the fund are raised through the compulsory and voluntary contributions of States, international organizations, public or private bodies, individuals and through any interest due on the resources of the Fund, as well

285 Article 5.

286 Article 11.1.

287 Article 11.2.

288 WHC.17/01, Operational Guidelines for the Implementation of the World Heritage Convention, World Heritage Centre, 2017, §77(vii). – Internet resource – https://whc.unesco.org/en/guidelines/ [Last visit: 30.04.2023]

289 Ibid, §77(viii).

290 Ibid, §77(ix-x).

291 *Krohn* [n. 276], p. 132.

as from other sources.[292] Thus, the Convention foresees a non-market-based approach to the conservation of natural and cultural heritage.

The development of market-based approaches is generally not foreseen by the UNESCO Convention. However, the idea has not been explicitly excluded. It is worth noting that Article 5.1 of the Convention stipulates the necessity of promoting the role of cultural and natural heritage sites in the life of local communities. The application of this provision can include, inter alia, the establishment of market-based mechanisms that aim to improve local livelihoods and raise funds for the conservation of natural and cultural heritage sites. However, in general, the WHC does not make use of possibilities to promote market-based approaches to the conservation of World Heritage sites. Moreover, States are not obliged to invest the revenues earned from the environmentally friendly exploitation of natural heritage sites, into their conservation. On the other hand, it can be seen as logical to invest financial resources into the conservation of World Heritage sites that deliver certain economic and political benefits.[293]

It is worth noting that the obligatory contributions of each party to the World Heritage Fund does not exceed 1% of its contributions to the UNESCO. In 2014, the income of World Heritage Fund was 4,156,380 US$.[294] Consequently, the Fund cannot provide the financial resources necessary to cover all the costs and expenditures related to the conservation of World Heritage forests and the lost opportunity costs. Thus, the UNESCO Convention does not create enough economic incentives for the inclusion of forested areas on the World Heritage List. For instance, the WHC does not provide any remedies for situations whereby crude oil or other mineral resources are found beneath a World Heritage Forest. Instead, it promotes the adoption of more strict measures for the protection of already established national parks and nature conservation areas. In some cases, it can also promote the establishment of new ones. It is worth noting that Article 6.2 of the Convention reaffirms the obligation of States to support one another's conservation efforts. However, this provision refers more to scientific, technical, and artistic assistance. Concerning the financial aspect, it should be understood only in the light of each parties' obligations towards the Fund. It cannot justify

292 Article 15.

293 *Goodwin* [n. 281], pp. 171 ff.

294 UNESCO, Report on the execution of the budget for the biennium 2014–2015 and preparation of the budget for the biennium 2016–2017, Paris, 15 May 2015. – Internet resource – http://whc.unesco.org/archive/2015/whc15-39com-15-en.pdf [Last visit: 30.04.2023]

any additional claims for financial support, which would exceed the general conventional framework.[295] However, it can be argued that parties can be ready to provide further financial assistance due to political benefits.[296]

It is also worth noting that the Convention can be appreciated for the particular attention that it pays to the conservation of ecosystems and biodiversity in World Heritage forests. As the Convention is deemed to promote the preservation of natural heritage sites for future generations, protection standards and requirements are much higher than in many other international environmental legal regimes, meaning that economic activities in the relevant areas should be either prohibited or restricted.[297] Theoretically, the need for these strict conservation measures, combined with the already mentioned lack of financial incentives, should keep States from including their natural sites and areas in the List. However, the very act of preserving World Heritage Sites can improve a State's international image and attract visitors, so that it can enjoy direct and indirect benefits, including economic and political ones.[298]

As has already been mentioned, many natural sites included on the World Heritage List already have the status of *national park* and are therefore already protected by national legislation. Consequently, their inclusion in the World Heritage List has only a limited positive impact on their protection, even though it helps to attract international attention to their conservation.[299] Arguably, additional funding, the monitoring mechanisms of the UNESCO Convention and enhanced international attention can improve the conservation status of national sites.

Due to the lack of a clear, unique, and all-inclusive definition of forests in international law and scientific literature, the estimation of the number of forest areas protected under the UNESCO Convention can be a challenging task. However, for this purpose, one commonly used definition can be applied. It identifies a "World Heritage Forest" as a natural heritage site, a substantial part of which is covered with forests, according to the data provided by the State in which it is located.[300] A Site also falls under this category if the forest contributes to the

295 See also *Schulte zu Sodingen* [n. 233], p. 144.

296 *Goodwin* [n. 281], pp. 167 ff.

297 *Krohn* [n. 276], p. 132.

298 *Goodwin* [n. 281], pp. 167 ff.

299 Ibid.

300 *Ripley, Steven/Patry, Mark*, World Heritage Forests: Leveraging Conservation at the Landscape Level, in: Proceedings of the 2nd World Heritage Forest Meeting, 9–11 March 2005, World Heritage Reports, No. 21, 2007, p. 15; *Thorsell, Jim/Sigaty, Todd*, A Global Overview of Forest Protected Areas on the World Heritage List, a Contribution

Sites' Outstanding Universal Value.[301] From this point of view, as of July 2006, 91 World Heritage Sites, the total area of which amounts to approximately 75 million ha, can be considered as the *World Heritage Forest*.[302] However, as certain parts of many "World Heritage Forest" Sites are covered by natural formations and ecosystems other than forests, the total forest area that is protected under the World Heritage Convention amounts to some 63.7 million hectares.[303]

The World Heritage Committee coordinates the implementation of the Convention and makes legally non-binding recommendations in this regard. Although not directly empowered by the Convention to do so, the Committee also gives these recommendations without consulting Parties, which are directly affected by the decision in question.[304] Additionally, the Committee is also able to make formal requests to improve the protection status of World Heritage Sites and even to apply sanctions, which would mean the temporary or permanent loss of the economic and politic benefits granted by membership in the Convention.[305] It is worth noting that States are obliged to report every six years on the implementation of the Convention to the General Conference of the UNESCO. All these elements make up the compliance control system of the Convention.[306] It is submitted that the compliance and enforcement mechanisms of the WHC are stronger and more robust than the relevant mechanisms of most environmental treaties. It has already proven to be able to effectively tackle the cases of non-compliance by encouraging links between the fulfillment of the duties according to the Convention and the promotion of national interests of States, as it happened in the case of Ecuador.[307] The WHC has an interesting example of a supra-national compliance control system, over which no individual State can exercise any power. It can be argued that the sanction mechanism is not effective enough in order to remedy situations whereby the potential benefits of the economic exploitation of a natural heritage site are higher than the political and economic benefits of collaboration with the Committee. However, the enforcement

---

to the Global Theme Study of the World Heritage Natural Sites, Natural Heritage Programme, IUCN, 1997, p. 2.

301 *Ripley/Patry* [n. 300], p. 15; *Thorsell/Sigaty* [n. 300], p. 2.

302 *Ripley/Patry* [n. 300], p. 15.

303 For the current list of UNESCO heritage sites see: http://whc.unesco.org/en/list/?type=natural#note1 [Last visit: 30.04.2023]

304 *Ripley/Patry* [n. 300], p. 15.

305 *Goodwin* [n. 281], pp. 171 ff.

306 Ibid, p. 172.

307 Ibid, pp. 172, 176 ff.

of the current compliance control system is unlikely, due to the scarce resources of the Committee. Moreover, most compliance and enforcement tools can be applied to national heritage sites only if they are included in the World Heritage List. Overall, one should evaluate positively the contribution of the WHC to the protection of representative samples of forest ecosystems of scientific, conservationist or aesthetic importance, as well as those of outstanding universal value, including unique habitats. However, its positive impact on the preservation of global forests is limited by its insufficient financial resources and by its inability to extend the protection regime of the convention to all existing natural forests.

IV. Convention on International Trade in Endangered Species of Wild Fauna and Flora (CITES)

Another important international legally binding document that should be further highlighted here is the Convention on International Trade in Endangered Species of Wild Fauna and Flora (CITES).[308] It was adopted on 3 March 1973, almost immediately in the aftermath of the Stockholm Conference. The CITES entered into force on 1 July 1975.

The CITES is dedicated to the regulation of international trade in the most endangered species of fauna and flora. The species protected by the Convention are included in one of three Appendixes to the Convention. Each Appendix presupposes a different level of protection. Appendix I includes all endangered species, international trade in which can affect adversely their survival.[309] For this reason, commercial trade in these species is forbidden.[310]

---

308 Convention on International Trade in Endangered Species of Wild Fauna and Flora, adopted on 3 March 1973 and entered into force on 1 July 1975, 993 UNTS 243; see also: *Hutton, Jon/Dickson, Barbara*, Endangered Species, Threatened Convention: the Past, Present and Future of the CITES, the Convention on International Trade in Endangered Species of Wild Flora and Fauna, 2000, pp. 29 ff.; *Wijnstekers, Willelm*, The evolution of CITES: A reference to the Convention on International Trade in Endangered Species of Wild Fauna and Flora, 11th edition, International Council for Game and Wildlife Conservation, 2011, pp. 31 ff.; *Favre, David S.*, International Trade in Endangered Species: A Guide to CITES, 1989, pp. 1 ff.

309 Article 2.1.; see also *Hutton, Jon/Dickson, Barbara*, Endangered species, threatened convention: the past, present and future of CITES, the Convention on International Trade in Endangered Species of Wild Flora and Fauna, 2000, pp. 29 ff.; *Schulte zu Sodingen* [n. 233], p. 145; *Krohn* [n. 276], p. 100.

310 *Schulte zu Sodingen* [n. 233], p. 145; *Krohn* [n. 276], p. 100.

Appendix II includes species that are not currently under threat but may be threatened by extinction if international trade in them is not restricted.[311] Additionally, it also includes species whose international trade should be regulated in order to effectively control the trade of potentially endangered species, as per Appendix I.[312] The trade in species of Appendix I is permitted upon the presentation of an export permit or a re-export certificate.[313]

Appendix III includes those species, the trade in which is regulated and/or restricted by the national legislation of one of the parties to the CITES, and whose survival requires international cooperation.[314] An export permit for Appendix III species is to be granted only when it has been obtained in accordance with the internal legislation of the State of the species' origin, and when the transport conditions do not represent any threat to living specimens.[315] The importation of specimens of an Appendix III species is subject to the prior granting and presentation of a certificate of origin. In the case, when a specimen is imported from a State that added this particular species to Appendix III, an export permit is required.

It is worth noting that the purpose of the CITES is not to establish a comprehensive protection regime for endangered species or their habitats, but to promote their protection by regulating and/or restricting the international trade therein. This approach can be criticized insofar as it does not promote effective measures for the conservation of species on a national level.[316] As its scope is limited to the regulation of trade in endangered species, the CITES can have only a limited impact on forest ecosystems, as it does not take into account the multiple functions of forests. Many tree species that are crucial for the survival of healthy forest ecosystems are currently not recognized as being endangered. Despite commercial exploitation resulting in deforestation and forest degradation, the conservation status of many tree species that play a critical role in forest ecosystems does not allow them to be considered as threatened. Additionally, it is not the international tropical timber trade that is the leading cause of deforestation and forest degradation in many regions, but forest conversion and unsustainable

---

311 Article 2.2(a).

312 Article 2.2(b).

313 Article 4.2.

314 Article 3.3.

315 Ibid.

316 See also *Skala-Kuhmann, Astrid*, Rechtliche Instrumente zur Förderung der nachhaltigen Nutzung und zum Schutz der forstlichen Ressourcen im Internationalen Zusammenhang, 1996, p. 17; *Schulte zu Sodingen* [n. 233], pp. 146 f.

agricultural practices.[317] Arguably, an international agreement endeavoring to tackle the driving forces of deforestation would need to regulate and restrict the trade in agricultural crops, such as palm oil, cacao and soya beans, as the unsustainable production thereof fosters forest conversions in many areas. However, the regulation of the trade therein goes beyond the scope of the CITES.

Under the current conditions, the CITES can contribute to forest conservation by regulating or restricting international trade in certain animal and plant species, especially endangered tree species. By requiring an export permit for the export and import of Appendix I species, it could potentially facilitate the combating of illegal trade in endangered timber tree species. From the very beginning, predominantly only ornamental and houseplants were included in the Appendixes, while States were reluctant to include commercial timber tree species and claimed to have an exclusive sovereign responsibility for the regulation of the international timber trade. The inclusion of new species is particularly complicated by the fact that any amendments made to Appendices I or II must be approved by a two-thirds majority.

Although a number of commercial timber tree species were already included in the Appendixes in 1976, in the mid-90s the parties realized that the Appendixes did not consider the current international trade situation and existing threats to timber species.[318] Consequently, the protection status of several different species under the CITES had to be changed. Several species had to be excluded from the Appendixes, whereas a great number of others had to be added. However, this process was hindered by a lack of consent from certain export States. Additionally, many parties were generally opposed to the applicability of the CITES to timber species, especially in the tropics. For instance, at CITES COP 9 in 1994, several tropical States argued that the ITTA already regulates the trade therein and that its provisions should be applied because this treaty represents a classical *lex specialis* in this context.[319] For this reason, a time-lapse working group in charge of examining the possibility of extending the protection regime of the CITES on internationally traded timber species (the

---

317 See Subsection B.I of Chapter 1.

318 *Schulte zu Sodingen* [n. 233], p. 147.

319 *Kelso, Bobbie Jo*, Ninth Meeting of the Conference of the Parties to CITES, in: Traffic Bulletin, Volume 15, Issue 2, 1995, p. 71; *Sand, Peter H.*, Das Waschingtoner Artenschutzabkommen (CITES) von 1973, in: *Gehring, Thomas/Oberthür, Sebastian* (eds.), Internationale Umweltregime, Umweltschutz durch Verhandlungen und Verträge, 1997.

Timber Working Group) was created.[320] On the basis of the results presented by the Timber Working Group, COP 10 in 1997 agreed that both the CITES and ITTA could be simultaneously applied to trade in tropical timber species, as they regulate different aspects of trade.[321] Accordingly, the ITTA is a mix of a commodity and environmental treaty, and environmental issues are subordinated to its main goal, that is to regulate international tropical timber trade. In contrast, the CITES is an environmental treaty. Therefore, it is endeavored to regulate and/or restrict international trade in endangered species of flora and fauna.[322] As a result, States were encouraged to consider both current threats to the survival of internationally traded timber species, as well as the economic significance thereof.[323] Parties that wanted a specific timber species to be included in the Convention's Appendixes were encouraged to contact at least four major international forest-related organizations, including the African Timber Organization, the Asian-Pacific Timber Organization, the Center for International Forestry Research, the FAO, the WWF, and others. During these consultations, parties have to gather and verify information concerning the environmental status of the timber species in question, as well as international trade therein.[324] The verified data must then be transferred to the CITES Secretariat, that will forward it to other member States. The CITES Secretariat itself should consult the ITTO, FAO, and the IUCN on the issue.[325]

---

320 *CITES DOC 9.52*, Interpretation and Implementation of the Convention, Trade in Plant Specimens, Implementation of the Convention for timber species, 9th meeting of the CITES COP, 1994, Paragraph 9. – Internet resource – https://www.cites.org/sites/default/files/eng/cop/09/doc/E9-Doc-52-53.pdf [Last visit: 30.04.2023]; *CITES COP 9* Decisions of the Conference of the Parties, Decisions directed to the Standing Committee, 1994, Paragraph 4. – Internet resource – https://www.cites.org/sites/default/files/eng/cop/09/E9-Dec.pdf [Last visit: 30.04.2023]

321 CITES DOC 10.52, Interpretation and Implementation of the Convention, Trade in Plant Specimens, Implementation of the Convention for timber species, 10th meeting of the CITES COP, 1997. – Internet resource – https://www.cites.org/sites/default/files/eng/cop/10/doc/E10-52to54.pdf [Last visit: 30.04.2023]

322 CITES, Preamble and Article 2; *Schulte zu Sodingen* [n. 233], p. 147.

323 CITES COP Conf. 10.13, Implementation of the Convention for timber species, in: CITES COP 10, Resolutions of the Conference of the Parties, 1997, Paragraph 1(j) – Internet resource – http://www.ciesin.columbia.edu/repository/entri/docs/cop/CITES_COP010_res013.pdf [Last visit: 30.04.2023]; See also *Krohn* [n. 276], p. 103.

324 CITES COP Conf. 10.13, Paragraph 1(a,b).

325 Ibid.

Fulfilling of all these requirements is time-consuming and prevents endangered species from being immediately added to the Appendices. However, it allows for the controversial opinions of many different actors, including non-governmental organizations, to be considered.[326] Arguably, it makes the process of inclusion more legitimate and transparent, and also facilitates the negotiation of the inclusion of new species in Appendices at the COP. It is important to note that the above-mentioned organizations also represent the interests of the developing timber-exporting States. Thus, consultations facilitate the achievement of a consensus on the inclusion of specific species.

Determining the current survival status of specific species is a further problem. The data concerning the current population size and distribution of specific species is difficult to compile and is even less reliable than international trading estimates.[327] The volume of internationally traded timber species can be underreported, due to a lack of monitoring and because of illegal trading. Estimating the current population status of a species can be particularly challenging, as certain tropical species cannot be easily identified and their natural habitat can be difficult to allocate.[328] For this reason, CITES COP 9 agreed to establish a new classification system for endangered species that is based on the precautionary principle.[329] According to this principle, scientific uncertainty about the ecological status of species should not impede their inclusion in Appendices. This new system has reduced the need for proof of a species' vulnerability that is needed for the inclusion of a new timber species into the Appendices.[330]

The story of the inclusion of the Mahogany species *Swietenia Macrophylla* into Appendix II demonstrates how difficult achieving of the consensus in CITES bodies can be.[331] The first unsuccessful attempt to include this Mahogany species in Appendix II was undertaken in 1992.[332] At CITES COP 9 in 1994, El

---

326 *Krohn* [n. 276], p. 103.

327 *Krohn* [n. 276], p. 104.

328 Ibid.

329 *Krohn* [n. 276], p. 104; *Eikermann* [n. 234], pp. 69 f.

330 *Krohn* [n. 276], p. 104.

331 See *Barbosa, Luiz C.*, Guardians of the Brazilian Amazon Rainforest: Environmental Organizations and Development, 2015, pp. 48 ff.; *Blundell, Art*, A review of the CITES listing of big-leaf mahogany, in: ORYX, Volume 38, Number 1, 2004, pp. 1 ff.

332 CITES COP 8, Proposals 8.94 and 8.95, in: Amendments to the Appendices I and II of the Convention, Other Proposals, 1992, pp. 111 ff. – Internet resource – https://cites.org/sites/default/files/eng/cop/08/prop/E08-Prop-94_Swietenia.pdf [Last visit: 30.04.2023]

Salvador and the Netherlands submitted a proposal to include several Mahogany species in Appendix II.[333] Several States, including Malaysia, Cameroon, Brazil and the Democratic Republic of Congo were decisively opposed to this initiative, arguing that national legislative measures were sufficient in order to ensure the survival of the species.[334] The USA and Bolivia's proposal to include *Swietenia Macrophylla* in Appendix II was rejected again during CITES COP 10 in 1997.[335] Little progress was achieved until the COP 11 held in 2000, where only several Parties to the CITES expressed their readiness to implement the CITES system of granting export permits for *Swietenia Macrophylla*.[336] In the end, the consensus needed was achieved following long and controversial negotiations. At CITES COP 12, held in 2003, Parties agreed to add commercially tradable bigleaf mahogany (*Swietenia macrophylla*) to Appendix II.[337]

The understanding of the necessity of prior consultations with export States has facilitated the reaching of a consensus on the inclusion of new species. For example, during CITES COP 17 in 2016, States agreed to add all commercially tradable rosewood species under the genus *Dalbergia* and three Bubinga species (*Guibourtia demeusei*, *Guibourtia pellegriniana*, and *Guibourtia tessmannii*) to Appendix II of the CITES.[338] As of 2 January 2017, dozens of timber species have

333 *CITES COP* 9, Proposal 9.101, in: Amendments to the Appendices I and II of the Convention, Other Proposals, 1994, pp. 101 ff. – Internet resource – https://cites.org/sites/default/files/eng/cop/09/prop/E09-Prop-101_Swietenia.PDF [Last visit: 30.04.2023]; *Schulte zu Sodingen* [n. 233], p. 147.

334 *Barbosa* [n. 331], p. 59.

335 CITES COP 10, Proposal 10.69., in: Consideration of Proposals for the Amendment of Appendices I and II of the Convention, Other Proposals, 1997, pp. 1 ff. – Internet resource – https://cites.org/sites/default/files/eng/cop/10/prop/E-CoP10-P-06.pdf [Last visit: 30.04.2023]

336 *Barbosa* [n. 331], pp. 59 ff.

337 CITES COP 12, Amendments to Appendices I and II of the Convention, adopted by the Conference of the Parties at its 12th meeting, Santiago, Chile, 3 to 15 November 2002, 2002. – Internet resource – https://cites.org/sites/default/files/eng/cop/12/Adopted_Amendments.pdf [Last visit: 30.04.2023]

338 CITES COP 17, Amendments to Appendices I and II of the Convention, adopted by the Conference of the Parties at its 17th Meeting, Johannesburg (South Africa), 24 September – 4 October 2016, 2016. – Internet resource – https://cites.org/sites/default/files/notif/E-Notif-2016-063.pdf [Last visit: 30.04.2023]; See *Saunders, Jade*, CITES Conference Marks Major Breakthrough in Battle Against Endangered Rosewood Trade, Chatham House, the Royal Institute of International Affairs, 20 October 2016. – Internet resource – https://www.chathamhouse.org/expert/comment/

been added to the Appendixes; 7 to Appendix I, 53 to Appendix II and 10 to Appendix III.[339] In total, more than 600 tree species are currently included in the Appendixes to CITES. Great progress has been made, considering that in 1975 this number amounted to but 18 tree species.[340]

Although the addition of new tree species to the Appendixes of the CITES is a time-consuming and complicated process, it should not be forgotten that the Convention can contribute to the protection of forests by regulating and/or restricting international trade in animal and non-tree plant species. In fact, forest ecosystems do not consist only of trees, as other wildlife species can also play a major role in its functioning. For example, all species Hummingbirds (*Trochilidae*), which has been included in Appendix II to the Convention since in 1987,[341] plays an important role as a pollinizer in the tropical rainforests of Central America.[342] As forest ecosystems are highly complex natural formations, the role of non-tree species should not be underestimated.

The CITES is unlikely to create incentives to protect forests due to the lack of provisions it has at its disposal with which to deal with economic issues. Unlike the ITTA, the CITES does not endeavor to promote trade in legally and sustainably harvested timber. As it has been shown above, economic ratio plays a great role, and export States are very reluctant to extending the scope of the CITES to the protection of commercial tree species. States, on the other hand, can be interested in the inclusion of timber species in Appendixes I or II. For example, international trade in illegally harvested tree species can cause significant damage and losses to national economies, and the CITES delivers a potentially effective, internationally recognized trade control system. If a State implements a long-term strategy of sustainable development, it will be also interested in the inclusion of endangered commercial tree species in the CITES Appendices, if

cites-conference-marks-major-breakthrough-battle-against-endangered-rosewood-trade [Last visit: 30.04.2023]

339 US Fish and Wildlife Service, Current CITES Listings of Tree Species. – Internet resource – https://www.fws.gov/international/plants/current-cites-listings-of-tree-species.html [Last visit: 30.04.2023]

340 *Schulte zu Sodingen* [n. 233], p. 147.

341 *CITES COP 6*, Amendments to the Appendices I and II of the Convention adopted at the sixth meeting of the Conference of the Parties in Ottawa, Canada, 12 to 24 July 1987, 1987, p. 99. – Internet resource – https://cites.org/sites/default/files/eng/cop/06/E06-Amendments-to-Appendices.pdf [Last visit: 30.04.2023]

342 See *Beauty of Birds*, Emerald-chinned Hummingbirds. – Internet resource – https://www.beautyofbirds.com/emeraldchinnedhummingbirds.html [Last visit: 30.04.2023]

their population density has significantly reduced. This measure can facilitate the recovery of a species' population and also allow for its commercial use in the future. However, the CITES does not create direct short-term incentives for the protection of forests.

In addition, the volume of the annual illegal trade in flora and fauna species, which amounts to 10–20 billion $, indicates that there are serious problems with the implementation of the CITES, and that a robust and comprehensive system for monitoring international trade in flora and fauna species has not yet been established.[343] Inter alia, the national authorities of many Member States encounter difficulties while monitoring the legality of harvesting and the scope of imports and exports.[344] For instance, in the case of *Swietenia Macrophylla*, a system for monitoring of all the stages of the international trade in it, from the harvesting process to the export stage, is required. The establishment and maintenance of these systems requires significant technical and financial resources, as well as the collaboration of trading companies.[345] Additionally, the identification of specimens of a species protected under the CITES may represent a further challenge. In many cases, customs officers and other public officials do not receive adequate training in order to be able to recognize specific specimens.[346] The identification of an endangered specimen can be further complicated, as its trade name can refer to the scientific name of a different species, only some of which are actually endangered. At the same time, one scientific name can correspond to many different trade names. This circumstance allows merchants to illegally buy and sell specimens of an endangered species by means of an uncommon trade name.[347] Moreover, as in the case of *Swietania Macrophylla*, in order to facilitate the identification of a specimen protected under the CITES, the control of trade in tropical timber is often limited to a few types of timber products, like logs, sawn timber, veneer and plywood.[348]

343 *Eikermann* [n. 234], pp. 69 ff.

344 *Blundell* [n. 331], p. 5; *Rodan, Bruce D./Blundell, Arthur G.*, Can Sustainable Mahogany Stem from CITES Science?, in: Bioscience, Volume 53, 2003, pp. 619 ff.

345 *Rodan/Blundell* [n. 344], p. 620.

346 Ibid.

347 Ibid, p. 621.

348 Ibid.

1. Compliance Mechanism of the CITES

Inter alia, the contribution of the CITES to the protection of forests depends on the effectiveness of its compliance mechanism. This mechanism is of great importance, given that the CITES possesses one of the most innovative tools to ensure compliance with its provisions. Thus, the possibility of developing compliance mechanisms based on the experience of the CITES in other forest-related international environmental treaties should be examined.

Unlike most other international environmental treaties, which foresee, *inter alia*, the limitation of transgressors' rights in the Convention bodies or the reduction of international financial assistance, CITES is now able to apply trade sanctions to the State which violates its provisions.[349] It is worth noting that the Convention's bodies have been able to develop a comparatively effective and robust compliance mechanism, although the Convention itself does not foresee trade sanctions as a tool for ensuring compliance.[350] Sanctions are applied in the form of trade embargoes, meaning that any international trade in those species listed in the CITES Appendices is suspended for any State that violates the provisions of the Convention. This mechanism is codified in the Guidelines on Compliance with the Convention, which were adopted in 2007.[351] Interestingly, it can be applied to both parties and non-parties of the CITES.[352]

Since 1981, the Technical Expert Committee, and later the CITES Standing Committee and the Secretariat, were entrusted with the identification of measures for the enforcement of the CITES provisions.[353] In the following decades,

---

349 *Treves, Tullio/Tanzi, Attila/Pineschi, Laura/Pitea, Cesare/Ragni, Chiara/Jacur, Francesca Romanin,* Non-Compliance Procedures and Mechanisms and the Effectiveness of International Environmental Agreements, 2009, pp. 50 ff.

350 *Koester, Veit,* Compliance Committees Within MEAs and the Desirability and Feasibility of Establishing Special Compliance Bodies under CITES, SC54 Doc.Inf.3, 2004, pp. 1 ff.

351 CITES COP 14 Document 23, Interpretation and implementation of the Convention, Compliance and enforcement issues: Guidelines for Comliance with the Convention, 2007. – Internet resource – https://cites.org/sites/default/files/eng/cop/14/doc/E14-23.pdf [Last visit: 30.04.2023]

352 CITES Resolution Conf. 14.3, CITES Compliance Procedures, Annex: Guide to CITES Compliance Procedures, 2007 – Internet resource – https://www.cites.org/eng/res/14/14-03C15.php [Last visit: 30.04.2023]; see also: *Payne, Cymie,* Introductory Note to Convention on International Trade in Endangered Species of Wild Flora and Fauna (CITES) Compliance Procedures, in: ILM, Volume 46, 2007, p. 1174;

353 *Reeve, Rosalind*, Policing International Trade in Endangered Species: The CITES Treaty and Compliance, 2002, pp. 49 f.

the problem of enforcing the provisions of the Convention grew significantly in importance. International reports have revealed that illegal trade in flora and fauna species constitutes a major threat to the survival of endangered species and serves as an important source of finance for criminal groups.[354] Subsequently, the issue of promoting compliance with the CITES and other international environmental agreements has become even more important. Nevertheless, the option of establishing a mechanism of sanctions vis-à-vis most international environmental agreements has been generally perceived as being unrealistic, as the necessary provisions have been absent in treaty texts. The potential positive effect of these mechanisms has been disputed and even their general necessity often denied.[355] On the other hand, the CITES provides an interesting example, which proves that the possibility of developing a sanction mechanism still exists.

Article XIV.1(a) of the CITES has served as the legal basis for the development of the embargo mechanism under the Convention. According to this Article, Member States are authorized to implement "stricter domestic measures" in order to regulate the trade, possession and otherwise use of specimens of species protected under the Convention, or completely prohibit them. To be more specific, Article XIV.1(a) authorizes States to implement unilateral sanctions in the form of trade restrictions and trade bans directed against other States that have violated the provisions of the Convention.[356] However, the compatibility of these measures with other international legal regimes, including international trade law, should be further considered.[357]

It can be problematic that this procedure is based on the unilateral interpretation of international environmental treaties, and that a national authority is entrusted with verifying whether or not any international environmental conventions have been violated. To avoid any possible infringements of international law, national trade sanctions should be also legitimized by the CITES' bodies, including the COP and/or Standing Committee. Notably, the Supreme Court of the United States has clarified that the decision to apply trade sanctions can be

354 *UN Office on Drugs and Crime,* Transnational Organized Crime in East Asia and the Pacific: A Threat Assessment, UNODC, 2013.

355 See *D'Amato, Anthony*, The Coerciveness of International Law, in: German Yearbook of International Law, Volume 52, 2009, pp. 437 ff.; *Beyerlin, Ulrich/Stoll Peter-Tobias/ Wolfrum, Rüdiger*, Ensuring Compliance with Multilateral Environmental Agreements: A Dialogue between Practitioners and Academia, 2006, pp. 1 ff.

356 *Sand, Peter H.* Enforcing CITES: The Rise and Fall of Trade Sanctions, in: RECIEL, Volume 22, Issue 3, 2013, p. 253.

357 Ibid.

taken by the American administration even without further consultations with the CITES bodies or in defiance of their opinions.[358] However, such unilateral practices could impede the development of a unified international legal strategy, one that could promote compliance with the CITES.[359]

To tackle this issue, the COP and the Standing Committee developed rules for the application of collective measures in response to violations of the CITES, that were consolidated and codified in Resolution 14.3 during the COP CITES 14 in 2007.[360] The collective measures were based on Article XIII of the Convention. Accordingly, if the Secretariat finds out that an endangered species included in Appendices I or II is being affected adversely by international trade, or that the provisions of the CITES are not being implemented correctly, it should inform the Party/Parties concerned.[361] The Party/Parties informed should deliver all relevant information to the Secretariat and propose remedial measures.

In practice, if all the attempts to resolve a situation of non-compliance via communication between the CITES bodies and the Party that has violated the provisions of the Convention are not successful, the Conference or the Standing Committee can recommend as an *ultima ratio* a collective trade embargo. This should, however, be approved by the national authorities of Member States.[362] The embargo can include not only the species directly threatened by the lack of compliance with the CITES rules, but all the species of Appendices I and II. As a result, the threat of being excluded from the flora and fauna species market can force the Parties to comply with the Convention.[363] This noteworthy sanction mechanism can also be applied to those non-Parties failing to comply with the documentation standards established for trade with third States, as in accordance

358 Japan Whaling Association v. American Cetacean Society (30 June 1986), in: 478 US 221; See *Setear, John K.*, Can Legalization Last? Whaling and the Durability of National (Executive) Discretion, in: Virginia Journal of International Law, Volume 44, Issue 3, 2004, p. 753.

359 See also *Sand 2013* [n. 356], p. 253.

360 CITES Resolution Conf. 14.3, CITES Compliance Procedures (2007), Annex: Guide to CITES Compliance Procedures, 2007; See *Payne* [n. 352], p. 1174.

361 CITES, Article 13.

362 Ibid.

363 See *Sand, Peter H.*, Sanctions in Case of Non-compliance and State Responsibility: Pacta Sunt Servanda – Or Else?, in: *Zaelke, Durwood/Kaniaru, Donald/Kružíková, Eva*, (eds.), Making Law Work: Environmental Compliance and Sustainable Development, Volume 1, 2005, p. 259

with Article X of the CITES.[364] This issue has meanwhile lost its importance, though, as the CITES have achieved almost universal partnership.

2. Legitimacy of Trade Sanctions under the CITES

The question could be asked as to whether or not the CITES sanction mechanisms are legally impeccable. For instance, it can be seen problematic that trade embargoes are developed and applied by the Governing Bodies of the CITES without the specific authorization of a treaty text. However, this question of legality loses its relevance, once one takes into account that trade embargoes need to be approved by national or, in case of the EU, supranational authorities.[365]

A more serious problem arises in relation to the compatibility of the sanction mechanism with the GATT/WTO regime.[366] The most important GATT provision in this regard is Article XX. This article authorizes GATT members to apply measures in order to protect a number of critical intangible goods and values, including human, animal, plant life or health. However, these measures should not be applied in a manner that would constitute a means of "arbitrary or unjustifiable discrimination between countries where the same conditions prevail, or a disguised restriction on international trade". As ambivalent as this provision is, it can be argued that because of the almost universal CITES membership, the described sanction mechanism can hardly violate the provisions of Article XX of GATT.[367] The CITES provides a platform for trade sanctions, the legality of which would be far more questionable if applied unilaterally. It should be noted that several States have already contested the legality of sanctions that have been applied by referring to the GATT provisions.[368] However, the States in question have never raised a complaint before the WTO dispute settlement body. Without

364 See CITES Resolution Conf. 9.5 (Rev. CoP13), Trade with States not Party to the Convention, 1994. – Internet resource – https://cites.org/eng/cop/09/doc/index.php [Last visit: 30.04.2023]; *Wijnstekers* [n. 308], pp. 342 f.

365 *Sand 2013* [n. 356], pp. 255 ff.

366 Marrakesh Agreement establishing the World Trade Organization (with final act, annexes and protocol), Concluded at Marrakesh on 15 April 1994, in: 1867 UNTS 3; see: *Cameron, James/Robinson, Jonathan*, The Use of Trade Provisions in International Environmental Agreements and Their Compatibility with the GATT, in: Yearbook of International Environmental Law, Volume 2, Issue 1, 1991, pp. 8–12.

367 *Yeater, Marceil/Vasquez, Juan*, Demystifying the Relationship between CITES and the WTO, in: RECIEL, Volume 10, Issue 3, 2001, p. 271; *Beyerlin, Ulrich/Marauhn, Thilo*, International Environmental Law, 2011, p. 434.

368 *Reeve* [n. 353], pp. 62 ff.

a relevant WTO decision, any discussions on the topic remain predominantly hypothetical.[369] Overall, it seems possible to assume that there is a silent international consensus on the compatibility of CITES trade embargoes with the WTO regime.[370]

The admissibility of trade sanctions as countermeasures constitutes another important legality issue. It should be specified, that the recommendations of the Governing Bodies of the CITES to apply trade embargoes against treaty violators does not aim to suspend the treaty in whole or in part in relation to a specific State, as is generally allowed in accordance with Article 60 of the 1969 Vienna Convention on the Law of Treaties,[371] but rather to restrict its several trade-related rights and obligations.[372] Due to the lack of international legally binding treaties regulating the admissibility of sanctions, the Articles on the Responsibility of States for Internationally Wrongful Acts is the main point of reference in this regard.[373] Part III/Chapter II of this legally non-binding document, which, however, chiefly represents international customary law,[374] deals with the legality of countermeasures in international relations. Despite the intentional exclusion of retorsions, such as trade embargoes, from the scope of the Articles due to the lack of the State practice,[375] this document, particularly its Articles 49–54, provides a key yardstick for the analysis of legality of trade embargoes under the CITES regime.[376] According to these trade embargoes, countermeasures have

---

369 *Yeater/Vasquez* [n. 367], p. 271.

370 *Beyerlin/Marauhn* [n. 367], p. 434

371 Vienna Convention on the Law of Treaties (with annex), adopted on 23 May 1969, 1155 UNTS 331.

372 *Rosenne, Shabtai*, Breach of Treaty, 1985, pp. 35–44; *Sand 2013* [n. 356], p. 258.

373 United Nations A/RES/56/83, Resolution adopted by the General Assembly on 12 December 2001, Responsibility of States for internationally Wrongful Acts, in: Report of the International Law Commission on the Work of Its Fifty-Third Session, in: Yearbook of the international Law Commission, Volume 2, Part 2, 2001, pp. 31 ff. Annex, Articles 43 f.

374 *Klein, Eckart/Schmahl, Stefanie*, Die Internationalen und Supranationalen Organisationen, in: *Proelß, Alexander/Graf Vitzthum, Wolfgang* (Hrsg.), Völkerrecht, 2016, p. 338.

375 United Nations, UN Doc. A/56/10, Introductory commentary by Special Rapporteur J. Crawford on Part III/Chapter II of the Draft Articles on State Responsibility adopted by the International Law Commission ('ILC Draft Articles'), Report of the International Law Commission on its 53rd Session, 29, 2001, p. 326.

376 See *Sand 2013* [n. 356], p. 259; for the necessity of the lawfulness of countermeasures in international relations see also *Koskenniemi, Mariti*, Solidarity Measures: State

to be consistent with the protection of human rights and with the general principles of international law as embodied in the UN Charter, including the rejection of the threat or use of force, and other documents.[377] Additionally, they have to be consistent with the dispute settlement procedure existing between the States engaged, and with the international law of diplomatic relations.[378] These countermeasures are deemed to induce compliance with a State's obligations under international law and should not create unnecessary obstacles for the resuming of the fulfillment of a Party's duties under the CITES.[379] Moreover, countermeasures should be temporal and proportional to the violation of the international legal provisions that induced their application.[380] Importantly, specific procedural requirements have to be fulfilled, including prior notification vis-à-vis possible countermeasures and the offer of resolving the issue through negotiations.[381] Sanctions should never be applied or should be immediately cancelled, if the State has resumed the fulfillment of its obligations or the case has been brought to the attention of the relevant international court. To sum up, sanctions should be temporal, proportional, and applied only as *ultima ratio*. It can be stated that the current sanction mechanism of the CITES as regulated by its Compliance Procedures satisfies all these requirements.[382]

3. *Trade Sanctions under the CITES and the Protection of Forests*

The sanction mechanism of the CITES has proven itself to be so effective, that the treaty does not need a special compliance control body.[383] The efficiency of the described sanction mechanism has been proven by the fact that trade embargoes and bans could always be lifted during the first year after having been applied, as those States to which they were applied always returned to compliance with the Convention.[384] It was done, *inter alia*, by adopting necessary national

---

Responsibility as a New International Order?, in: British Yearbook of International Law, Volume 72, Issue 1, 2002, pp. 337 ff.

377 Article 50.1(a–d).

378 Article 50.2(a–b).

379 Article 50.1,3.

380 Articles 51, 53.

381 Article 52.

382 See also *Sand 2013* [n. 356], p. 259.

383 *Koester* [n. 350], p. 1 ff.

384 Analyze of different cases of CITES trade embargo application at *Reeve* [n. 353], pp. 91–180.

legislative acts, submitting overdue implementation reports and complying with action plan requirements.[385] However, its effectiveness in relation to the promotion of forest protection is limited by the generally insignificant role of the CITES in this area.

The sanction mechanism of the CITES can serve as a basis for the development of sanction mechanisms to be included in forest-specific treaties and mechanisms, such as REDD+, but only if the documents in question regulate the trade in wood and non-wood forest goods, other forest-related commodities, and international financial transfers. In order to avoid legality concerns, the respective provisions should be included in the original treaty text. However, as these provisions can alter the reaching of a consensus, a compliance mechanism may also be developed by the Governing Bodies of any international environmental treaty dealing with forests, as in the case of the CITES.

V. International Tropical Timber Agreement

The International Tropical Timber Agreement (ITTA)[386] represents another major international forest-relevant convention adopted in the aftermath of Stockholm Conference in 1983. However, the text was subsequently revised, and new versions of the treaty were adopted in 1994[387] and 2006.[388] The version of the agreement that was released in 2006 is referred to as ITTA3 and has been in force since 2011. As the ITTA can be described as being a certain mix of a commodity treaty and an environmental agreement, one placing a great deal of emphasis upon the exploitation of forest timber resources, its contribution to conservation of forests is controversial.[389] Nevertheless, an attempt was undertaken to reconcile the ecological and economical aspects of the industrial utilization of tropical forests in the treaty text.[390] It can be seen as a positive development that since the

385 Ibid.

386 International Tropical Timber Agreement, adopted on 18 November 1983 and entered into force on 1 April 1985, in: 1393 UNTS 67.

387 International Tropical Timber Agreement, adopted on 26 January 1994 and entered into force on 1 January 1997, in: 1955 UNTS 81.

388 International Tropical Timber Agreement, adopted on 27 January 2006 and entered into force on 7 December 2011, in: 2797 UNTS 75.

389 *Beyerlin/Marauhn* [n. 367], pp. 201 ff.; for alternative view considering the ITTA as a purely commodity agreement with several environmentally relevant provisions, see: *Schulte zu Sodingen* [n. 233], p. 163; *Krohn* [n. 276], p. 93.

390 *Wassermann, Ursula*, International Tropical Timber Agreement, in: Journal of World Trade, Volume 18, Issue 1, 1984, p. 89.

treaty text was revised in 1994 and 2006, States have tended to further accentuate environmental issues. It can be argued that the 1983 ITTA should be perceived in the light of general post-Stockholm developments in international environmental law, whereas the modern version of the agreement should be analyzed as part of a post-Rio system of global environmental governance. This distinction is important in order to compare the ITTA with the UNFCCC, the CBD, and other documents effectively.

The main objective of the 2006 ITTA includes environmental goals. It states that the Convention aims to promote both international trade in legally harvested tropical timber from sustainably managed forests and sustainable forest management in the tropics.[391] Environmental issues are also considered in the list of means designed to achieve this objective. It is interesting to note how the environmentally relevant provisions of the ITTA have evolved over time. The 1983 ITTA seeks to promote forest-related scientific research and to encourage tropical timber reforestation, although the last provision can also be interpreted as the promotion of artificial monoculture tree plantations.[392] Moreover, Article 1(h) of the ITTA1 from 1983 encourages the development of national policies aimed at the promotion of sustainable forestry as well as the conservation of forests and their genetic resources, by contributing to the maintenance of an environmental balance in the tropics. Article 1(m) of the ITTA3 has almost completely inherited this provision, except for its reference to the genetic resources of forests, which was removed in order to avoid possible overlaps with the CBD, which deals specifically with these issues.

Other provisions included in the 2006 ITTA also deal with environmental issues. The promotion of tropical timber trade and of sustainable forestry in the tropics may contribute to sustainable development and poverty alleviation.[393] In contrast to the 1983 ITTA, the modern version of the treaty seeks to promote scientific research not only in relation to timber production issues, but also concerning the conservation and enhancement of non-timber forest values.[394] Moreover, the Agreement seeks to promote the better understanding of the contribution of non-timber forest products and environmental services of forests to sustainable forest management.[395] The Parties should consult on this issue

391 Article 1.
392 Ibid.
393 Article 1(c).
394 Article 1(f).
395 Article 1(q).

other relevant international institutions,[396] which opens the door to collaboration with the United Nations Forest Forum (UNFF), the UNFCCC and the CBD. Members are also encouraged to promote tropical reforestation, the rehabilitation and restoration of degraded lands, all with due regard to the interests of forest-dependent ILCs, and to recognize their role in achieving sustainable forest management.[397] The treaty also aims to promote trade in sustainably cultivated and legally harvested goods, by strengthening the consumer's awareness. This last provision implicitly refers to forest certification schemes, which are directly addressed in Article 1.(o).[398] Furthermore, it aims to promote the enforcement of forest law and to combat both illegal logging and trade in illegally harvested timber.[399] Importantly, the modern version of the ITTA mentions the issue of illegal logging for the first time, whereas the 1983 ITTA does not address this issue and the 1994 ITTA contains only a vague reference to "undocumented trade".[400] It is submitted that the Forest Law Enforcement and Governance (FLEG) process has contributed to the inclusion of the issues in question in the ITTA's text.[401] The inclusion of these issues can be seen as a great success, as initially the ITTO refused to acknowledge the very existence of the problem of illegality.[402]

Thus, the ITTA includes a comprehensive list of goals aiming to increase the efficiency of tropical timber harvesting and trade and to promote sustainable forestry. However, the ITTA does not stipulate any commitments or obligations that must be met and does not deliver any instruments or tools that would promote the achievement of its objectives. Only Article 29 stipulates that ITTA members have to make efforts in order to promote the achievement of the objectives of the convention and should refrain from any actions contrary thereto.[403]

396 Ibid.

397 Article 1(j,r).

398 Article 1(k,o).

399 Article 1(n).

400 See also: *Gulbrandsen, Lars H./Humphrayes, David*, International Initiatives to Address Tropical Timber Logging and Trade, A Report for the Norwegian Ministry of the Environment, Fridtjof Nansens Institutt, 2006, p. 24.

401 Ibid.

402 Ibid.

403 *Schulte zu Sodingen* [n. 233], p. 162; *Krohn* [n. 276], pp. 93 ff.; *Nagtzaam, Gerry*, The International Tropical Timber Organization and Conservationist Forestry Norms: A Bridge Too Far, 2008. – Internet resource – https://citeseerx.ist.psu.edu/viewdoc/download?doi=10.1.1.880.8364&rep=rep1&type=pdf [Last visit: 30.04.2023]

For a number of reasons, the ITTA was dedicated only to tropical-timber related issues, thus excluding other types of forests from the scope of the agreement. It was argued, among other things, that tropical timber constitutes a major part of international timber trade, whereas timber originating from temperate and boreal forests is usually designated for internal consumption and/or industrial processing. Moreover, rapid forest loss in the tropics was the international community's main concern at that time.[404] It was argued that boreal and temperate forests are already sustainably managed, and that industrial pollution and acid rain represent their main threats. This argument can no longer be maintained, as recent research has confirmed that significant parts of forests in northern latitudes are managed unsustainably and overexploited.[405] Additionally, the North-South controversies have significantly influenced the negotiation process of the new ITTA. As in other cases, certain industrial States were opposed to any international regulations that would impose restrictions on international trade in their timber products, establish a legally binding obligation for them to conserve temperate and boreal forests, or otherwise restrict their national sovereignty.[406] At the same time, many developing forest-rich States have been motivated to join the ITTA by the possibility of improving the regulation of the international tropical timber market and the maintenance of stable and fair prices on tropical timber.[407]

Despite its general focus on the regulation of tropical timber harvesting and trade, and the exclusion of non-tropical timber from the scope of the convention, the ITTA still delivered a number of environmentally relevant provisions. Furthermore, the text of the ITTA3 indicates that the agreement should be used as a platform for the elaboration of policies concerning non-tropical timber as well. In this regard, Article 1(a) stipulates that the agreement should provide an effective framework for consultations and policy developments in relation to all

404 *Humphreys, David*, Forest Politics: The Evolution of International Cooperation, 1996, pp. 95 ff.

405 *Dudley, Nigel/Jeanrenaud, Jeau-Paul/Sullivan, Francis*, Bad harvest? The Timber Trade and the Degradation of the World's Forests, 1995, pp. 8 f.; *Gagnon, Jennifer*, A Global Conservation Assessment of Temperate Forests: Status and Protection, University of New Orleans: Theses and Dissertations, N. 45, 2003, p. 48; *Dudley, Nigel*, Forests in Trouble: A Review of the Status of Temperate Forests Worldwide, 1992, pp. 15 ff.; *de Gouvenain, Roland C/Silander, John*, Temperate Forests, in: *Roitberg, Bernard D./Cotter, Paul D.* (eds.), Reference Module in Life Sciences, 2017, pp. 1 ff.

406 *Dudley 1992* [n. 405], pp. 1 ff.

407 Ibid.

forests. Despite the fact that most of the provisions of the Agreement refer explicitly to tropical timber, it can be seen as positive that some developed States have also undergone voluntary commitments to promote sustainability in their forests.[408]

The adoption of environmentally relevant decisions under the framework of the ITTA is often complicated by the prevalence of tropical timber export States in the Governing Bodies of the ITTO.[409] On the other hand, these States should arguably be interested in the conservation of their timber producing forests. The growing acknowledgement of sustainable forest management and the carrying out of preliminary negotiations with tropical forest-rich States can facilitate the reaching of a consensus that is needed for the elaboration of conservationist policies.[410]

The International Tropical Timber Organization (ITTO), founded in 1985, serves as the main Governing and Executive Body of the Convention. The ITTO carries out its functions through the International Tropical Timber Council (ITTC), which includes three Committees, the Executive Director, and the Secretary. According to Article 7, the ITTC is authorized to adopt rules and regulations that are necessary for the implementation of the Agreement and for the achievement of its goals. These regulations include those dealing with the management of financial resources. It is argued that Article 19.2 indicates that every decision made by the ITTC must be implemented in the national legislative systems of its Member States, as appropriate.[411] However, the wording of Article 19.2 can hardly support this assumption. It only stipulates that Parties must undertake measures in order to accept and implement the Decisions of the Council and must refrain from any actions contrary thereto. It can be argued that States are encouraged to implement the Decisions of the Council, but that the extent of implementation thereof is left to their discretion.

However, although not required by the treaty, the ITTC's decisions have to be implemented in the national law of Member States in order to achieve the consistent and effective implementation of the treaty. The Decisions in question often suffer from the lack of concreteness. Consequently, it is the responsibility of Member States to "fill in the gaps" by adopting national laws and regulations.[412]

408 *EUROFOR*, Europe and the Forest, External Study carried out for European Parliament, European Parliament, 1997.

409 *Schulte zu Sodingen* [n. 233], p. 162; *Krohn* [n. 276], pp. 93 ff.

410 EUROFOR [n. 408].

411 *Krohn* [n. 276], pp. 93 ff.

412 *Eikermann* [n. 234], pp. 82 ff.

This approach can be criticized as it can easily lead to the insufficient implementation of ITTC's decisions, but it also facilitates the reaching of a consensus needed for the adoption of the ITTC's decisions.[413]

Another downside of the ITTA is its week compliance mechanism. States are neither requested to submit reports on the implementation of the ITTA, nor has a monitoring mechanism been established. On the other hand, the ITTC is authorized to organize and conduct independent research in order to analyze the state of the implementation of sustainable forest management practices in tropical timber export States, but only with the allowance of the State concerned. One example is the ITTC mission that was active in the Sarawak State of Malaysia from 1989 to 1990. In May 1990, the mission presented the results of its research activities in a final report, outlining a dramatic rate of deforestation in Sarawak and a lack of ILCs' involvement in the process of decision-making. Inter alia, the mission recommended ensuring the protection of different types of ecosystems, in order to promote the conservation of biodiversity, to facilitate the access to decision-making processes for ILCs, and to enhance the socio-economic and environmental benefits of natural parks. However, neither the mission report nor the further ITTO decisions on the topic requested to reduce significantly the rate of deforestation in the region. The report treated the industrial-scale logging in the Sarawak forests as an admissible practice and stipulated that only the State-dominated process of interest-balancing should determine the amount of tropical timber gained. With this mission, ITTO engagement in Sarawak began, which included both consultation on and financial support of a number of projects.

During the ITTC meeting in Bali in 1990, the parties involved committed themselves to ensure that as of the year 2000, all tropical timber designated for international trade would come from sustainably managed sources. The achievement of this goal (the so-called Target 2000 or "Bali commitment") was unrealistic, because in the precedent period only 1% of loggings found place in more or less sustainably-managed forests. For this reason, at the meeting in November 2000, the ITTC only outlined that the full transition to the timber trade from sustainably managed sources would have to be achieved "as [rapidly] as possible" without specifying any deadlines further.[414]

413 *Krohn* [n. 276], p. 95.

414 Article 1 of ITTC Decision 2 (XXIX). – Internet resource – http://www.itto.int/sessions_detail/contents_type=184 [Last visit: 30.04.2023]

It is worth noting that the ITTO has also developed voluntary guidelines for the sustainable management of natural tropical forests.[415] Accordingly, SFM should be based on strong forest governance and secure forest tenures with clearly defined access and usage rights (Principle 1). Additionally, a science-based approach to land use and forest management planning, as well as to the constant monitoring of forests is required (Principle 2) in order to maintain and enhance the socio-economic and environmental functions of forests. For every forest management unit (FMU), each plan should include clearly stated management objectives and the measures required for the achievement thereof (Principle 5). The ability of forests to provide multiple benefits to society should be promoted with special regard to ecological balance and climate change issues (Principles 3 and 4). Moreover, forest management should be based on participatory decision-making, with particular attention being given to the interests of ILCs and suitable conditions and favorable working environments for forest workers being created (Principle 6). Importantly, the voluntary guidelines in question recognize that a strict interdependence exists between the promotion of SFM and the provision of additional substantial financial resources. However, the ITTO does not accentuate the need for further provision of forest finance from donor States. Instead, States are encouraged to promote economic evaluation of the environmental services provided by forests and to ensure the equitable distribution of the costs and benefits derived from SFM (Principle 7).

It is worth noting that the ITTO was a pioneer in the development of the Criteria and Indicators for sustainable forest management in the tropics.[416] The C&I represent a set of tools that serve to assess and monitor any progress made towards the implementation of SFM. They include the size and conditions of forests; forest ecosystem health and resilience; forest production; forest biological diversity; soil and water protection, economic, social and cultural aspects.[417] Theoretically, the Voluntary SFM Guidelines and the C&I can also be applied to non-tropical forests.

---

415 ITTO, Voluntary guidelines for the sustainable management of natural tropical forests, 2015. – Internet resource – http://www.itto.int/files/user/pdf/PR%20and%20Communications%20material/ITTO%20Voluntary%20Guidelines%20BROCHURE%20Web.pdf [Last visit 30.04.2023]

416 *Schulte zu Sodingen* [n. 233], p. 162; *Krohn* [n. 276], pp. 93 ff.

417 ITTO, Criteria and indicators for the sustainable management of tropical forests. ITTO Policy Development Series No. 21, 2016, p. 15.

By developing the Voluntary SFM Guidelines and the C&I, the ITTO sought to provide scientific assistance to timber-producing States while fulfilling their sustainability goals. However, once a forest-rich State commits itself to the implementation of the said documents, it may apply for the provision of further training and financial support.[418] The dependence of the achievement of the ITTA's objectives on financial support from the international community is also recognized in the ITTA's Preamble (s) and in Article 1(g) of the ITTA. In order to meet this need for additional sources of finance, Article 18 of the ITTA ordered the establishment of the Special Account and Bali partnership Fund, so as to manage the voluntary contributions of Member States. Thus, the ITTA facilities are used to both financially and technologically support a number of projects in tropical timber producing States.

It is noteworthy that the concept of SFM as promoted by the ITTO needs to be revised under the consideration of new studies that have demonstrated that sustainable industrial logging in the tropics is often not compatible with the natural lifecycles of endangered tree species.[419] The ITTA's impact on forest conservation is also relativized by the fact that this commodity agreement is unable to combat the main driving forces behind deforestation, such as forestland conversions, the spread of monoculture plantations and unsustainable agricultural practices.[420] To sum up, there is still a room for improvement vis-à-vis empowering the ITTO with further tools and instruments that would permit it to consider multiple forest functions and to combat multiple driving forces of deforestation. However, the positive outcome that these amendments to the ITTA might have for the forest conservation would be limited by the main scope of the agreement.

C. The Rio Summit in 1992 and Its Impacts on the International Legal Regime for Forests

The United Nations Conference on Environment and Development, that was held in Rio-de-Janeiro, Brazil, on 3–14 June 1992, was an important milestone in

418 *Schulte zu Sodingen* [n. 233], p. 162; *Krohn* [n. 276], pp. 93 ff.

419 *Barber, Victor Charles*, 25 Years of So-Called "Sustainable Forest Management" in Sarawak, World Resources Institute,19 November 2015. – Internet resource – https://www.wri.org/insights/25-years-so-called-sustainable-forest-management-sarawak [Last visit: 30.04.2023]

420 *Young, Tom*, Malaysian Palm Oil Destroying Forests, Report Warns, in: The Guardian, 2 February 2011. – Internet resource – https://www.theguardian.com/environment/2011/feb/02/malaysian-palm-oil-forests [Last visit: 30.04.2023]

the development of international environmental law.[421] The very fact that almost every State, each represented by a head of State, participated in the conference, shows how great a success it was. Political, economic, and cultural controversies were, inspired by a common understanding of the urgent need to combat environmental crises, largely put to one side.[422] During the conference, two important international, legally binding agreements were adopted: the UNFCCC and the CBD. To this list should be added the United Nations Convention to Combat Desertification in those Countries Experiencing Serious Drought and/or Desertification, Particularly in Africa (UNCCD). The UNCCD was only adopted in 1994, but the negotiations thereof were launched at the Rio Conference. Additionally, a number of non-legally binding documents of high political importance came out of Rio: the Rio Declaration on Environment and Development,[423] the Agenda 21,[424] and the Non-Legally binding Authoritative Statement of Principles for a Global Consensus on the Management, Conservation, and Sustainable Development of All Types of Forests (Rio Forest Principles).[425] Although all these documents play a relatively significant role in international forest regimes, the latest document deserves particular attention. Among other things, its status as non-legally binding but still authoritative deserves particular attention.[426] Arguably, the choice of wording was inspired by the States' dissatisfaction with the document. In fact, the Forest Principles are regarded as the main fiasco of the Rio conference, as on the eve of the conference States were negotiating a legally binding forest convention, but during the meeting were unable to reach the consensus needed for its adoption.[427] Actually, the Forest Principles differ from other soft law documents adopted in Rio in that the Rio

---

421 *Beyerlin/Marauhn* [n. 367], pp. 13. ff.; *Schulte zu Sodingen* [n. 233], pp. 185 ff.

422 *Schulte zu Sodingen* [n. 233], pp. 185 ff.

423 Rio Declaration on Environment and Development, United Nations Conference on Environment and Development, Rio de Janeiro, Brazil, 3–14 June, 1992, in: ILM, Volume 31, 1992, pp. 874 ff.

424 Agenda 21, United Nations Conference on Environment and Development, Rio de Janeiro, Brazil, 3–14 June, 1992. – Internet resource – https://sustainabledevelopment.un.org/content/documents/Agenda21.pdf [Last visit: 30.04.2023]

425 Non-Legally Binding Authoritative Statement of Principles for a Global Consensus on the Management, Conservation and Sustainable Development of All Types of Forests, adopted on 14 June 1992, in: ILM, Volume 31, 1992, pp. 881 ff.

426 *Schulte zu Sodingen* [n. 233], pp. 191 ff.

427 *Schulte zu Sodingen* [n. 233], pp. 191 ff.; *Krohn* [n. 276], pp. 168 ff.

Declaration on Environment and Development and Agenda 21 deal with general issues of environmental conservation. Instead, specific forest-related topics are at the focus of the Forest Principles. That such a generalized subject area is covered by the Rio Declaration and Agenda 21 is a reason for their non-legally binding character. If these documents were meant to establish legal obligations, they would evolve into some kind of "Universal Environmental Agreement", the time for which has not yet come at the time of Rio because of sovereignty issues. The negotiations of this document would also go beyond the mandate of the Rio Conference.[428] These documents were destined to serve as a general guideline on the implementation of environmental policies and as the manifestation of the existing political consensus on the issues in question. At the same time, the Forest Principles deals with a specific and precisely defined environmental problem, the regulation of which presupposes the adoption and implementation of detailed, legally binding provisions as being the perfect outcome. Thus, a legally binding form should be preferred for this document.

The failure of the Rio Conference to adopt a legally binding forest convention is often explained by a number of factors.[429] As in the case of the ITTA whose scope is limited only to tropical forests, the issue of North-South controversies is often cited in this regard. Developing States expected that developed ones would undergo the same restrictions concerning the utilization of forest resources and forestland conversions.[430] Otherwise, they would have regarded the legally binding agreement as a disguised attempt of arbitrary discrimination in the global timber trade.[431] Mahathir Mohammad, the Prime Minister of Malaysia, a State that unofficially lead a group of developing States in opposing the adoption of a legally binding forest convention in Rio, can be cited in this regard.[432]

---

428 *Laitos, Jan/Okulski, Julian*, Why Environmental Policies Fail?, 2017.

429 *Davenport, Deborah S.*, An Alternative Explanation for the Failure of the UNCED Forest Negotiations, in: Global Environmental Politics, Volume 5, Issue 1, February 2005, pp. 107 ff.

430 Ibid, p. 108.

431 Ibid.

432 At the 2nd Ministerial Conference of developing countries States on environment and development that was held on 27 April 1992, Mahathir Mohammad stated his position on the forest-related negotiations: "I sincerely hope that the historic opportunity in Rio will not be wasted. Rio can be the occasion to take important historic steps for a true global partnership. The South has suffered enough. It is wrong that we should be made scapegoats for the past sins of the North. The South cannot remain the repository of the resources for the North including locking up its forest to serve as the

It must also be mentioned, that developing forest-rich States were expecting developed ones to provide substantial financial resources to not only cover the costs of forest protection but also any lost development opportunities.[433] However, developed States were unwilling to do this. They were also extremely reluctant to make any legally binding commitments to preserve forest ecosystems within their territories. Moreover, as in the case of the ITTA, it was argued that developed States' forests were already managed sustainably, with forest health and vitality both preserved and promoted (which was actually not the case).[434] However, the North-South controversies cannot be sufficient in order to explain the failure of negotiations of a legally binding forest convention during the Rio Conference, as well as during subsequent international meetings.[435] The above-mentioned North-South controversies can be applied to all the thematic elements of the Rio conference, including the negotiation of the UNFCCC, which however did not hinder its adoption. Arguably, the UNFCCC, and the subsequently adopted Kyoto Protocol and Paris Agreement, possess a higher potential to restrict the national sovereignty of developed States and the development policies of developing States then the Forest Convention itself. Legally binding goals for the reduction of emissions impose restrictions on the industry and transportation branches, not only on the forestry sector. At the same time, legally binding goals for the reduction of deforestation and/or forest degradation rates can have a comparatively smaller impact on the economic development. Both industrial and developing States have already sacrificed a part of their sovereignty and decisions-making discretion vis-à-vis economic issues by adopting the UNFCCC. A legally binding forest convention, which could assume a form of a framework treaty – meaning that no precise and verifiable targets would have to be included in it –, would not make matters much worse for them. The unwillingness of developed States to provide assistance and substantial financial

global green lung and its genetic resource laboratory. The North must help the South to develop for it is in their environmental, economic and security interest that they do so". Source: Speech of Mahathir Mohammad, held in Kuala Lumpur, on 27 April 1992 can be found in the Internet at: http://www.mahathir.com/malaysia/speeches/1992/1992-04-27.php [Last visit 30.04.2023].

433 *Scholz, Imme,* A Forest Convention – Yes or No?, Position Paper for the BMZ on the State and the Perspectives for Creating a Legally Binding Instrument for International Forest Policy, Deutsche Institut für Entwicklungspolitik, Bonn, 2004; *Schulte zu Sodingen* [n. 233], pp. 191 ff.

434 See Subsection B.I of Chapter 1.

435 *Davenport* [n. 1715], p. 105.

resources to assist forest-rich States to protect their forests cannot be explained by a pure reluctance to the provision of additional funding, as the UNFCCC, the CBD, and the CCD also require significant financial resources for their implementation. It can be suspected that if the negotiation of a forest convention were successful, it would be possible to avoid the inclusion of legally binding financial goals in its text. Instead, a mechanism for fundraising could be established. The unwillingness of developed States to commit themselves to the enhancement and improvement of the quality of the forests under their jurisdiction is not clear from the economic point of view, because the emission reduction commitments they have agreed to under the UNFCCC and Kyoto Protocol already represent a clear self-limitation of their economic ambitions. One could argue that tropical deforestation has been given less attention at international negotiations since Rio, receiving less media coverage and public attention than climate change and loss of biodiversity. In reality, however, people in developed States were already concerned about the survival of tropical forests before Rio, with a number of European States introducing import bans on tropical timber in the 1980s and 90s.[436] Consequently, the question remains open as to why the progress that has been made on such crucial aspect of environmental conservation as that of combatting, halting and reversing deforestation and forest degradation has been so insignificant. One should also consider that the forest conservation is one of the most cost-effective approaches to tackling climate changes, biodiversity loss and desertification.

One of reasons can consist in the technical obstacles related to the measuring and verifying of deforestation rates reduction goals. Moreover, any more-or-less effective legally- binding commitments specifically designed in order to promote a reduction of deforestation rates should be of a global nature and permit to avoid leakage, otherwise they would contradict the very scope of a hypothetical convention. It is noteworthy that the market-based mechanisms are supposed to play a major role in the promotion of environmental protection. As the future development of the REDD+-mechanism has demonstrated, the development of a market-based mechanism is more likely, if an investor is allowed to use the reduction of the net CO_2-emissions that have been achieved through the forest conservation to comply with their own commitments on the mitigation of climate change.[437] However, this scheme would likely have been rejected by developing States in Rio, as, in their view, climate change mitigation would only

436 *Schulte zu Sodingen* [n. 233], pp. 369 ff.
437 Ibid.

have their developed counterparties to implement those measures and policies necessary.[438] It is worth noting that it is currently impossible to link the multiple forest functions to any single market in order to evaluate properly the positive impacts of forest conservation.

It cannot be proven whether or not the participants of the Rio Earth Summit also took the forthcoming considerations into account. However, it should be noted that the amount of financial resources to be provided in order to compensate the lost development opportunities linked to the forest conservation in forest rich States would be difficult to estimate.[439] Converted forestlands can be used for low-revenue farming activities as well as highly profitable mining and industrial operations. Developing States' claims to this significant amount of resources could be in contradiction with the principle of common but differentiated responsibility,[440] which presupposes that all States strive together to resolve global environmental problems according to their available capabilities and resources, whereby developing States can rely on financial help from developed ones.[441] However, this principle, which plays an important role in international environmental law, does not imply that the entire responsibility for the protection of the world's forests be assumed by developed States alone.[442]

Additionally, a focus on financial help from abroad would undermine the intrinsic value of forests as well as their substantial economic benefits, which can be retrieved from non-timber forest goods and the ecological functions of forests. It was not taken into consideration, that in a number of cases deforestation and forest degradation did not contribute to the improvement of local livelihoods but have actually aggravated the socio-economic conditions of local populations (as in Haiti and Madagascar).[443] Some forest-rich States were able to improve local livelihoods in the long term, while still conserving their forests, as in the cases of Costa Rica and Dominican Republic.[444]

Another major reason why an international forest convention has never been adopted could also be a lack of will to focus on environmental issues. A legally

---

438 Ibid, p. 370.

439 *Schulte zu Sodingen* [n. 233], p. 192.

440 *Proelß, Alexander*, Raum und Umwelt im Völkerrecht, in: *Proelß, Alexander/Graf Vitzthum, Wolfgang* (eds.), Völkerrecht, 7. Auflage, 2016, pp. 423 ff.

441 *Schröter, Jessica*, Strukturprinzipien des Umweltvölkerrechts und ihr Beitrag zur Eindämmung des Klimawandels, 2015, p. 248.

442 Ibid.

443 See Subsection A.II of Chapter 1.

444 See Subsection A.II of Chapter 1.

binding agreement on forests would require an integrated approach that would consider a number of environmental, economic, socio-cultural, and political issues. As for other environmental agreements, a common understanding of the driving forces and consequences of deforestation, as well as of the ways to combat the forest loss with due regard to local circumstances, is required.[445] Differing understandings of crucial forest-related issues have hampered the achievement of a wide-ranging consensus.

The absence of a legally-binding forest convention can be also explained by a lack of effort on behalf of developed States to steer negotiations in such a way that they would cause developing States to change their negotiating positions. Arguably, at the preparatory meetings for the Rio Conference, developing States showed on several occasions their readiness to cooperate on the achievement of a consensus. However, developed States lacked the will to apply their existing resources in order to facilitate the negotiation of the Forest Convention.[446]

As has been shown, the absence of a legally-binding Forest Convention has been determined by the difficulties associated with the raising of funds, the lacking recognition of both non-timber benefits and the environmental services delivered by forests, the high complexity of forest-related issues, and by developed States' lack of interest in steering the negotiation process in a direction that would make the adoption of a Forest Convention feasible. As ongoing forest negotiations are not currently leading towards the adoption of an effective forest-specific convention in the near future, forest-related Rio documents should be analyzed at the current stage of the research, with the exclusion of the UNFCCC and the CBD, which are examined in the next chapters.

I. UNCCD and the Protection of Forests

The UNCCD[447] was adopted on 17 June 1994 and came into force on 26 December 1996. Although certain States are not affected by desertification directly, this disastrous phenomenon has had adverse impacts on water and food supplies and thus could have a negative impact on sustainable development and security

445 *Scholz* [n. 433], p. 18.

446 See Subsection D.I of Chapter 6.

447 Convention to combat desertification in those countries experiencing serious drought and/or desertification, adopted on 17 June 1994 and came into force on 26 December 1996, in: 1954 UNTS 3.

worldwide.[448] With this in mind, the UNCCD enjoys almost universal membership, with 197 States currently being Party to the Convention. The negotiations and adoption of the UNCCD were established by Chapter 12 of Agenda 21 ("Managing fragile ecosystems: combating desertification and drought").[449] It is worth noting that when the negotiations were launched, there was a lack of a consensus on the necessity of the Convention.[450] Developing States argued that the CO_2-emissions originating from the industrial North are the main driving force of desertification.[451] From this point of view, an international treaty combating desertification is needed, meaning that the Convention in question can be considered as a logical extension of the UNFCCC and the CBD.[452] At the same time, developed States argued that deforestation, unsustainable forestry and agricultural practices and other internal factors are the main causes of desertification in developing States. With this in mind, they denied the necessity of the convention on desertification, while arguing that developing States are solely responsible for desertification within their territories.[453] However, they finally abandoned their position and consented to the adoption of the UNCCD in a vain hope of pursuing the achievement of a consensus on the adoption of the forest convention. One of the reasons why the UNCCD did not contribute to the forest negotiations is that those States prone to desertification had already cleared the most of their forests, and had less political power at meetings than forest-rich States in Latin America and Asia.[454] Moreover, the adoption of the UNCCD could not facilitate the negotiations of a forest convention because it focuses only on one environmental problem related to deforestation, whereas an effective forest convention would have to take into account numerous forest-related issues and properly evaluate the ecological benefits of forests. Thus, the adoption of the UNCCD did not facilitate the negotiation of a forest convention.

448 *Kürzinger, Edith/Schipule, Peter Hans*, Desertifikationskonvention: Ein Lehrstück für den Rio-Folgeprozeß? in: Entwicklung und Zusammenarbeit, Band 37, Heft 1, 1996, pp. 8 f.

449 Paragraph 40 of Chapter 12 of Agenda 21.

450 *Kürzinger/Schipule* [n. 448], p. 8; *Burns, William C.*, The International Convention to Combat Desertification: Drawing a Line in the Sand?, in: Michigan Journal of International Law, Volume 16, Issue 3, 1995, pp. 831, 849 ff.

451 *Burns* [n. 450], pp. 831 ff.

452 Ibid.

453 Ibid.

454 Ibid.

Nevertheless, the adoption of the UNCCD can be seen as an important achievement, particularly under the consideration of immense damage caused by the desertification worldwide.[455] Despite the graveness of the desertification problem, the combating thereof played only a marginal role in international activities until the adoption of the UNCCD.[456]

Combating of desertification, land degradation, and drought (DLDD) are in the focus of the UNCCD. The Convention could not avoid mentioning the problem of deforestation, as deforestation and forest degradation belong to the leading causes of DLDD. Article 1(a) identifies desertification as a process of *land degradation* due to climate changes and human activities. Article 1(f) indicates further that one of the major causes of *land degradation* is the long-term loss of natural vegetation. Consequently, the UNCCD recognizes that there is a deep interdependence between desertification and deforestation. According to its provisions, States should develop national action programs in order to identify the actions that are needed in order to combat desertification and promote their implementation.[457]

The Annexes to the Convention include a vast list of activities that should be included in National Action Programmes. In total, the UNCCD has five Annexes, each of which deals with desertification-related issues in five specific regions: Africa, Asia, Latin America and the Caribbean, the North Mediterranean and Central and Eastern Europe. Article 8.3(b)(i) of the Annex for Africa determines that the National Action Programmes in this region should promote the conservation of natural resources by assuring the sustainable management of natural resources, including forests, natural vegetation, and wildlife. States in Latin America and the Caribbean should also include the promotion of sustainable agricultural and forestry in their National Action Programmes. However, the forest-related aspects of the UNCCD do not go beyond these provisions. In total, forests play only a marginal role in this treaty. However, the Convention bears the potential to foster forest conservation activities in the process of its implementation.[458]

---

455 See *King, Ed,* Desertification Crisis Affecting 168 Countries Worldwide, Study Shows, in: The Guardian, 17.04.2013. – Internet resource – https://www.theguardian.com/environment/2013/apr/17/desertification [Last visit: 30.04.2023]

456 *Honey, Wahl,* Vereinte Nationen und Umwelt, UN Basis-Informationen, 1993, p. 6.

457 Article 2 UNCCD.

458 *Schulte zu Sodingen* [n. 233], p. 257.

The UNCCD particularly emphasizes the importance of active public participation in the process of its implementation.[459] While referring to national governments on this issue, the Convention not only invites rural residents to promote the gentle use of land resources, but also delivers an international legal framework for the stable improvement of their living conditions. Both the capacity building[460] and the income diversification are foreseen.[461] All these issues play a major role in forest protection, because unsustainable agricultural and forestry activities that are strictly related to poverty and unemployment represent a major source of hazards for forest conservation.

Despite its almost universal membership, political and financial support for the implementation of the Convention remained insignificant for a long time.[462] In order to combat the insufficient implementation of the treaty, UNCCD COP 8 in 2007 adopted a Strategic Action Plan for the period of 2008–2018.[463] This document contained four strategic objectives, including the improvement of local livelihoods, the restoration of degraded ecosystems, the promotion of global benefits, and resource mobilization. It was expected that after the ten-year period an adequate international response to the problem of DLDD would be achieved. The synergies with climate change issues and biodiversity protection should also be promoted. The engagement of civil society and the scientific community in the implementation of the UNCCD should be strengthened. It is important to note that the list of indicators used to evaluate the success achieved in the process of implementing of the Strategic Action Plan in question included Indicator S-7, which refers to the areas of forest under SFM.

The Outcome Document of Earth Summit 2012 in Rio (Rio+20), "Future we want",[464] reaffirmed the commitment of its participants to achieving the goals of the UNCCD Strategic Action Plan. It further underlined the necessity to halt

---

459 *Beyerlin, Ulrich*, Umweltvölkerrecht, 2000, p. 258.

460 Article 12 UNCCD.

461 Article 10.3(d,e); Article 8.3(a) of Annex for Africa, UNCCD.

462 *Birnie, Patricia/Boyle, Alan/Redgwell, Catherine*, International Law and the Environment, 2009, p. 694.

463 UNCCD, The 10-year strategic plan and framework to enhance the implementation of the Convention (2008–2018), UNCCD, 2008. – Internet resource – http://www2.unccd.int/sites/default/files/relevant-links/2017-01/Strategy-leaflet-eng.pdf [Last visit: 30.04.2023]

464 United Nations A/RES/66/288, Resolution adopted by the General Assembly on 27 July 2012: The future we want, 2012. – Internet resource – https://undocs.org/A/RES/66/288 [Last visit: 30.04.2023]

and reverse land degradation. Subsequently, DLDD-related issues have gained more international attention, with the post-2015 development agenda being elaborated upon by UN General Assembly,[465] which replaced the Millennium Development Goals. This document includes a lot of forest-related issues. The voluntary commitments to stop and reverse land degradation (Goal 15) and to strive towards the achievement of land degradation neutrality until 2030 are particularly important in this regard. In order to promote the achievement of these goals, the UNCCD has launched a number of activities, including the land degradation neutrality (LDN) project, within the framework of which a number of States receive scientific, technical and financial assistance for the promotion of land degradation neutrality. Moreover, in September 2017, the new land degradation fund was established during UNCCD COP 13 in Ordos, China. With an initial target of 300 million dollars, the fund is supposed to raise additional financial resources to promote land degradation neutrality in developing States, which also include forest conservation and reforestation.

Despite all these achievements, funding for the DLDD remains scarce.[466] The UNCCD can be referred to as a "people's convention" because of its predominantly bottom-up approach, meaning that important decisions are taken exactly at the levels where they have to be implemented.[467] Despite certain benefits, this situation results in the underestimation of Convention's value both on the national and international levels.[468]

Arguably, the contribution of the UNCCD to forest protection is likely to be insignificant on a global scale because those States affected by DLDD rarely have vast forest ecosystems within their territories. Moreover, the measures that have been taken by the UNCCD are more likely to include afforestation rather than the

465 United Nations A/RES/70/1, Resolution adopted by the General Assembly on 25 September 2015: Transforming our world: the 2030 Agenda for Sustainable Development, 2015. – Internet resource – https://undocs.org/en/A/RES/70/1 [Last visit: 30.04.2023]

466 *Beyerlin, Ulrich*, Desertification, in: *Wolfrum, Rüdiger* (ed.), Max Planck Encyclopedia of Public International Law, Online edition, 2009. – Internet resource – https://opil.ouplaw.com/view/10.1093/law:epil/9780199231690/law-9780199231690-e1571?rskey=aAFzV1&result=1&prd=MPIL [Last visit: 30.04.2023]; See *Eikermann* [n. 234], p. 129.

467 *Eikermann* [n. 234], p. 129.

468 Ibid; See also UNCCD, Evaluation of the effectiveness of national action programmes to implement the United Nations Convention to Combat Desertification, UNCCD Publications, 2017, p. 8. – Internet resource – https://www.unccd.int/sites/default/files/relevant-links/2017-01/NAP%20evaluation_0.pdf [Last visit: 30.04.2023]

preservation of particularly precious natural forests, because those areas covered by LDN-projects and similar activities are predominantly converted forestlands. However, the UNCCD can play a major role on a regional level by supporting the conservation of particularly fragile forest ecosystems. Additionally, despite the fact that forests are not always included in the activities implemented under the umbrella of the UNCCD, any activity under the umbrella of this treaty can either directly or indirectly contribute substantially to their protection, as the combating of DLDD and the restoration of degraded lands can help to reduce agricultural pressure on existing forest ecosystems.

II. Rio Forest Principles

The Forest Principles underline the socio-economic role of forests and the sovereign right of States to exploit their natural resources in accordance with their own environmental legislation and practices.[469] Like other environmental documents, the Forest Principles include one of the most prominent rules of international environmental law, according to which every State has to avoid causing any harm to the environment of the territories beyond its national jurisdiction by conducting activities under its jurisdiction.[470] As it was already pointed out above, this rule is not substantially codified in international treaties, but has been accepted as customary international law.[471] The reference to this rule in the Forest Principles can be perceived as a declaratory one. Thus, no sanctions or other mechanisms can be applied in order to enforce compliance with the principle. However, a State that has suffered damage caused by dangerous environmental activities can call the International Court of Justice (ICJ) as long as it either effectively subjected itself to the Court's jurisdiction by unilateral declaration, or ratified any treaty which contains a clause that established jurisdiction

469 Principle 1.

470 Trail Smelter Arbitration (United States v. Canada), Arbitral Judgements of 16.04.1938 and 11.03.1941, Rewards of International Arbitral Awards, Volume III, p. 1965.

471 Legality of the Threat or Use of Nuclear Weapons, Advisory Opinion, ICJ Reports 1996, pp. 241–242, Paragraph 29; Pulp Mills on the River Uruguay (Argentina v. Uruguay), Judgment of 20.04.2010, ICJ Reports 2010, pp. 55 f., Paragraph 101; *Shelton, Dinah L./Kiss, Alex*, Strict Liability in International Environmental Law, in: *Ndiaye, Tafsir Malick/Wolfrum, Rüdiger* (eds.), Law of the Sea, Environmental Law and Settlement of Disputes, Brill Academic Publishers, 2007, p. 1132 ff.; See Section B of this Chapter.

of the Court.[472] Otherwise, it can use other legal means to settle international disputes.[473] As the "Whaling in the Antarctic" case showed, a State can take legal action even when its own territory and State interests are not affected.[474] However, this practice is generally accepted by the ICJ only in cases of the violations of obligations *erga omnes partes*.[475] Such obligations could be established in relation to the forests by a legally binding forest conservation, which has not been adopted so far.

Moreover, in practice, a claim can be raised only when extensive and significant environmental harm has been done, otherwise every economic activity would be illegal from the point of view of international environmental law.[476] Additionally, a State can be exempt from responsibility by proving that it was acting with due diligence and that all necessary precautionary measures, including the conducting of environmental impact assessments, were taken.[477] Without these reservations, most economic activities would be deemed illegal.[478] Moreover, the responsibility of a State only becomes relevant if and to the extent to which a direct link between its activities and the environmental damage that has been caused to the territory beyond its jurisdiction can be established.[479] With regard to the environmental harm caused by deforestation, it can be particularly difficult to provide this evidence.[480] Although the depletion of genetic resources as a result of deforestation represents an enormous loss for humanity as a whole, as is the case with any public good, it is impossible to determine which State has been directly affected and to express the damage caused in monetary terms. Nevertheless, as the already cited "Whaling Case" has

472 Legality of the Threat or Use of Nuclear Weapons, Advisory Opinion, ICJ Reports 1996, pp. 241–242, Paragraph 29.

473 Ibid.

474 Whaling in the Antarctic (Australia v. Japan: New Zealand intervening), Judgement of 31.03.2014, ICJ Report 2014, pp. 226–230.

475 Questions relating to the Obligation to Prosecute or Extradite (Belgium v. Senegal), Judgment of 20 July 2012, ICJ Reports 2012, pp. 449 f., Paragraphs 68–70.

476 See *Dederer, Hans Georg*, Staatenverantwortlichkeit („State responsibility") und Haftung („liability") im Bereich der „ultrahazardous activities", in: *Hecker, Bernd/Hendler, Reinhard/Proelß, Alexander/Reiff, Peter* (eds.), Verantwortlichkeit und Haftung für Umweltschäden, 28. Trierer Kolloquium zum Umwelt- und Technikrecht vom 6. bis 7. September 2012, 2013, pp. 15 f.

477 *Shelton/Kiss* [n. 471], p. 1138.

478 Ibid.

479 Ibid.

480 See Section B of this Chapter.

demonstrated, a State still can be charged for the breach of its obligation to manage responsibly its forest resources. Even a State that has not been directly affected by deforestation within another State's territory can raise a claim against it in the cases when it can justify that the environmental damage caused represents a breach of an *erga omnes partes* obligation.[481] Despite this, such obligations in relation to forests can hardly be derived from any existing treaty. Even if a State were able to justify the existence of the violation of an *erga omnes partes* obligation, in the case of damage being caused to forest ecosystems, a court would face major challenges while forming its judgement. Unlike the "Whaling Case",[482] States' obligation to protect and manage forest ecosystems sustainably cannot be derived from existing international law.[483] Due to their non-legally binding nature, the Forest Principles cannot establish such commitments.

It is worth noting that the Rio Forest Principles refer to all the world forests, without any differentiation between tropical, temperate, northern, natural, or artificial forests.[484] The preamble of the document in question recognizes the socio-economic significance of forests and their role in the development of human civilization. In fact, economic issues occupy a greater place in the document than environmental ones.[485] For example, the Rio Forest Principles stipulate the necessity to facilitate international timber trade and to cancel timber trade bans and restrictions.[486] In this way, it takes into account the interests of developing forest-rich States. If environmental issues were to prevail in the Forest Principles, and if their scope were to be restricted to tropical forests, such a constellation, combined with the absence of a financial stimulus, would likely result in the failure of negotiations of the document in question.[487] The environmental provisions that finally found their way into the Forest Principles include the recognition of the ecological significance of forests[488] and the promotion of the sustainable management of forest resources.[489]

The provisions on the protection of natural forests are vague. States are only supposed to include "the protection of ecologically viable representative or

---

481 Questions relating to the Obligation to Prosecute or Extradite (Belgium v. Senegal), Judgment, I.C.J. Reports 2012, pp. 449 f., Paragraphs 68–70.

482 Whaling in the Antarctic (Australia v. Japan: New Zealand intervening), Judgement of 31.03.2014, ICJ Report 2014, pp. 226–230.

483 See Subsection C.II of Chapter 6.

484 *Schulte zu Sodingen* [n. 233], p. 191; *Krohn* [n. 276], p. 168.

485 *Schulte zu Sodingen* [n. 233], p. 191.

486 Principles 13–14.

487 *Schulte zu Sodingen* [n. 233], p. 191.

488 Principle 4.

489 Principle 2(b).

unique examples of forests, including the primary/old-growth forests, cultural, spiritual, historical, religious and other unique and valued forests of national importance" in their national forest policies.[490] Obviously, no measurable and verifiable targets concerning the reduction of deforestation rates and in relation to the promotion of reforestation are included in the document. Each State is free to decide which forest ecosystem should be designated as a protected area, and which conservation measures should be implemented in the relevant regions.[491]

In an attempt to enhance the legal value of the Forest Rio Principles, the word "Authoritative" was added to its title.[492] In a similar vein, paragraph (d) of the preamble also points out that the said document represents the "first global consensus on forests", and that States will strengthen their efforts to promote its full and effective implementation. It is further stated that the Forest Principles are to be subject to further development and that subsequent international cooperation is going to take place.[493] However, unlike Chapter 11 of Agenda 21, this document does not explicitly foresee the elaboration of new forest-related international agreements in the future.

III. Forests in the Agenda 21

Agenda 21 is the most comprehensive document to come out of Rio. It represents a non-legally binding political strategy, one that should serve as a guideline for the promoting of main environmental policies, elaborated in Rio, with particular accent made on sustainable development.[494]

The second of Agenda 21's four sections are dedicated to the conservation and management of natural resources and deals with such issues as ecosystem management, atmosphere protection and the promotion of sustainable agricultural activities. Almost every chapter of this section refers to forest issues at some point.[495] In this way, Agenda 21 considers the fact that forest conservation is

---

490 Principle 8(f).

491 *Schulte zu Sodingen* [n. 233], p. 193; *Krohn* [n. 276], p. 170.

492 *Krohn* [n. 276], p. 170.

493 Preamble (d).

494 *Schulte zu Sodingen* [n. 233], p. 193; *Krohn* [n. 276], p. 170; *Robinson, Nicholas A.*, AGENDA 21: Earth's Action Plan, IUCN Environmental Policy and Law Paper No. 27, 1993, pp. 1 ff.; *Hohmann, Harald*, Ergebnisse des Erdgipfels von Rio-Weiterentwicklung des Umweltvölkerrechts durch die UN-Umweltkonferenz von 1992, in: Neue Zeitschrift für Verwaltungsrecht, 12. Jahrgang, Heft 4, 1993, pp. 311, 315.

495 *Schulte zu Sodingen* [n. 233], p. 193; *Krohn* [n. 276], p. 170.

essential for the achievement of almost all environmentally relevant goals and is playing a crucial role in ecological sustainability.

Additionally, Chapter 11 of Agenda 21 is specifically dedicated to the combating of deforestation. This Chapter includes four program areas which describe different activities that should constitute a guideline for the promotion of forest conservation and sustainable forest management. Each program area includes its own basis for actions, objectives and activities, as well as for the means of their implementation.[496]

Program area A is particularly dedicated to the promotion of multiple forest and forest land functions, which should be achieved through the reinforcement of national forest-related national institutions, while both supporting activities related to the management, conservation, and sustainable development of forests as well as promoting the sustainable utilization and production of forests' goods and services.[497] It is important that Agenda 21 underlines the necessity of international assistance towards the accomplishment of every program area.[498] Of course, Agenda 21 does not stipulate any legal obligations for the donors to provide this financial support.

Program area B envisages the strengthening of conservation efforts, sustainable management and the protection of forests, as well as the rehabilitation of degraded areas, through both afforestation and forest restoration. This program area intends to combat the uncontrolled conversion of forests to other types of land, as promoted by increased human pressure on forests, agricultural needs, a lack of adequate forest-fire control and illegal logging.[499] The document further outlines that the destruction of forests manifests itself in the form of negative impacts upon the environment, reduces the chances of sustainable development, and that urgent actions are therefore needed.[500] These actions should include the maintenance of existing forests and the implementation of national forest action plans that have been designed with due regard to the multiple functions of forests, by way of the sustainable management and, as appropriate, conservation of forests.[501] The provision on the expansion of artificial forest plantations in this program area should not be regarded as a shortcoming of Chapter 11, because

---

496 *Schulte zu Sodingen* [n. 233], p. 193; *Krohn* [n. 276], p. 170.
497 Paragraph 11.2(a).
498 Paragraph 11.6.
499 Paragraph 11.10.
500 Paragraph 11.11.
501 Paragraph 11.13(a).

it should serve primarily to combat deforestation and to reduce human pressure on natural forests.[502] Agenda 21 foresees the possibility of new international arrangements on forests being elaborated upon in the future without pointing out if a legally binding form should be preferred for these documents or not.[503]

Program area C includes measures for the facilitation of the efficient utilization of forests and of the assessment of forest-related goods and services in order to determine their full value. The program points out that the full extent to which forests and forestlands might be able to contribute to sustainable development have not been ascertained yet.[504] Against this backdrop, it is necessary to promote understanding of the multiple benefits of forests, including ecological ones, and to recognize the adverse impacts of deforestation.[505]

The next point made by this program area is primarily of economic relevance. It consists of the promotion of "efficient forest-based processing industries", which would permit value-added secondary wood products to be produced in forest-rich States.[506] The next measure, which foresees the promotion of a more effective and sustainable use of forest resources to extract firewood and for other energy purposes, can be criticized from the following point of view.[507] Namely, although firewood represents the most important source of energy for hundreds of millions of people, the use of it should be reduced rather than promoted. In this regard, it can be pointed out that the use of firewood in Haiti has resulted in tremendous adverse environmental impacts with negative consequences for socio-economic development there.[508]

Finally, this program area calls for the promotion of eco-tourism in forest areas. Eco-tourism can contribute to forest protection because it allows for the pressure on forests to be reduced, generates further incomes, and creates incentives for forest protection, while ecotourism is only supposed to flourish in sustainably managed forests.[509] However, due to a lack of public awareness about

502 Paragraph 11.13(b)
503 Paragraph 11.13(e).
504 Paragraph 11.20.
505 Paragraph 11.21(a).
506 Paragraph 11.21(b).
507 Paragraph 11.21(c).
508 See Subsection B.II of Chapter 1.
509 *Novelli, Marina/Gebhardt, Bwana Tucke-Tucke, Kathrin*, Community Based Tourism in Namibia: "Reality Show" or "Window Dressing"?, In: Current Issues in Tourism, Volume 10, Issue 5, 2009, pp. 443–497; *Das, Madhumita/Chatterjee, Bani*, Ecotourism: A Panacea or a Predicament?, in: Tourism Management Perspectives, Volume

the significance of natural ecosystems, even forests with reduced biodiversity or with a great number of artificially introduced species can attract tourists. Additionally, a lack of infrastructures can make access to natural forests relatively difficult for tourists.[510] Many natural forest ecosystems are situated in remote or mountainous areas. The creation of a tourist infrastructure is associated with additional environmental risks. Thus, despite all of its benefits, eco-tourism can contribute to the degradation of the environment, even when visitor levels are low.[511] The adverse impacts of ecotourism include soil erosion, the loss of habitats, a reduction in biodiversity, water, air and noise pollution, and litter.[512] For example, in Nepal, ecotourism largely depends on the utilization of forest wood for fuel, which results in the increased anthropogenic pressure on forests.[513] Agenda 21 fails to consider all these issues.

Programme area D is dedicated to the establishment and/or enhancement of those capacities required for the planning, assessment and monitoring of forests and forest-related programmes and activities, including the trade in timber and non-timber forest products. This programme area aims to improve the monitoring systems used to assess the availability and quality of forest resources and those land areas available for afforestation, with a particular regard to land tenure issues.[514] These systems should be integrated into a continuing research process, one that would allow for their constant improvement, in order for the best possible outcome to be achieved.[515]

In addition, it is worth noting that the measures included in Chapter 11 do not only aim to promote the sustainable utilization of forest resources, but also the enhancement of forest conservation.[516] Additionally, Chapter 11 promotes the cross-sector approach. It foresees that the decision makers, which are not directly engaged in the forest sector, are supposed to take forest-related issues into

---

14, 2015. pp. 3–16; *Das, Madhumita/Chatterjee, Bani*, Ecotourism and Empowerment: A Case Analysis of Bhitarkanika Wildlife Sanctuary, Odisha, India, in: IIM Kozhikode Society & Management Review, Volume 4, Issue 2, 2015, pp. 136 ff.

510 *Das/Chatterjee* [n. 509], pp. 7 f.

511 Ibid.

512 Ibid.

513 Ibid.

514 Paragraph 11.30(a).

515 Ibid.

516 *Schulte zu Sodingen* [n. 233], p. 198; *Krohn* [n. 276], p. 174.

account while developing their own policies.[517] It is submitted that this provision incites countries to combat such drivers of deforestation as mining activities and the construction of dams.

Despite these positive aspects, Chapter 11 does not include any incentives for its implementation. Furthermore, the raising of the financial resources needed for the implementation of Convention was an unrealistic goal.[518] On the other hand, Chapter 11 has served as a basis for the development of new forest-related documents and policies.[519] It ought to be assessed positively that multiple forest-related environmental and socio-economic issues were for the first time mentioned in one single document. However, the holistic approach towards forest conservation did not come into being. In other words, there was a need for the further development and amendment of the measures and policies outlined in Chapter 11 and in other forest-related documents. It is important in this regard, that Chapter 11 leaves a certain amount of freedom of choice for decision makers. The criteria and guidelines for forest management and conservation, national action programs and other forest policies can be developed and subsequently amended in accordance with the newest scientific findings. Chapter 11 does not address the problem of scientific uncertainties, but it underlines the necessity of assessing the environmental impact of the activities implemented in forest areas.[520]

D. Development of the International Legal Regime for Forests after the Rio Conference

I. Intergovernmental Panel on Forests

The outcome document of the third meeting of the Commission on Sustainable Development (CSD)[521] outlined the urgent need for strengthening international

---

517 *Hönerbach, Frank*, Verhandlungen einer Waldkonvention: Ihr Ansatz und ihr Scheitern, WZB Discussion Paper, No. FS II 96–404, 1996, pp. 58 ff.

518 *Krohn* [n. 276], p. 174; *Eikermann* [n. 279], pp. 85 ff.

519 *Schulte zu Sodingen* [n. 233], p. 198.

520 Paragraphs 11.22(a), 11.23(b), 11.31(c).

521 CSD was created by ECOSOC with its decision 1993/207 on 12 December 1993 with the aim to promote the implementation of Agenda 21. The decision 1993/207 is to be found in: United Nations, Resolutions and decisions of the Economic and social Council, E/1995/95 Official records, New York, 1996, pp. 93 ff. – Internet resource – https://documents-dds-ny.un.org/doc/UNDOC/GEN/N94/932/99/PDF/N9493299.pdf?OpenElement [Last visit: 30.04.2023]

efforts dedicated to the implementation of the Rio Forest Principles and the forest-related provisions of Agenda 21.[522] In accordance with these recommendations, the ECOSOC founded the Intergovernmental Panel on Forests (IPF) by way of adopting decision 1995/226.[523] At its final session 5 held in 1997, the IPF presented to the CSD its proposals for action, which were approved with the consequence that a number of States committed themselves to their implementation. The IPF's proposals included an extensive number of recommendations dedicated to the implementation of the Forest Principles.[524] The main strategy promoted by the IPF's proposals is SFM, and the main instrument for its achievement are the national forest and land-use programs, that should deal with the underlying causes of deforestation and take into account the multiple benefits of forests.[525] Consequently, traditional forest-related knowledge and skills should be protected.[526] The proposals give particular attention to the need for further actions dedicated to the protection of those fragile forest ecosystems affected by desertification and drought and affected by airborne pollution.[527] The application of the internationally accepted C&I for SFM is recommended but not explicitly required.[528] The proposals deal with the needs of the States with low forest cover in particular, underlying the need for further action aimed at conserving existing forests, promoting afforestation and forest restoration, and assuring the long-term availability and sustainability of the multiple benefits of forests.[529] The IPF's proposals also address the need to enhance international financial assistance and the transfer of technology without stipulating any obligations to provide the funding required. In order to improve the effectiveness of

---

522 United Nations, Commission on Sustainable Development, Report on the Third Session (11–28 April 1995), E/CN.17/1995/36, Economic and Social Council, Official Records, Supplement No.12, United Nations, New York, 1995, Annex 1, Paragraph 1. – Internet resource – https://undocs.org/E/CN.17/1995/36 [Last visit: 30.04.2023]

523 United Nations, Resolutions and decisions of the Economic and social Council, E/1995/95, Official records, 1995, New York, 1996, p. 73. – Internet resource – https://undocs.org/E/1995/95 [Last visit 30.04.2023]

524 United Nations, Report of the Ad Hoc Intergovernmental Panel on Forests, E/CN.17/1997/12, Economic and Social Council, New York, 20 March 1997. – Internet resource – https://undocs.org/E/CN.17/1997/12 [Last visit: 30.04.2023]

525 Ibid, Paragraphs 17(a–i), 27(a)–31(c).

526 Ibid, 40(a–r).

527 Ibid, Paragraphs 46(a–g), 50(a–e).

528 Ibid, Paragraph 89(a–h)

529 Ibid, Paragraph 58(a–e).

funding, the document in question urges recipient States to manage the provided financial resources sustainably and to promote the generation of further finance directly from forests under SFM.[530] The participation of the private sector should also be promoted in this process.[531] In summary, the main message of the IPF's proposals is that actions made on a national level should play a dominant role in the achievement of SFM, but that the implementation of these actions presupposes extensive international cooperation.

II. Intergovernmental Forum on Forests and United Nations Forum on Forests

As the IPF did not manage to develop detailed recommendations for the implementation of its provisions, the CSD created a new advisory body, the ad hoc open-ended Intergovernmental Forum on Forests (IFF) with the aim to promote the IPF's Proposals.[532] The IFF's final report was presented at its fourth Session in 2000 and is called the IFF Proposals for action.[533] The IFF underlined the role of national forest programs in the implementation of forest strategies.[534] National forest programs should be based on a holistic approach. The IFF's proposals foresee the promotion of access to forest-related information and the protection and utilization of traditional forest-related skills with the particular consideration being given to the need for equitable sharing of the benefits derived from its utilization.[535] Another positive incentive coming from the IFF's proposals is that States are strongly encouraged to promote the expansion of protected forest areas.[536] However, this provision does not stipulate any measurable or verifiable targets.[537] The IFF's Proposals also stress the need for enhanced economic incentives for the promotion of environmentally sound forest use.[538] However, the

530 Ibid, Paragraphs 67(a)–68(c).
531 Ibid, Paragraph 69(a–e).
532 *Schulte zu Sodingen* [n. 233], p. 231.
533 United Nations, Report of the Intergovernmental Forum on Forests on its Fourth Session, E/CN.17/2000/14, Economic and Social Council, 20 March 2000. – Internet resource – https://undocs.org/E/CN.17/2000/14 [Last visit: 30.04.2023]
534 Ibid, Paragraphs 56, 66 ff.
535 Ibid, Paragraphs 96(a)–98(c).
536 Ibid, Paragraphs 84–90.
537 Ibid, Paragraphs 84–90.
538 Ibid, Paragraph 85(f).

IFF's proposals also fail to deliver positive incentives for the promotion of their implementation.

Based on the IPF/IFF's proposals for action, the ECOSOC decided to adopt the International Arrangement on Forests (IAF) with the aim of catalyzing international forest-related cooperation.[539] One of the main provisions of the IAF consisted in the creation of the United Nations Forum on Forests, which has played a major role in forest-related negotiations since 2000. One of the first major achievements of the UNFF has been the adoption of a Non-Legally Binding Instrument on All Types of Forests, which constitutes the most significant step forward in the development of the international legal regime for forests since the Rio Conference.

III. Non-Legally Binding Instrument on All Types of Forests

The Non-legally Binding Instrument on All Types of Forests (NLBI) represents an important milestone in the elaboration of international, forest-specific policies. 15 years of numerous consultations, meetings, and negotiations were needed in order to develop this new comprehensive international document dedicated to forests, while the Rio Forest Principles have remained the most insightful embodiment of an international consensus on forests throughout this period. The IPF/IFF's action proposals did not go far beyond the listing of policies and procedures, which should be adopted in the forest sector, without containing further provisions for the promotion of their implementation. The International Arrangement on Forests as embedded in ECOSOC Resolution 2000/35 is a largely subsidiary document that refers to previous international forest-related documents and establishes organizational procedures for the UNFF.[540] The NLBI was negotiated and adopted during a general decline in global efforts concerning forest protection.[541] This decline was reflected in the Millennium Development

539 United Nations E/CN.17/2000/20, Report on the eighth session of the Commission on Sustainable Development, Economic and Social Council, Official Records, 2000, Decision 8/2, Paragraph 3. – Internet resource – https://undocs.org/E/CN.17/2000/20 [Last visit: 30.04.2023]

540 ECOSOC Resolution 2000/35, Report on the forth Session of the Intergovernmental Forum on Forests, Economic and Social Council, United Nations, 2000. – Internet resource – https://www.un.org/esa/forests/wp-content/uploads/2013/09/2000_35_E.pdf [Last visit: 30.04.2023]

541 *Schneider, Thomas W.*, A Non-Legally Binding Instrument as an Alternative to a Forest Convention, 2006.

Goals (MDG), which were developed primarily in order to catalyze the reduction of extreme poverty and with forest-related issues playing only a marginal role in it.[542]

Despite all these developments, the participants of the 5th Session of the UNFF in May 2005 accepted the fact that the current IAF could not satisfy the need for the promotion of sustainable management of forests and of forest conservation to an appropriate extent, and that new documents should be developed in order to enhance the effectiveness of the UNFF.[543] However, the attempt to start negotiations on new documents failed because of the spirit of misunderstanding that dominated the 5th Session.[544] The proposition that was made by the EU, Canada, Switzerland, Mexico, Norway, and South Korea to formulate measurable and time-bound targets was bogged down by the resistance from the USA, Brazil and India.[545] Subsequently, the session was closed after only some informal consultations on the Chairman's Draft Text of a new International Arrangement on Forests had taken place.[546]

Negotiations were resumed during the 6th Session of the UNFF, which was held in February 2006. The future of the IAF and the options for elaborating a legally binding framework on all types of forests in accordance with the ECOSOC Resolution 2000/35 and the UNFF's multi-year program of work were on the meeting agenda.[547] It was also expected that parties would reconsider the negotiation of the Chairman's Draft Text in accordance with decision 5/2 of the 5th Session of the UNFF.[548] Thus, the main topics that were negotiated upon during the UNFF-6 consisted of the main structural elements of a new agreement and

542 United Nations A/RES/55/2, United Nations Millennium Declaration, General Assembly Resolution, 18 September 2000. – Internet resource – https://undocs.org/A/RES/55/2 [Last visit: 30.04.2023]

543 United Nations E/CN.18/2005/18, United Nations Forum on Forests, Report of the fifth Session (14 May 2004 and 16 to 27 May 2005), Economic and Social Council, Official Records, New York, 2005, pp. 13 ff. – Internet resource – https://undocs.org/E/CN.18/2005/18 [Last visit: 30.04.2023]

544 *Schneider* [n. 541], p. 7.

545 Ibid.

546 Ibid.

547 United Nations E/CN.18/2006/18, United Nations Forum on Forests, Report of the sixth Session (27 May 2005 and 13 to 24 February 2006), Economic and Social Council, Official Records, New York, 2006, pp. 1 ff. – Internet resource – https://undocs.org/E/CN.18/2006/18 [Last visit: 30.04.2023]

548 United Nations E/CN.18/2005/18 [n. 543], pp. 2 ff.

a new instrument on all types of forests. Negotiations were concluded with the adoption of four Global Objectives on Forests, which were developed in order to incentivize global efforts in the sphere of forest protection and sustainable forest management.[549] The lack of a consensus on the adoption of a legally binding framework on forests resulted in the abandonment of this issue. Instead, the decision was made to dedicate further efforts to the negotiation of a non-legally binding instrument on all types of forests.[550] The reason for this was that key States opposed the adoption of a legally binding forest convention. Major donor States, which were reluctant to undergo any commitments to provide additional financial resources, were unwilling to change their position.[551] However, the decision was made to reconsider the possibility of adopting a legally binding document at the UNFF-11 in 2015.[552] It is important to note that in 2015 the option of a legally binding document was rejected and that the decision was made to reaffirm the validity of the NLBI with particular emphasis on the new developments under the Rio Conventions but without changing the non-legally binding character of the said document.[553]

The numerous propositions for the development of the NLBI that were submitted by parties and different expert groups stipulated that under conditions that were unfavorable for a legally binding forest convention, any new document should at least create enough incentives to catalyze forest-related international efforts, by promoting new and additional forms of financial assistance.[554] At the same time, it was necessary to ensure that the highest possible number of States would join the NLBI and make the necessary political commitments. Considering the non-legally binding nature of the document, the suggestion was made for States to join the document without any ratification or other form of approval being necessary. It meant that the document would be able to become operational immediately after its adoption at the international conference. However,

549 *Schneider* [n. 541], p. 8.

550 United Nations E/CN.18/2006/18 [n. 547], Paragraphs 26 ff.

551 *Schneider* [n. 541], p. 8.

552 United Nations E/CN.18/2006/18 [n. 547], Paragraph 32.

553 United Nations E/CN.18/2015/14, United Nations Forum on Forests, Report on the eleventh Session (19 April 2013 and 4 to 15 May 2015), Economic and Social Council, Official Records, New York, 2015, p. 12. – Internet resource – https://www.undocs.org/en/E/CN.18/2015/14 [Last visit: 30.04.2023]

554 *Kunzmann, Katharina*, The Non-legally Binding Instrument on Sustainable Management of All Types of Forests Towards a Legal Regime for Sustainable Forest Management?, in: German Law Journal, Volume 09, Number 08, January 2008, pp. 981 ff.

in the course of the final negotiations this idea was rejected, as forest-rich States were opposing any possible political commitments.[555]

Following the meeting of the Ad Hoc Expert Group in December 2006, the NLBI was finally adopted at the seventh session of UNFF in April 2007.[556] Despite the non-legally binding character of the NLBI, it was negotiated in a detailed, precise, and at the same time restrained manner, implying that it would be able to establish political commitments while at the same time demonstrating the high political value that they attached to the document.[557] Since the very beginning of the negotiations, it was clear that the achievement of a consensus presupposes that any main critical issues be negotiated in a package, including the definition of SFM, the elaboration of measurable and time-bounded targets, and a mechanism for the promotion of NLBI implementation, including the financing thereof.[558]

The EU Member States expected that the above-mentioned document would include much stronger commitments in comparison to previous forest-related documents, that more forest-related topics would be covered, that it would establish certain monitoring and compliance procedures for the monitoring of its implementation and that it would contribute to the strengthening of national forest-related legislation.[559] On the other hand, a group of States, which included both forest-rich developing and developed States led by the USA and Brazil, rejected any stronger commitments and more precise wording by opposing any developments that might lead to the establishment of new rules of customary international law, including provisions on transparency and good governance.[560] No single State was ready to give up the smallest part of its national sovereignty unless others were to take the same step. For this reason, many proposals that could have strengthened the NLBI were rejected, including the EU's proposition to establish a Committee of Experts for the promotion of NLBI's implementation.[561]

---

555 *Kunzmann* [n. 554], pp. 990 ff.; *Schneider* [n. 541], p. 9.

556 United Nations E/CN.18/2007/8, United Nations Forum on Forests (24 February 2006 and 16–27 April 2007), Economic and Social Council, Official Records, 2007. – Internet resource – https://www.undocs.org/en/E/CN.18/2007/8 [Last visit: 30.04.2023]

557 *Kunzmann* [n. 554], pp. 991 f.

558 *Kunzmann* [n. 554], pp. 991 f.; *Schneider* [n. 541], p. 9.

559 *Kunzmann* [n. 554], pp. 992 f.

560 Ibid.

561 Ibid.

The opposition of the group of forest-rich States to any stronger commitments was as strong as the reluctance of donor States to commit themselves to the provision of new and additional financial resources. It was argued that the abundant funding had already been raised and that instead of applying for new financial assistance, developing States should use their available resources and promote the income-generation from SFM.[562] The question arises as to why the opposition to fundraising was so strong, if at the same time donor States were ready to undergo costly climate change mitigation commitments.[563] One possible explanation for this may be the lack of tools for ensuring the proper utilization of financial resources in recipient States, a lack of understanding of the role of forests in climate change mitigation/adaptation policies, and a readiness to discuss forest-related issues only in the light of already existing international policies, including the MDGs.[564] In this regard, it should be mentioned that the absence of strong commitments to the reduction of greenhouse gases found in the UNFCCC and the Kyoto Protocol (KP) did not favor the enhancement of forest funding. In addition, the negotiations of REDD+ were only in their initial phases.[565]

The title of the NLBI itself is worth further consideration. From a linguistic point of view, it is decisive that the wording "Non-Legally Binding" was chosen over "Non Legally binding" or "Legally Non-Binding", which signalizes that the NLBI does not create legal rights and obligations but can still be a source of strong political commitments.[566] In addition, the adoption of the NLBI as an annex to the GA resolution indicates that it can potentially serve as a basis for the

562 Ibid, p. 993.

563 It is difficult to determine the actual costs of climate change mitigation, due to current weak economic models, which are incapable of predicting expenditures like industrial modification or energy supply schemes, and unexpected revenues from these activities. It is submitted that climate change mitigation policies are inspired by concerns about global climate and are not driven by the economic ratio behind them. Cf.: *Rosen, Richard A./Guentherb, Edeltraud*, The Economics of Mitigating Climate Change: What Can We Know?, in: Technological Forecasting and Social Change, Volume 91, February 2015, pp. 93 ff.

564 *Kunzmann* [n. 554], pp. 992 f.

565 See Subsection D.I of Chapter 4.

566 Non-Legally Binding Authoritative Statement of Principles for a Global Consensus on the Management, Conservation and Sustainable Development of All Types of Forests, adopted on 14 June 1992, in: ILM, Volume 31, 1992, pp. 881 ff.; See *Kunzmann* [n. 554], pp. 993 f.

development of new rules of customary international law.[567] Moreover, it means that the document can be applied universally to all the UN Member States, without any divisions between subscribers and non-subscribers. In this way, the properly chosen title and the inclusion of the text as part of the GA resolution gave higher legal relevance to the document, which was otherwise weakened by the omission of stronger commitments.

Notably, the title of the NLBI refers to "sustainable forest management", which reaffirms the adherence of international forestry policies to this principle, one that was steadily prioritized in international forestry policies at the cost of other possible approaches, such as the placement of major forest areas under protection. It is also important that the title of the document in question explicitly indicates that, like the Rio Forest Principles, the document is to be applied to "all types of forests".

The purpose of the NLBI is to contribute to the implementation of SFM and to promote the achievement of shared global objectives concerning forests, which also deal with the promotion of SFM.[568] At the same time, the contribution of forests to the achievement of the MDGs should be increased, especially concerning the eradication of poverty and the preservation of environment.[569] The NLBI can contribute to all these developments by providing a framework for activities carried out on a national and international level. Consequently, as one can also see from the whole text of the NLBI, SFM plays a crucial role in this document. Consequently, the NLBI fails to consider the shortcomings of SFM. One of these shortcomings materializes in the incompatibility of the industrial forest logging, which is presumed to be sustainable, with forest vegetation life cycles.[570]

Despite the lack of a consensus on the issue, negotiators were finally able to include a definition of SFM into the NLBI, even although it is characterized by vague and ambiguous wording.[571] §4 of the NLBI stipulates that "sustainable forest management, as a dynamic and evolving concept, aims to maintain and enhance the economic, social and environmental values of all types of forests, for the benefit of present and future generations". This rather anthropocentric definition indicates the main scope of SFM, while contributing in this way to the

567 *Kunzmann* [n. 554], pp. 993 f.

568 NLBI, Paragraph 1.

569 Paragraph 1(b).

570 See Subsection B.I of Chapter 1.

571 *Barbara L. Zimmerman/Cyril F. Kormos*, Prospects for Sustainable Logging in Tropical Forests, in: BioScience, Volume 62, Number 5, 2012, p. 485.

future elaboration of its comprehensive definition.[572] However, the NLBI's definition does not strike a balance between socio-economic and ecological issues and does not give any insight on how the natural forests can be better conserved through SFM. Moreover, it is hardly applicable in those situations, in which the promotion of the ecological functions of forests is not compatible with their economical exploitation. The undermining of these critical issues by the definition in question could be seen as harmful for international forest conservation efforts. However, States obviously avoid agreeing on any definition that could be a source of any strong political commitments.

With the adoption of the NLBI, the list of Global Objectives on Forests that was already agreed upon by parties at the sixth session of UNFF was adopted.[573] The first of the objectives deals with the reversal of the loss of forests all around the world, which should be achieved through SFM, and also includes the protection and restoration of forests.[574] It should be noted that this objective does not provide any insight in the role played by the conservation of natural forests. The second objective deals with the enhancement of the socio-economic and ecological benefits of forests, which, in accordance with the MDGs, should occur simultaneously with the improvement of local livelihoods and income generation/diversification.[575] The next objective foresees "the enhancement of areas of protected forests worldwide as well as the areas of other forests under SFM". The number of products that originate from sustainably managed forests should also be enhanced.[576] Arguably, the wording of the third objective is not completely clear; the wording "other forests under SFM" indicates that protected forests should be regarded as forests under SFM. However, as has already been pointed out above, the NLBI's definition of SFM does not give any insights on how natural forests can be protected through SFM. In any case, the wording of the third objective indicates that the promotion of SFM also presupposes the expansion of protected forest areas. The fourth objective takes into account the concerns of forest-rich developing States, deals with the decline of official development assistance, and calls for the mobilization of significant financial resources for the promotion of SFM.[577]

572 *Kunzmann* [n. 554], pp. 994 f.
573 Paragraph 5.
574 Ibid.
575 Ibid.
576 Ibid.
577 Ibid.

Paragraph 6 of the NLBI describes measures that every State should undertake in order to promote the implementation of the instrument. Similar to previous forest-related documents, the NLBI stipulates that national forest programs, as well as other strategies concerning SFM, should be developed, in order to identify the forestry policies needed and to ensure the implementation thereof.[578] The seven thematic forestry elements, as proposed by the NLBI[579], should serve as a reference framework for SFM, on whose base specific forest-related elements should be identified in order to serve as criteria and indicators for SFM.[580] Consequently, the NLBI does not deliver detailed C&I for SFM. Instead, it contains general regulations for the development of C&I on a national level. Thus, the document in question can be criticized for its failure to deliver the detailed uniformed and internationally agreed C&I for SFM.

States should promote the environmental impact assessments of projects implemented in forest areas and implement these projects in an environmentally sound way.[581] Moreover, they should support SFM by promoting numerous forest goods and services and also by encouraging waste recycling.[582] The NLBI also stresses the need for the protection and more effective utilization of traditional forest-related knowledge and practices (TFRK&P) with the consent and participation of the holders of such knowledge, and promote the fair and equitable distribution of those benefits derived from the utilization of such knowledge and practices.[583] This provision is particularly important, as TFRK&P play an important role in the promotion of environmentally sound forest management, and its application can facilitate forest-related scientific research. Forest-rich States should also promote the financing of SFM with particular emphasis on the involvement of the private sector and the generation of financial resources from the environmentally sound use of forests, inter alia, by promoting the monetary evaluation of the ecological functions of forests.[584] Like Agenda 21, the NLBI also calls for the enhanced cooperation and cross-sectoral coordination aimed

578 Paragraph 6(a).

579 According to the text of NLBI, these elements are (a) extent of forest resources; (b) forest biological diversity; (c) forest health and vitality; (d) productive functions of forest resources; (e) protective functions of forest resources; (f) socio-economic functions of forests; and (g) legal, policy and institutional framework.

580 Paragraph 6(b).

581 Paragraph 6(c).

582 Paragraph 6(d,e).

583 Paragraph 6(f).

584 Paragraph 6(h–j).

at integrating forest-related issues into national decision-making processes. It is important that among the threats to a forest's health and vitality, which should be addressed on a national level, the NLBI mentions not only pollution, fire, and pest outbreaks, but also threats from invasive alien species.[585]

Networks of protected forest areas should be expanded, their conditions should be assessed, and their management should be improved.[586] Finally, the NLBI stipulates the need to facilitate public access to forest resources and relevant markets, with specific regard to forest-dependent people. In doing so, the improvement of local livelihoods and income diversification should be achieved.[587] This provision can be criticized, however, for the lack of attention it pays to finding a balance between the interests of forest-dependent people and conservation measures. For example, pressure on forests could be reduced by promoting alternative income sources, inter alia, by establishing new energy supply systems in order to reduce people's dependence on forest wood for heating and cooking.

The NLBI includes a detailed list of measures that should be taken on an international level in order to promote SFM. Paragraph 7 in particular underlines the need for international assistance, including the provision of new and additional financial resources, in order to implement SFM, and links the combating of deforestation to the eradication of poverty. However, the allocation of resources from other thematic topics of the ODA to SFM is also foreseen.[588] This provision will likely serve to meet the expectations of donor States, which urged the recipients of funding to manage better the already provided finance. Furthermore, the international community should create positive incentives for developing States to promote the conservation and restoration of their forests and widespread application of SFM.[589] The promotion of these incentives represents a critical issue, one that is directly related to combating the underlying causes of deforestation.

One could argue that the said incentives should combat those economic phenomena that favor deforestation and forest degradation. This aspect is not explicitly mentioned in the NLBI. Instead, the document underlines that developed States should support the economically, socially, and environmentally sound measures serving as incentives for forest protection in developing States.[590]

---

585 Paragraph 6(o).
586 Paragraph 6(p, q)
587 Paragraph 6(y).
588 Paragraph 7(a-c).
589 Paragraph 7(d).
590 Paragraph 7(e).

Due to strong resistance on the part of forest-rich States, only weak provisions on the monitoring and assessment of its implementation are included in the document. In this regard, the NLBI simply indicates that States should monitor and assess their achievement of the NLBI's aims. They are also expected to submit reports about the progress they have made vis-à-vis the implementation of the NLBI to the UNFF, but only on a voluntary basis and in accordance with national circumstances.[591] The process of the implementation of the NLBI will be analyzed by the UNFF, meaning that the UNFF will serve as the main governing body of the NLBI.[592] It is problematic, however, that States were reluctant to empower further the UNFF. For this reason, the UNFF was not authorized to create subsidiary bodies with regard to the implementation of the NLBI. The relevant option was discussed during negotiations but rejected afterwards.[593] One of the reasons for this is that the creation of such a subsidiary body was not laid out by the ECOSOC Resolution 2000/35,[594] which authorized the UNFF to create certain subsidiary bodies. If the UNFF had extended its initial mandate to the creation a specific body for the NLBI, it would arguably have acted *ultra vires*.[595] The possibility of extending the powers of the UNFF via a new ECOSOC Resolution was not discussed.

In conclusion, the final version of the NLBI represents a non-legally binding agreement on the smallest common denominator, just as in the case of the Forest Principles and the subsequent forest-related documents. Its main value consists in that the NLBI serves as the main manifestation of the existing global political consensus on forests, and that for the first time this consensus was enshrined into a GA resolution. In terms of its environmental relevance, the NLBI constitutes a significant step forward in comparison to the Forest Principles. The last document tends to underline the economic role of forests, whereas the NLBI is characterized by its focus on SFM, which seeks to reconcile environmental and socio-economic issues related to forests.[596] At the same time, those approaches to forest management, which would prioritize the conservation of forests over

591 Paragraphs 8–9.

592 Paragraph 10.

593 *Kunzmann* [n. 554], pp. 1004 ff.

594 ECOSOC Resolution 2000/35, Report on the forth Session of the Intergovernmental Forum on Forests, Economic and Social Council, United Nations, 2000. – Internet resource – https://www.un.org/esa/forests/wp-content/uploads/2013/09/2000_35_E.pdf [Last visit: 30.04.2023]

595 Ibid.

596 *Eikermann* [n. 234], p. 55.

their utilization and seek to minimize the anthropogenic impact on forests by replacing timber forest products through other technologies and materials, are not mentioned in the NLBI. Natural forests are not mentioned at any point in the NLBI. Certain issues, such as the utilization of forests' genetic resources, are not addressed by the NLBI. Other issues are treated only sketchily, as, for example, the C&I for SFM. It has been reaffirmed that positive incentives should be created for forest-rich States to encourage them to implement SFM, but the NLBI does not include further details on how these incentives could be created. Financial provisions are both weak and superficial and the document in question does not ensure that new and additional substantial financial resources will be provided. Consequently, the NLBI does not provide a comprehensive international framework for forest protection or the environmentally sound use of forest resources. These and other gaps would not have been remedied even if the NLBI had been adopted as a legally binding international treaty. From this point of view, and in light of on-going deforestation, which rarely contributes to human development in forest-rich States, there was a need for further negotiations and for the elaboration of new, international legal forest-related documents and strategies.

IV. International Arrangement on Forests Beyond 2015

An independent Ad-Hoc Expert Group was created in April 2013 in order to support the mandate of the UNFF at its 11th session by providing advice and input.[597] At the first meeting of the ad-hoc expert group in February 2014, it was suggested that international arrangements on forests be dedicated to the implementation of the NLBI, to the promotion of four Global Objectives on Forests and to the achievement of the post-2015 development agenda. At the same time, it was suggested that the possibility of adopting a legally binding forest agreement be reconsidered.[598]

---

597 United Nations E/CN.18/2013/18, United Nations Forum on Forests, Report on the tenth Session (4 February 2011 and 8 to 19 April 2013), Economic and Social Council, Official Records, New York, 2013, p. 16. – Internet resource – https://undocs.org/E/CN.18/2013/18 [Last visit: 30.04.2023]

598 United Nations E/CN.18/2015/11, United Nations Forum on Forests, eleventh Session, (4–15 May 2015), Report on the first meeting of the Open-ended Intergovernmental Ad Hoc Expert Group on the International Arrangement on Forests, Item 3 of the provisional agenda, Economic and Social Council, New York, 2015, pp. 11 f. – Internet resource – https://undocs.org/E/CN.18/2015/11 [Last visit: 30.04.2023]

In January 2015, the second meeting of the Ad-Hoc Expert Group, proposed four options for future International Arrangement on Forests. The *first option* consisted of the adoption of a strengthened International Arrangement on Forests, which would catalyze the evaluation and amendment of the NLBI, and which would allow for the creation of new forest-related bodies as part of the UN, with a particular attention being given to the establishment of the UN forest assembly, which would replace the UNFF.[599] The creation of a new scientific body called "UN-Forest", which would both promote and coordinate scientific research in the sphere of forest management and facilitate the international exchange of forest-related scientific information, was also proposed. A special UN envoy on forests would be appointed, the existing UNFF Trust Fund would be developed and new trust funds and sub-accounts would be created under its framework.[600] The *second option* is based on the first one, with the main difference being that it urges the further development of the "UN-Forest's" body with a view to create a new international institution.[601] The *third option* also foresees the possibility of Member States undergoing unilateral, legally binding commitments vis-à-vis the achievement of SFM.[602] The *fourth option* would also include regional legally binding agreements as an important part of international forest policies. The adoption of these regional agreements would be promoted under the umbrella of the NLBI, building upon existing international conventions and agreements.[603] During the meeting, controversial views were expressed on the possibility of adopting a universal, legally binding agreement on forests. Not only was the possibility of such an agreement being reached regarded as unrealistic, but doubts were also expressed as to whether a legally binding document really would contribute to the implementation of SFM.[604] Already at this point it should be argued that these doubts are justified, because even if such a legally binding document had been adopted and States had given up part of their sovereignty on forest issues, forest conservation and SFM would not have been promoted unless positive incentives had been delivered. Consequently,

---

599 Ibid, p. 6.

600 Ibid.

601 Ibid.

602 Ibid.

603 Ibid, p. 9.

604 *Humphreys, David*, Negotiating the future under the shadow of the past: the eleventh session of the United Nations Forum on Forests and the 2015 renewal of the International Arrangement on Forests, in: International Forestry Review, Volume 17, Issue 4, 2015, p. 390.

elaborating upon a comprehensive international legal definition of SFM, creating an all-inclusive international framework for its implementation, and thinking up incentives for the achievement of SFM, can be seen as more important issues than the adoption of a legally binding document itself. However, such an approach does not take into account that different stakeholders that are engaged in the promotion of forest conservation and SFM require transparency and legal certainty that can be delivered by a legally binding treaty.

However, on the eve of the eleventh Session of the UNFF, a group of 20 States led by China, Malaysia, Costa Rica, and Singapore declared their support for a forest convention. During the eleventh session, Russia, supported but Switzerland, became the most active supporter of a legally binding convention, arguing that it would bring more clarity and transparency to international decision-making processes.[605] This initiative faced strong opposition from a group of states led by the USA, Canada, Australia, and Japan.[606] As a result, China, Costa Rica, and a number of other States gave up any attempts to resume the negotiation of a legally binding agreement.[607] As the supporters of this agreement wanted to leave at least some hope for its adoption in the future, the final text of the International Arrangement on Forests has set a task for the UNFF to reconsider the possibility of adopting a forest convention during the mid-term evaluation of the IAF in 2024.[608] The provisions in question simply repeat the wording of the resolution 10/2 UNFF in relation to examining the possibility of adopting a legally binding agreement during the eleventh session of the UNFF.[609] Therefore, the UNFF-11 did not bring forward the negotiations of a legally binding forest convention.

Once again, two of the main controversial issues that hindered the negotiation of a forest convention were a lack of funding and a desire to set specific time-bounded and quantifiable targets that would be accepted by forest-rich States. The Group of 77, that at the time of the negotiations already had 134 members, required the creation of a global forest fund to assist forest-rich States in their attempts to achieve SFM.[610] Developed nations were reluctant to make any promises to transfer a predefined amount of money on a regular basis,

605 Ibid, p. 390.
606 Ibid.
607 Ibid.
608 United Nations E/CN.18/2015/14 [n. 553], Paragraph 42.
609 United Nations E/CN.18/2015/11 [n. 598], p. 9.
610 *Humphreys 2015* [n. 604], p. 390.

whereas the G77 insisted that any progress made with regard to the implementation of SFM could only be achieved if this were to happen.[611] However, no developed State supported this idea, with Finland arguing that the discussion of such fund goes beyond the authority of the UNFF.[612] Japan and the United States argued that if the fund was created, otherwise sources of finance for SFM would be reduced. For this reason, they advocated the promoting of voluntary markets, the strengthening of PES schemes, and fostering of income generation from SFM itself.[613] Switzerland stated that it would only support the creation of a fund, should a legally binding agreement be adopted.[614] Meanwhile, China was ready to support both the global forest fund as well as regional forest funds.[615]

Major developed States have also opposed strengthening the UNFF's role into a kind of a global mechanism for SFM implementation. Instead, it was decided that it should maintain its role as an agency for the promotion of dialogues between States and catalyzing political will.[616]

Overall, the progress that has been achieved with regard to forest finance and implementation issues has been rather modest so far. Instead of creating a standing committee on the implementation of commitments and technical advice, it was agreed that since 2015 the UNFF's meetings were to take place on an annual instead of a biannual basis, with every second meeting being dedicated to the implementation of commitments and technical advice. Every reference to the creation of special sub-accounts as a part of the Forum Trust Fund was rejected, but it was agreed that the operation of the Fund should be subject to subsequent revision.[617] At the same time, in order to meet the requirements of developing

611 Ibid, p. 390.
612 Ibid.
613 Ibid.
614 Ibid, p. 234.
615 United Nations E/CN.18/2015/9, Letter dated 6 January 2015 from the permanent representative of China to the United States addressed to the Secretary-General, Economic and Social Council, 21 January 2015. – Internet resource – http://undocs.org/E/CN.18/2015/9 [Last visit 30.04.2023].
616 United Nations E/CN.18/2015/14 [n. 553], Paragraph 78; *Blaser, Juergen/Chipeta, Mafa Evaristus/Illueca, Jorge/Lobovikov, Maxim/Umali, Ricardo Martinez*, Independent Assessment of the International Arrangement on Forests (IAF), Report of the Team of Independent Consultants, 2014, p. 68; United Nations E/CN.18/AEG/2014/INF/1, Compilation of Views and Proposals by Member States and Other Relevant Stakeholders on the International Arrangement on Forests, Economic and Social Council, New York, 12 February 2014, p. 262. – Internet resource – https://undocs.org/E/CN.18/AEG/2014/INF/1 [Last visit: 30.04.2023]
617 United Nations E/CN.18/AEG/2014/INF/1 [616].

States, it was decided that the UNFF Facilitative Process should be renamed the "Global Forest Financing Facilitation Network", and serve as a "clearing house on existing, new and emerging financing opportunities". The creation of the "Global Forest Financing Facilitation Network" represents one of the most important achievements of the UNFF-11 with regard to promoting forest finance. It was also agreed that the mid-term evaluation of the IAF in 2024 should include the consideration of a number of options, including the possibility of the creation of a "Voluntary global forest fund". It is worth noting that it was the first time that the option of the creation of such fund found its way into an official UN document. Additionally, it was decided to encourage the creation of a new focal area of finance dedicated to SFM operating under the supervision of the GEF.[618]

Despite the attempts that were made to raise the profile of forests in the UN, any proposals on this regard were adamantly rejected during the consultations preceding the UNFF-11, including the idea to create a UN Forest Assembly and the suggestion to introduce a special envoy for forests under the guide of the UN's Secretary General. This means that both the UNFF's status and the role of the Collaborative Partnership on Forests have remained unchanged.[619]

The proposal to create a new body for scientific cooperation and the implementation of SFM – so-called UN-Forest, was also rejected at the very beginning of negotiations because of the lack of political will.[620] Nevertheless, the separate members of the CPF tried to strengthen scientific cooperation. Indeed, the Center for International Forestry Research (CIFOR), the International Union of Forest Research Organization (IUFRO) and the World Agroforestry Center (ICRAF) made a joint statement on the creation of an inter-organizational science-research platform and invited the participants of the UNFF-11 to consider the possibility of including such an option into the IAF.[621] However, this proposition did not find support either. It seems that developed States were reluctant to strengthen the political mandate of the UNFF, and that forest-rich States were not ready to commit themselves to following the recommendations of the major international scientific body on forest issues in their national decision-making processes. Overall, the final document confirmed that sharing of forest-related knowledge, "including… science-policy interference", represents one of the core

618 United Nations E/CN.18/2015/14 [n. 553], Paragraph 15.

619 *Humphreys 2015* [n. 604], p. 394.

620 Ibid.

621 *Earth Negotiations Bulletin*, Summary of the 11th session of UNFF, in: Earth Negotiations Bulletin, Volume 13, Number 199, 2015, p. 12.

functions of the UNFF.[622] Additionally, it was outlined that one of the core functions of the CPF is to "promote scientific and technical advice on forests".[623]

Neither the Swiss proposition to develop the revised IAF into a base for the establishment of the "Global stewardship on forests" nor the concept of "Mother Earth" that was introduced by Bolivia and supported by Ecuador in the course of the Ad-Hoc Expert Group's meeting found its way into the UNFF 11's final decisions.[624] The first proposition was rejected by developing States, as they saw in it as an attempt to interfere with their national sovereignty over forest resources without adequate compensation from developed States.

At the same time, the concept of "Mother Earth", which recognizes nature as a subject endowed with legal rights similar to humans, ran counter to liberal market principles, existing development strategies, prevailing religious and philosophic traditions, and the general anthropogenic character of international environmental law.[625] It can be argued, inter alia, that the ability of any legal subject to acquire legal rights presupposes its capacity to assume legal obligations, meaning that nature itself could be legally obliged to provide humans with certain resources. Thus, in order to promote forest conservation, it is necessary to create positive incentives for the environmentally friendly exploitation of forest resources and for the renunciation of those revenues associated with industrial logging, forest clearance and other damaging activities. However, the very act of recognizing nature's intrinsic rights could promote a shift to the environmentally sound management of forest resources. The State of Ecuador, for example, enshrined the right of integral respect for nature ("Pacha Mama") into its constitution in 2008.[626] This is exemplified by Ecuador's recent push to reduce its

---

622 United Nations E/CN.18/2015/14 [n. 553], Paragraph 6.

623 United Nations E/CN.18/2015/14 [n. 553], Paragraph 20.

624 United Nations E/CN.18/2015/12, Note verbale dated 23 March 2015 from the Permanent Mission of Switzerland to the United Nations addressed to the Secretary-General, Economic and Social Council, New York, 4–15 May 2015, Paragraph 16. – Internet resource – https://undocs.org/en/E/CN.18/2015/12 [Last visit: 30.04.2023]; *Earth Negotiations Bulletin*, Summary of the eleventh Session of the United Nations Forum on Forests, Volume 13, Issue 199, 2015, pp. 7, 10. – Internet resource – https://enb.iisd.org/download/pdf/enb13189e.pdf [Last visit: 30.04.2023]; See *Humphreys 2015* [n. 604], p. 394.

625 See *Humphreys 2015* [n. 604], p. 394.

626 Constitution of the Republic of Ecuador 2008, Article 71.

deforestation rates. Once one of the highest in the world,[627] they have fallen rapidly, even although they remain above the reduction commitment levels.[628]

Some progress was achieved in relation to the promotion of the role of the NLBI. The suggestion to include time-bound, quantifiable and verifiable targets in the NLBI, made by an independent assessment team, was rejected. However, the role of the NLBI and of its four Global Objectives on Forests was reaffirmed. It was also decided to replace the reference to the "Millennium Development Goals" by the reference to the "Sustainable Development Goals" (SDGs) of the post-2015 development agenda.[629] In this regard, it should be noted that environmental issues play a significantly higher role in the SDGs when compared to the MDGs.[630] Additionally, the participants of the UNFF-11 reaffirmed the role of Global Objectives on Forests and extended the time period needed for their achievement until 2030.[631] It was also decided to change the name of the NLBI to the *United Nations Forum Instrument*, but the voluntary and non-legally binding character of the instrument was not revised.[632] It can be argued that the change of the Instrument's title reflected the increase in its international significance and signaled continuing political commitment to its implementation.

In conclusion, the progress achieved at UNFF-11 was modest; the outcome document represents an agreement on the less common denominator, with almost absent compliance and control mechanism and with only weak provisions for the promotion of its implementation. The meeting was hampered by contradictory interests of the participating States. Some pleaded for strengthened forest conservation, while others prioritized the utilitarian aspects of forest management. The main interest of the opposing coalition was to promote

---

627 Global Forest Atlas, Forest Governance – Ecuador. – Internet resource – https://globalforestatlas.yale.edu/amazon-forest/forest-governance/forest-governance-ecuador [Last visit: 30.04.2023].

628 *Gaworecki, Mike*, New Report Examines Drivers of Rising Amazon Deforestation on Country-by-Country Basis, in: Mongabay, 23 May 2019. – Internet resource – https://news.mongabay.com/2019/05/new-report-examines-drivers-of-rising-amazon-deforestation-on-country-by-country-basis/ [Last visit: 30.04.2023]

629 United Nations E/CN.18/2015/14 [n. 553], Paragraphs 12, 44(a).

630 United Nations A/RES/70/1, Resolution Adopted by the General Assembly on 25 September 2015: Transforming our World: the 2030 Agenda for Sustainable Development, United Nations, 21 October 2015, pp. 1 ff. – Internet resource – https://www.un.org/ga/search/view_doc.asp?symbol=A/RES/70/1&Lang=E [Last visit: 30.04.2023]

631 United Nations E/CN.18/2015/14 [n. 553], Paragraph 8.

632 Ibid.

international timber trade and to support forest-based economic development. Subsequently, in light of weak provisions on finance within the framework of the IAF, a number of forest-rich States turned their attention to the emerging REDD+-mechanism.[633]

V. United Nations Strategic Plan for Forests 2017–2030

Despite the little progress achieved during the previous forest-related negotiations under the UNFF's framework, the year 2017 suddenly brought a relative breakthrough. The adoption of a legally binding agreement did not appear on the agenda of the talks held by different international bodies. However, during the special session of the UNFF held in January 2017, the participants successfully elaborated and adopted the first-ever UN Strategic Plan for Forests.[634] The UN Strategic Plan for Forests (UNSPF) was subsequently adopted by the General Assembly Resolution 71/285 on 27 April 2017.[635]

The UNSPF includes a number of important innovations. First of all, the list of Global Objectives on Forests was extended from 4 to 6. Remarkably, the goals also include 26 targets to be reached by 2030, some of which are reportable and quantifiable. The universal, but voluntary, character of these targets is outlined in the document. First Four Forest Goals generally correspond to the 4 Global Objectives on Forests as adopted at the UNFF-6. The main difference is that some of the Goals include verifiable Targets. The Global Forest Goal 1 consists of the reversion of the global loss of forest cover "through SFM, including protection, restoration, afforestation, and reforestation". At the same time, forest degradation should be tackled and the contribution of forests to combating climate change should increase. It is important that Target 1 of this Goal foresees that the forest area should increase by 3% worldwide by 2030 as compared to the level of 2015.[636] Additionally, according to Target 2, the deforestation has to be halted by

---

633 See Subsection D.II of Chapter 4.

634 United Nations E/RES/2017/4, Resolution adopted by the Economic and Social Council on 20 April 2017: United Nations Strategic Plan for Forests 2017–2030 and Quadrennial Programme of Work of the United Nations Forum on Forests for the Period 2017–2020, Economic and Social Council, New York, 7 July 2017. – Internet resource – https://undocs.org/E/RES/2017/4 [Last visit: 30.04.2023].

635 United Nations A/RES/71/285, Resolution adopted by the General Assembly on 27 April 2017: United Nations Strategic Plan for Forests 2017–2030, General Assembly, New York, 1 May 2017. – Internet resource – https://undocs.org/E/RES/2017/4 [Last visit: 30.04.2023].

636 Ibid, Global Forest Goal 1.

2020, SFM should be promoted, the forest restoration, afforestation, and reforestation should be significantly enhanced[637]. This Target can be also understood in a different way: as a commitment to compensate deforestation by promoting afforestation with a view to achieve net zero forest loss. This interpretation can be counterproductive for forest protection, as planted forests cannot fulfil all the socio-economic and environmental functions that natural forests do.

Two new Global Forest Goals consist of the promotion of good governance and international cooperation. Developing States gave up their initial resistance and accepted the inclusion of targets on integrating SFM in their national development strategies and on strengthening national legislation on forest issues, particularly with a view to combat illegal logging. However, no quantifiable targets are included in the Goal 4, which deals with financial issues. The finance-relevant paragraphs of the UNSPF repeat the weak wording of previous forest-related documents. They reaffirm the "crucial" need for the mobilization of substantial new and additional financial resources from all sources. Yet they fall short of stipulating any measurable, verifiable and, at least, politically binding commitments. From this point of view, forest-rich States are invited to use the financial resources provided by the GEF in order to finance the focal areas on SFM, and to apply the financial resources raised under the REDD+-mechanism.[638] States are also encouraged to promote the utilization of new innovative funding schemes, including the payments for ecological services of forests.[639] The UNFF-Members have also failed to achieve a consensus on creating new funds and sub-accounts within the framework of the UNFF Trust Fund. States are only invited to make voluntary contributions to the UNFF Trust Fund without undergoing any political commitments.[640] However, the decision of the UNFF-11 to reconsider the option of establishing a voluntary global forest fund during the mid-term revision of the IAF in 2024 was confirmed. It is noteworthy that the UNSPF also foresees the possibility to reconsider this option at any session prior to 2024, if there are signs for achieving consensus on this issue.[641] The UNSPF did not enhance the role of the Global Forest Financing Facilitation Network and reaffirmed that it serves "as a clearing house and database on existing, new, and emerging financing opportunities".[642]

637 Ibid.
638 Ibid, Paragraph 57.
639 Ibid, Paragraph 57.
640 Ibid, Paragraph 64(b).
641 Ibid.
642 Ibid, Paragraph 60(c).

Once again, the attempt to establish a United Nations Scientific Body on forests failed and the UNSPF only invited the scientific community to present their research and analysis to the Forum.[643] The scientific cooperation should be based on the existing mechanisms of the CPF and the creation of new mechanisms is not expected.[644]

Subsequently, the adoption of the UNSPF, which can be appreciated for the inclusion of several quantifiable and verifiable targets, represents an important achievement. The UNSPF enhanced the role of the UNFF in international forest-related negotiations. However, many critical issues remain unresolved, in particular forest finance and international cooperation. The UNSPF itself underlines the role of other international mechanisms, which were developed outside of international forest-related negotiations, but can play a significant role in financing forest-related activities, including, inter alia, REDD+. This means that there is still a need for further activities within the scope of the UNFF and of other international environmental legal regimes.

E. Conclusions

Despite numerous decades of negotiations over forest-related issues at the international level, an effective, comprehensive, and robust international mechanism for combating deforestation remains out of reach. Immediately after the Stockholm Conference, the adoption of a legally binding forest convention did not come into question. At this time, the main approach to international environmental governance consisted of promoting endangered species protection or of representative samples of natural ecosystems. Only in the case of the Ramsar Convention were specific types of ecosystems protected – in this case, watersheds. A holistic approach could not be achieved. It was only by narrowing the scope of application of the treaty could any notable success at international negotiations be achieved.

The dominant approach to international environmental regulation consisted of the promotion of trade bans and restrictions. The incentivizing effect of relevant provisions was often neglected. Under these conditions, it was particularly difficult to reconcile environmental protection with socio-economic development.

643 Ibid, Paragraph 59.
644 Ibid, Paragraph 59.

Despite this, a specific treaty on forestry issues – the ITTA, was adopted. Although the ITTA is a mixture of a commodity agreement and an environmental convention, the main focus rests upon the promotion and diversification of trade with legally harvested tropical timber. Additionally, the promotion of sustainable forest management in tropics plays a certain role in the Agreement. As the policies developed under the ITTA's framework are subordinated to its main scope, its positive impact on forest ecosystem conservation tends to be insufficient in order to halt and reverse the forest loss worldwide.

The Rio Summit in 1992 had created favorable conditions for a legally binding forest convention be adopted. International environmental policies demonstrated a shift to a more holistic approach. An increased willingness to confront major international environmental problems, without limiting regulations to the protection of sole species or representative samples of ecosystems, emerged. Sustainable development was proposed as a concept to reconcile socio-economic development with the protection of nature.

However, the achievement of consensus on forest issues was hampered by a lack of common vision of the deforestation crisis. The forest-rich developing States expected that developed nations would undergo the same legally binding commitments to conserve their forests, and that adequate funding would be provided to compensate for lost development opportunities. Therefore, both groups of States prioritized economic issues over the need to conserve the world forests.

For these reasons, Rio forest talks became a major failure of the Rio Summit. The Rio Forest Principles were adopted as the main non-legally binding manifestation of the existing political consensus on forests. This consensus was further developed during the negotiations of forest issues under the umbrella of the IPF/IFF and the UNFF; which all have failed to adopt a legally binding treaty on forests.

The Rio Forest Principles, the NLBI and other forest-specific non-legally binding documents contain a detailed list of policies to be adopted in the forest sector. However, they fail to promote forest finance and to deliver incentives for the conservation and sustainable use of forests. Ongoing deforestation remains a major environmental problem. Moreover, the pertinent international legal documents fail to consider new scientific evidence on the ecological functions of forest ecosystems. These ecological functions include, *inter alia*, the influence on global and regional precipitation patterns. Moreover, the fact that industrial logging practices, which are presumed to be sustainable, are not compatible with the natural lifecycles of the critical tropical forest tree species was never mentioned in the relevant documents.

Against this backdrop, the contribution of two major Rio Conventions – the CBD and the UNFCCC – to the development of the international legal regime for forests will be analyzed in next chapters.

Chapter Three Role of the CBD and the Nagoya Protocol in the Protection of Forests

A. Introduction

Forest ecosystems are major terrestrial biodiversity pools and natural habitats. Consequently, they are likely to play a major role in the achievement of the CBD's objectives regarding the conservation of biological resources and sustainable use of their compounds, as well as the fair and equitable sharing of benefits arising from the utilization of genetic resources.[645] As the inclusion of the provisions dealing with the utilization of genetic resources, a bulk of which is gathered in forests, and the sharing of relevant benefits were the main prerequisites behind the CBD approval by developing States, the achievement of the relevant third objective of the Convention in question can potentially promote the protection of global forest ecosystems.

For a long time, the utilization of biodiversity resources in pharmaceutics, cosmetics, and food industry has been a common practice.[646] The development of new technologies has promoted the application of biodiversity resources in different branches of industry. By modifying, recreating, or by simply applying natural chemical substances or genetic sequences derived from flora and fauna species, manufacturers are able to develop numerous products. The role of biodiversity compounds in different branches is steadily growing.

Against this background, it is important to ensure access to biodiversity resources for developers and manufacturers. However, in practice, regulation of access can be challenging. The relevant flora and fauna species are often available in-situ in biodiversity-rich developing States that do not possess technologies and infrastructure needed in order to develop and produce biodiversity-based products. Consequently, researchers and manufacturers make attempts to obtain access to those biodiversity resources that are found in-situ in developing States, which, in turn, seek to derive economic benefits from the utilization of these resources by foreign companies and, ideally, to develop their own research and

---

645 CBD, Article 1.

646 See *Young, Robert N.*, Importance of Biodiversity to the Modern Pharmaceutical Industry, in: Pure and Applied Chemistry, Volume 71, Issue 9, 1999, pp. 1655 ff.

manufacturing capacities. By sharing benefits from the utilization of biodiversity resources with the States providing these resources, foreign investors can motivate them to protect natural habitats.

Forest ecosystems are important natural habitats. They are distinguished by high amounts of genetic resources available in-situ. As long as developing States, on whose sovereign territory biodiversity-rich forests are situated, are unable to conduct research and development activities on their own, and as long as the equal and fair sharing of benefits arising from the utilization of these resources by foreign corporations and research institutes is not ensured, these States do not have incentives for the conservation of forest genetic resources. Greater forest areas can be subjected to unsustainable logging, clearances, and forestland conversions for agricultural purposes. Once the fair and equitable sharing of relevant benefits with forest-rich States is ensured, a new option for sustainable forest use in the tropics may be available. As a natural habitat rich in genetic resources, forests can ensure their owners income even if the trees are not cut and forestlands are not converted.

It remains questionable, however, to what extent the exploitation of biodiversity resources can promote the conservation of forests. Admittedly, the relevant activities tend to accentuate the utilitarian approach to forest governance instead of promoting the intrinsic values of forests. Moreover, relatively small forest areas, some of which are already enjoying the status of protected areas, can deliver genetic resources of scientific and/or commercial interest. In such a case, no incentives will be delivered for the creation of new protected areas and for the sustainable management of forests outside of these areas. Therefore, the question arises how forest ecosystems not currently used for gathering biodiversity samples can benefit from the scientific and commercial exploitation of genetic resources.

Reconciling the controversial interests of owners and users of genetic resources and of other relevant stakeholders can be particularly challenging. Researchers, manufacturers, and forest owners are likely to engage themselves in the development and manufacturing of biodiversity-based products only if the international legal framework fulfils their expectations concerning the safeguarding of their rights, and ensuring legal certainty as well as transparency. More importantly, all relevant activities and pertinent legal provisions must be designed and implemented in such a way as to ensure the best possible outcome for the preservation of forest ecosystems and for the promotion of rights of forest-dependent people. As has already been outlined, the utilization of biodiversity requires international cooperation, which must be regulated by international legal provisions. What can be evaluated positively is the fact that with the adoption of the Nagoya Protocol to the CBD, such legally binding provisions dealing with access to genetic resources and with sharing of benefits from their utilization have been delivered. However, it remains questionable if the available legal provisions tackle those challenges arising from the utilization of biodiversity resources in an effective manner, and if the existing international legal regime for the conservation and sustainable use of biodiversity can contribute to the promotion of forest protection. In order to assess solutions to this problem at this stage, the main international legal documents that affect the use of genetic resources, namely the CBD and the Nagoya Protocol,[647] must be analyzed.

B. General Issues

I. Development of the CBD and Other Relevant Documents

1. Legal Status of Biodiversity before the Adoption of the CBD

Prior to the negotiation and adoption of the CBD at the Rio Summit in 1992, States and other relevant actors advocated two contradictory and antagonistic views on the legal status of biodiversity.[648] According to one opinion, forest biodiversity resources are fully subject to the national sovereignty of the State on whose territory they are found.[649] Correspondingly, those States would have the full and undeniable right to preserve, exploit, or even destroy these forest resources without considering the issues of intra- and intergenerational justice.

647 Nagoya Protocol on Access to Genetic Resources and the Fair and Equitable Sharing of Benefits Arising from their Utilization to the Convention on Biological Diversity, adopted on 29 October 2010 and entered into force on 12 October 2014, in: 3009 UNTS No. 30619.

648 *Lochen, Tobias*, Die völkerrechtlichen Regelungen über den Zugang zu genetischen Ressourcen, 2007, pp. 26. ff.; *Wolfrum, Rüdiger/Klepper, Gernot/Stoll, Peter-Tobias/Franck, Stephanie L.*, Genetische Ressourcen, traditionelles Wissen und geistiges Eigentum im Rahmen des Übereinkommens über die biologische Vielfalt, Schlussbericht des F+E Vorhabens „rechtliche Analyse des Übereinkommens über die biologische Vielfalt unter besonderer Berücksichtigung der Fragen des geistigen Eigentums“, 2001, pp. 48 f.; *Subrata Roy, Chowdhury* Permanent Sovereignty Over Natural Resources in International Law: Principle and Practice, 1984, pp. 8 ff.; *Tvedt, Morten W./Young, Tomme*, Beyond Access: Exploring Implementation of the Fair and Equitable Sharing Commitment in the CBD, in: IUCN Environmental Policy and Law Paper No. 67/2, 2007, p. 1; *Henne, Gudrun*, Genetische Vielfalt als Ressource: Die Regelung Ihrer Nutzung, 1. Aufl., 1998, pp. 119 f.

649 *Lochen* [n. 648], pp. 32 f.; *Tvedt/Young* [n. 648], pp. 121 ff.; *Henne* [n. 648], pp. 120 f.

According to another opinion, biodiversity resources represent a common heritage of humanity, even if they fall under the jurisdiction of a certain State.[650] Even though such an approach appears to be highly ethical, in practice it often serves as a justification for the appropriation and commercial use of biodiversity compounds, without paying necessary attention to the rights and interests of developing States and ILCs. Manufacturers from industrial States claimed for themselves an unlimited and undeniable right to have access to the genetic resources of forests found in-situ in developing States.[651] In view of the prevailing uncertainty about the legal status of biodiversity resources, biopiracy became a common practice of private investors, and the interests of original owners of genetic resources of forests were widely neglected. This problem was partially resolved when the CBD confirmed the sovereign rights of States over biodiversity resources under their jurisdiction, and established the access and benefit-sharing (ABS) regime in order to reconcile the interests of both the owners and users of genetic resources. This mechanism was subsequently developed and reinforced through the documents adopted by the meetings of the CBD COP held on a biannual basis. Of these documents, the Nagoya Protocol deserves particular attention.

2. Adoption of the CBD

The CBD can be seen as one of the greatest successes of the Rio Summit.[652] Participating States managed to finalize necessary negotiations and preparatory

---

650 *Birnie, Patricia/ Boyle, Alan/Redgewell, Catherine*, International Law and the Environment, 3rd ed., 2009, pp. 197–198; *Morgera, Elisa/Tsioumani, Elsa/Buck, Matthias*, Unraveling the Nagoya Protocol, A Commentary on the Nagoya Protocol on Access and Benefit-sharing to the Convention on Biological Diversity, 2014, pp. 5 f.

651 *Henne* [n. 648], pp. 120 f.; *Curci, Jonathan*, The Protection of Biodiversity and Traditional Knowledge in International Law of Intellectual Property, 2010, p. 8.

652 *Schulte zu Sodingen, Beate*, Der völkerrechtliche Schutz der Wälder, 2002, pp. 201 f.; *Henne* [n. 648], pp. 117 f.; *Sdunzig, Tobias*, Die UN-Konvention über Biodiversität und ihre Zusatzprotokolle: Verhandlungshistorie, Inhalt, Kritik sowie Analyse der rechtlichen Steuerungsfähigkeit aus völkerrechtlicher und europarechtlicher, 1. Auflage, 2017, pp. 65 ff.; *Krohn, Susan Nicole*, Die Bewahrung tropischer Regenwälder durch völkerrechtliche Kooperationsmechanismen, 2002, pp. 108 f.; *Till, Markus*, Erhaltung und nachhaltige Nutzung der Biodiversität, in: *Proelß, Alexander* (ed.), Internationales Umweltrecht, 2017, p. 334; *Wolff, Nina/Köck, Wolfgang*, 10 Jahre Übereinkommen über die biologische Vielfalt: eine Zwischenbilanz, 1. Auflage, 2004, p. 20; *Proelß, Alexander*, Raum und Umwelt im Völkerrecht, in: *Proelß, Alexander/Graf Vitzthum, Wolfgang* (eds.), Völkerrecht, 7. Auflage, 2016, pp. 451 ff.; *Beyerlin, Ulrich*, Umweltvölkerrecht,

activities before the actual begin of the conference. Subsequently, the text was officially adopted in Rio de Janeiro on 22 May 1992 and signed by 152 participants of the conference. The Convention entered into force on 29 December 1993. The document's unexpectedly high level of approval by the forest-rich developing States was inspired by the fact that the Convention has acknowledged their undeniable sovereign rights over biodiversity resources under their jurisdiction and aimed to provide them with tools and instruments for combating biopiracy. At the same time, environmental issues played only a secondary role for these States.[653]

In contrast to the environmental agreements adopted between the Stockholm Conference and the Rio Summit, whose scope was restricted to protecting a limited number of wildlife species or only designated natural habitats, the CBD represents the first document which develops a holistic approach to biodiversity conservation while promoting the conservation of forests and other natural habitats.[654]

Notably, the CBD includes a reference to the intrinsic value of nature, but only in its preamble.[655] The preamble, as an introductory part of any international treaty or a national legal act, usually does not establish legal commitments like the document's main part does, and rarely imposes any direct legal obligations.[656] However, the preamble is an important means of treaty interpretation.[657] The

---

2000, pp. 198 f.; *Kiss, Alexander/Shelton, Dinah*, International Environmental Law, Second Edition, 2000, pp. 299 f.; *Curci* [n. 651], pp. 9 ff.

653 *Schulte zu Sodingen* [n. 652], pp. 201 f.; *Henne* [n. 648], pp. 117 f.; *von Saint André, Susanne*, Genetische Ressourcen und traditionelles Wissen: Zugang, Teilhabe und Rechtsdurchsetzung, 2013, pp. 32 ff.

654 *Krohn* [n. 652], pp. 109 f.; *Till* [n. 652], pp. 334 f., *Beyerlin* [n. 652], p. 199.

655 First paragraph of the Preamble of the CBD; *Michelot, Agnes/Aseeva, Anna*, From Ecosystem Services to Ecological Solidarity, in: *Westra, Laura/Gray, Janice/Gottwald, Franz-Theo* (eds.), The Role of Integrity in the Governance of the Commons, Governance, Ecology, Law, Ethics, 2017, p. 41.; *Raustlala, Kal/Victor, David G.*, Biodiversity since Rio: the future of the Convention on Biological Diversity, in: Environment: Science and Policy for Sustainable Development, Volume 38, Issue 4, 1996, p. 18.

656 *Hulme, Max, H.*, Preambles in the Treaty Interpretation, in: University of Pennsylvania Law Review, Volume 164, 2016, pp. 1296 ff., 1305 ff.; For the cases of legally binding preambles of the national constitutions see: *Orgad, Liav*, The Preamble in Constitutional Interpretation, International Journal of Constitutional Law, Volume 8, Issue 4, October 2010, pp. 728 f.

657 Vienna Convention on the Law of Treaties (with annex), adopted on 23 May 1969, in: 1155 UNTS 331, Article 31.2.

Preamble can shed light on the meaning of the treaty text and on its context, as well as provide a key to the analysis of the treaty's main goal and purpose.[658] Arguably, the wording of the preamble recognizing the intrinsic value of biodiversity represents a certain step backwards in comparison to the Preamble of the United Nations' General Assembly World Charter for Nature, which substantiates this concept by underlining the uniqueness of every form of life that should be respected independent of its worth for man.[659] However, the reference to the intrinsic value in the CBD's Preamble has higher practical significance, as the World Charter for Nature is a legally non-binding document.

The objectives of the Convention as contained in its Article 1 are significantly more pragmatic in comparison to the Preamble. The Convention aims to achieve three interdependent goals. The first two of them concern the conservation of biological diversity and sustainable use of its compounds. The third objective consists in the fair and equitable sharing of benefits arising from the utilization of genetic resources.[660] Although it is argued that the conservation of biodiversity represents basically the main goal of the Convention with its other goals serving for the achievement of this main goal,[661] the issues related to the utilization of genetic resources play a major role in the treaty. It is remarkable that the objectives of the CBD are supposed to be achieved by providing the appropriate funding, by granting access to genetic resources and by promoting the transfer of relevant technologies, while all the rights of owners and users of genetic resources and of corresponding technologies should be respected.[662] Interestingly, the promotion of fair and equitable sharing of benefits arising from the utilization of genetic resources serves both as the objective of the CBD and as a mean to achieve its first two objectives.

Notably, the Draft version of the CBD included a more holistic formulation of its objectives. Accordingly, the convention was supposed to achieve the

---

658 *Hulme* [n. 656], pp. 1297 ff.

659 United Nations General Assembly Resolution A/RES/37/7, Resolution adopted by the General Assembly on 28 October 1982, World Charter for Nature, United Nations, New York, 1982, Preamble. – Internet resource – https://undocs.org/en/A/RES/37/7 [Last visit: 30.04.2023]; See *Wood, Harold W.*, The United Nations World Charter for Nature: The Developing Nations' Initiative to Establish Protections for the Environment, in: Ecology Law Quarterly, Volume 12, Issue 4, 1985, pp. 977 ff.; Cf. *Michelot/Aseeva* [n. 655], p. 41.

660 Article 1.

661 *Sdunzig* [n. 652], p. 133.

662 Article 1.

conservation of biological diversity not only for the benefit of humanity, but also for its intrinsic value.[663] The sustainable utilization of biodiversity resources was not regarded as one of the Convention's main objectives, but rather as a means to conserve "the maximum possible biological diversity".[664] It can be argued that if this draft version had been adopted, the Convention would have been more conservation-centered compared to the actual treaty text, which seeks to balance conservation and utilization issues. Moreover, the reference to the financial aid for developing States in Article 1 of the Draft Version was more extensive and indicated that the funding should not only be appropriate, but also "new and additional". In order to promote the funding of biodiversity conservation, it was supposed to create a special Fund and to establish the mandatory contributions to the Fund for each developed State, which ratified the Convention.[665] Finally, the requirement that funding should be "new and additional" has found its way into Article 20 of the actual CBD text, although no obligation to provide mandatory contributions to the fund was established.

As already pointed out, the CBD's rapid ratification and taking effect was made possible by recognizing States' sovereign rights over biodiversity resources on their sovereign territory,[666] although this was put into perspective by the obligation to avoid environmental harm to other States and territories beyond national jurisdiction.[667] The idea of extending the concept of "common heritage of mankind" to biodiversity resources was completely rejected. Instead, the Preamble of the CBD acknowledges that the conservation of biodiversity resources represents a "common concern of humankind".[668] The main difference between "common concern" and "common heritage" is that the first one does not contest

663 See Article 1 of the Fifth Revised Draft Convention on Biological Diversity in: UNEP/Bio.Div/N7-INC.5/2 in 20 February 1992, Seventh negotiating session/Fifth session of Intergovernmental Negotiating Comittee on CBD, Fifth Revised Draft Convention on Biological Diversity, Nairobi, 11–19 May 1992. – Internet resource – https://www.cbd.int/doc/meetings/iccbd/bdn-07-inc-05/official/bdn-07-inc-05-02-en.pdf [Last visit: 30.04.2023]

664 Ibid.

665 Ibid.

666 *Schulte zu Sodingen* [n. 233], pp. 201 f.; *Henne* [n. 648], pp. 117 f.

667 CBD, Article 3.

668 CBD, Preamble; *Schulte zu Sodingen* [n. 233], pp. 202 f.; *Proelß* [n. 652], pp. 451 f.; *Krohn* [n. 652], p. 277; *Curci* [n. 651], p. 59.

the sovereign rights of the States of origin over their biodiversity resources.[669] Consequently, all States have to contribute to the conservation and sustainable use of biodiversity, whereas the national sovereignty of the forest-rich States over genetic resources under their jurisdiction and their authority to regulate the access to these resources remains undisputable.[670]

3. *Adoption of the Nagoya Protocol*

The CBD is generally characterized by the vague and declaratory character of its provisions dealing with environmental protection, relativized by ambiguous wording that includes "as far as possible and as appropriate", "in accordance with particular conditions and capabilities" etc.[671] The provisions dealing with access to genetic resources, technology transfer, and distribution of benefits from biotechnology are characterized by a higher level of specificity and tangibility. Despite this fact, the CBD still does not provide a comprehensive legal framework for the implementation of the access and benefit-sharing regime (ABS), whose application was challenged by the lack of substantiated regulations.[672] Subsequently, States realized the need for further regulations. After long and time-consuming negotiations, the legally binding Nagoya Protocol was finally adopted on 29 October 2010 and entered into force on 24 October 2014. It has so far been ratified by 104 States.[673] The contribution of this international treaty, which further specified the pertinent provisions of the CBD in relation to the development and implementation of the ABS regime, to the conservation of forests, will be further examined in this chapter. At first, however, a brief analysis of essential terms and concepts will be conducted.

---

669 *Thomas, Zakir*, Common Heritage to Common Concern, Preserving a Heritage and Sharing Knowledge, in: Journal of World Intellectual Property, Volume 8, Issue 3, 2005, p. 249.

670 Ibid, p. 250.

671 See, CBD, Articles 6, 7, 8.

672 *Krohn* [n. 652], pp. 114 f.

673 *Proelß* [n. 652], pp. 346 f.; *Sdunzig* [n. 652], pp. 406 f.; *Morgera/Tsioumani/Buck* [n. 650], pp. 1 ff.

II. Basic Terms and Concepts

1. "Genetic Resources" versus "Biodiversity Resources"

The term "genetic resources" was intentionally chosen by Parties in order to restrict the access and benefit-sharing regime's scope of application. The main purpose was to exclude any utilization of biodiversity from the scope of this regime that does not include genetic material.[674] Such exclusion served to avoid unnecessary interferences between the CBD and the commodity agreements, such as the ITTA.[675] For instance, Article 15 of the CBD regulates only the access to genetic resources and does not refer to biodiversity resources as a whole.[676]

The terms "genetic resources" and "genetic material" cannot embrace the full range of options for the utilization of biodiversity compounds found in forests, as the term "genetic" refers solely to material obtained from biodiversity compounds containing functional units of heredity.[677] Therefore, living organisms as a whole, parts thereof, populations of organisms as well as chemical compounds generated in living organisms in the process of metabolism which do not contain functional units of heredity are not embraced by the definition of genetic resources. For this reason, a broader term exists which defines all biodiversity compounds mentioned above as biological resources, providing that they have some potential or actual use or value for humanity.[678] However, this last reference to the potential or actual use or value of biological resources should be perceived widely, as all biological resources can be utilized in the future when new technologies are available.[679]

The proper use of the terms "genetic resources" and "biodiversity resources" is of great significance as it can influence the interpretation of the CBD and the

---

674 *Oberthür, Sebastian/Rosendal, G. Kristin*, Global Governance of Genetic Resources – Access and benefit sharing after the Nagoya Protocol, 2015, p. 23; *von Saint André* [n. 653], p. 9.

675 *von Saint André* [n. 653], p. 9.

676 *UNEP/CBD/WG-ABS/9/INF/119*, The Concept of "Genetic resources" in the Convention on Biological Diversity and how it relates to a functional international regime on access and benefit-sharing", Ad-hoc Open-ended Working Group on Access and benefit Sharing, Ninth meeting, Cali, Colombia, 22–28 March 2010, pp. 19 f. – Internet resource – https://www.cbd.int/doc/meetings/abs/abswg-09/information/abswg-09-inf-01-en.pdf [Last visit: 30.04.2023].

677 CBD, Article 2.

678 Ibid.

679 Cf. UNEP/CBD/WG-ABS/9/INF/119 [n. 676], pp. 26 f.

Nagoya Protocol. Strictly speaking, as these documents explicitly mention only genetic resources of forests, they can be applied only to such cases of biodiversity utilization where functional units of heredity are directly used for research and development activities.[680] Such interpretation can have significant consequences, as, in fact, the majority of products and innovations developed based on biodiversity material does not involve functional units of heredity at any stage of their development and manufacturing.[681] None of the 50 most traded drugs worldwide has been developed on the basis of genetic properties of any biological species.[682] In fact, products of metabolism play a crucial role in pharmacological and chemical industries, yet not genetic sequences themselves.[683] Subsequently, following such restrictive interpretation, one risks excluding a significant number of cases involving commercial utilization of forest biodiversity from the scope of application of the CBD regime.

Despite this fact, the negotiators of the CBD agreed to use the term "genetic resources". This choice was inspired, among other things, by new technologies that permitted to create the first genetically modified plant (tobacco) in 1983, and to conduct the first successful field trials of genetically engineered cotton in 1990.[684] Arguably, developing States and developers as well as manufacturers of biodiversity-based products were also interested in the restriction of the scope of application of the ABS regime. The former of the two were seeking to combat biopiracy while securing their national sovereignty over biological resources. At

---

680 Ibid, Paragraph 8.

681 UNEP/CBD/WG-ABS/9/INF/119 [n. 676], p. 26; *Young, Tomme Rosanne*, Legal Issues Regarding the International Regime: Objectives, Options, and Outlook, in: *Carrizosa, Santiago/Brush, Stephen B./Wright, Brian D./McGuire, Partrick E.* (eds.), Accessing Biodiversity and Sharing the Benefits: Lessons from Implementing the Convention on Biological Diversity, IUCN Environmental Policy and Law Paper No. 5, 2004, pp. 282 f.; European Comission Press Release IP/04/21, Commission encourages international solidarity when utilizing exotic plants, Brussels, 7 January 2004. – Internet resource – https://ec.europa.eu/commission/presscorner/detail/en/IP_04_21 [Last visit: 30.04.2023]; *WGBAS-3*, Briefing Paper on the Scope of an International ABS Regime, prepared for WGABS-3, 14, Bangkok, 18 February 2005. – Internet resource – https://www.publiceye.ch/fileadmin/doc/_migration/Biodiversitaet/050214_On_the_Scope_of_an_International_ABS_Regime.pdf [Last visit: 30.04.2023].

682 UNEP/CBD/WG-ABS/9/INF/119 [n. 676], p. 26.

683 *von Saint André* [n. 653], p. 32; *Henne* [n. 648], pp. 117 f., 125 f.

684 U. S. Department of Defence, Advances in Biotechnology and Genetic Engineering: Implications for the Development of New Biological Warfare Agents, June 1996, p. A-5; *Oberthür/Rosendal* [n. 674], p. 23.

the same time, the latter of the two were particularly keen to ensure the maximum possible freedom to conduct research and development activities without being limited by the Convention. It was generally accepted that any utilization of biological resources of forests as commodities that does not envisage any direct utilization of genetic resources should be exempt from the scope of the CBD.[685] Otherwise, the CBD would turn into a commodity agreement.[686] At the same time, the inclusion of the utilization of the so-called "derivatives" in the CBD's scope represented a major controversial issue. The last term was not defined in the Convention itself, however, it was generally understood that it is to be applied with regard to those biochemical substances which are the products of metabolism in living organisms, and which do not contain functional units of heredity.[687] According to another opinion, this term can also refer to the information derived from the genetic resources, and to any gene sequences obtained through artificial manipulation of natural genetic material.[688] As the CBD text did not further clarify the meaning of the term, it can be assumed that the treaty was not developed to be applied to the utilization of derivatives, but that States are able to extend the scope of the treaty to such cases by adopting specific provisions on a national level.

At any time, as the utilization of biodiversity compounds that does not include functional units of heredity becomes a common practice, the necessity arises to extend the interpretation of the term "genetic resources" also to the utilization of derivatives.[689] Nowadays, it is difficult to consider genetic resources as separate from the biodiversity compounds in which they are found, unless these compounds are used purely as commodities.[690] From this point of view, it can be argued that the CBD and the Nagoya Protocol can also be applied to the cases of the utilization of chemical substances developed in the process of metabolism in living organisms, taking into account that every metabolic reaction is regulated by "functioning units of heredity".[691]

685 *von Saint André* [n. 653], p. 32; *Henne* [n. 648], pp. 117 f., 125 f.

686 *von Saint André* [n. 653], p. 32; *Henne* [n. 648], pp. 117 f., 125 f.

687 *von Saint André* [n. 653], p. 9; *Henne* [n. 648], p. 139.

688 *Morgera/Tsiuomani/Buck* [n. 650], pp. 65 f.; UNEP/CBD/WG-ABS/7/2, Annex to the Report of the Meeting of the Group of Legal and Technical Experts on Concepts, Terms, Working Definitions and Sectorial Approaches, Paris, 2–8 April 2009, Paragraphs 19 f.

689 UNEP/CBD/WG-ABS/9/INF/119 [n. 676], p. 31, p. 33.

690 Ibid.

691 Ibid.

It is important to note that the Nagoya Protocol includes the definition of the term "utilization of genetic resources" which embraces "the research and development on the genetic and/or biochemical composition of genetic resources".[692] However, it is not completely clear if the cases of utilization of derivatives are also embraced by the term "research and development on the biochemical compositions of genetic resources". This issue becomes even more controversial as any reference to derivatives was deleted from the operational part of the final text of the Nagoya Protocol, and the term itself is mentioned only in its Article 2 defining the main terms. What is noteworthy is the fact that the Nagoya Protocol has also become the first international document to introduce the already existing scientific term "derivative" into the international legal terminology, defining it as a "naturally occurring biochemical compound resulting from the genetic expression or metabolism of biological or genetic resources, even if it does not contain functional units of heredity".[693] Combined with the reference to biotechnology – the term whose definition in the CBD Article 2 includes also the utilization of derivatives –, these provisions of the Nagoya Protocol can also be interpreted in a manner favorable to the extension of its scope to the use of derivatives, which do not contain functional units of heredity. Therefore, it can be affirmed that the controversy about the possibility of applying the ABS regime to the commercial utilization of derivatives was finally resolved affirmatively. However, many issues related to the utilization of derivatives remain unresolved and needs to be further clarified in national legislation.

With regard to possible doubts that may arise in the context of interpreting the phrase "genetic resources", it can be argued that the broad interpretation of this term permits extending the application area of international biodiversity legal regime also to the utilization of derivatives. At the same time, adhering to this term provides for purely commercial use of biological compounds, as commodities are not covered by the relevant provisions of the CBD and the Nagoya Protocol.[694]

2. *Bioprospecting and Biopatenting*

Based on the CBD COP and the WTO definitions, the process of gathering biodiversity samples with the subsequent exploration of their useful properties as well as the examination of their suitability for industrial and commercial exploitation

692 Article 2 of Nagoya Protocol.

693 Ibid.

694 Cf. *von Saint André* [n. 653], p. 32; *Henne* [n. 648], pp. 117 f., 125 f.

can be defined as "bioprospecting".[695] Consequently, the concept of bioprospecting[696] embraces the commercial exploitation of forest biodiversity resources, whereas these resources can be also acquired for purely scientific purposes. This means that not all aspects of utilization of genetic resources are included in the term "bioprospecting".[697] However, further legal controversies concerning benefit-sharing and ownership over genetic resources of forests and the products developed from them have arisen in relation to the commercial utilization of biodiversity resources. Scientific research related thereto can be effectively regulated within the general framework of international scientific cooperation.

However, genetic resources of forests which were initially acquired for purely scientific purposes as well as the results of scientific research can subsequently be transferred to a third party that will commercialize them. Such cases should be specifically regulated by international law. Consequently, as will be demonstrated further, efforts dedicated to the elaboration of an international legal framework for ABS have legitimately focused on the commercial aspect of the utilization of biodiversity resources of forests and of other ecosystems.

One major issue related to commercial exploitation of biodiversity resources consists of recognizing and respecting intellectual and commercial rights of both researchers and manufacturers.[698] The necessity of promoting such rights, while representing a controversial issue,[699] is currently recognized by international

695 *von Saint André* [n. 653], pp. 26 f; UN-Doc. UNEP/CBD/COP/5/INF/7, United Nations, Conference of the Parties to the Convention on Biological Diversity, Progress Report on the Implementation of the Programmes of Work – Information on Marine and Coastal Genetic Resources including Bioprospecting, Paragraph 7; TRIPS, CBD and Traditional Medicines: Concepts and Questions, Report of an ASEAN Workshop on the TRIPS-Agreement and Traditional Medicine, Jakarta, 13–15 February 2001. – Internet resource – https://grain.org/en/article/2025-recommendations-to-asean-on-ipr-trad-l-medicine [Last visit. 30.04.2023].

696 For the definition of "bioprospecting", see also *von Saint André* [n. 653], pp. 26 ff.; See also: *Gehl Sampath, Padmashree*, Regulating Bioprospecting: Institutions for Drug Research, Access and Benefit-Sharing, 2005, pp. 1 ff.; *Stoll, Peter Tobias*, Gestaltung der Bioprospektion unter dem Übereinkommen für biologische Vielfalt durch international verbindliche Verhaltensstandards: Hintergründe, Möglichkeiten und Inhalte, 2000, p. 3.

697 Cf. *Von Saint André* [n. 653], p. 26.

698 *Von Saint André* [n. 653], pp. 28 f.

699 Cf. *Royal Society*, Keeping Science Open: The Effects of Intellectual Property on the Conduct of Science, 2003; *Azoulay P, Ding W, Stuart T.*, The Impact of Academic

community.[700] As for the moment, so-called intellectual property rights can be protected through the system of internationally recognized patents.[701] The extended applicability of patents to products developed from biodiversity resources, a process which can be defined as biopatenting, was initially denied until the Supreme Court of the United States of America recognized it, as it is stated in its decision in the *Diamond v. Chakrabarty* case in 1980.[702] The Court confirmed the application of patents in relation to the protection of intellectual property rights on genetically modified plants and even on simple genes. Since this verdict, biopatenting has become a commonly acceptable practice in the USA and other developed States.[703] Considering the commercial value of products of research and development based on biodiversity compounds found in forests and other ecosystems as well as the widely disseminated practice of biopatenting, this legal mechanism required a unified international legal base for its regulation. The absence of such a regulatory framework had impeded the mutual recognition of patents and their application outside the jurisdiction of the State of issue for a long time.[704] The unified international legal framework for biopatenting was finally delivered with the adoption of the Agreement on Trade-Related Aspects of Intellectual Property Rights (TRIPs), which entered into force in January 1995.[705] TRIPs is an international agreement that establishes a legal framework for the standardization of the procedure of granting patents and their basic content all around the world.[706] According to the agreement, intellectual property rights concerning any invention, including both products and processes, should be protected by patents. The necessary requirement for

---

Patenting on the Rate, Quality and Direction of (Public) Research Output, in: Journal of Industrial Economics, Volume 57, 2009, pp. 638 ff.

700 *Oldham, Paul/Hall, Stephen/Forero, Oscar,* Biological Diversity in the Patent System, in: PLoS ONE, Volume 8, Issue 11, 2013, e78737.

701 Ibid.

702 *Von Saint André* [n. 653], pp. 28 f.; US. *Supreme Court,* Diamond, Commissioner of Patents and Trademarks v. Chakrabarty, 447 U.S., 1980, p. 330.

703 *Von Saint André* [n. 653], pp. 28 f.

704 Ibid.

705 TRIPS Agreement: Agreement on Trade-Related Aspects of Intellectual Property Rights, adopted on 15 April 1994 and entered into force on 1 January 1995, Marrakesh Agreement Establishing the World Trade Organization, Annex 1C, in: 1869 UNTS 299.

706 *Curci* [n. 651], pp. 36 ff.

patentability is that the product is new, inventive, and that its industrial application is possible.[707]

However, according to TRIPs, some product categories can be exempted from patentability by national authorities. This option is mainly intended for those cases where the exclusion of a certain innovation from patentability is needed in order to protect public order, morality, health, and life of people, animals and plants, or to preserve the integrity of the environment.[708] This provision's application is subject to the interpretation of the terms "public order" and "morality" as well as to the assessment of risks to society generated by patenting a certain innovation. Additionally, innovations can be excluded from patentability when they play a critical role in the treatment of human or animal illnesses,[709] and also with regard to "plant and animal species other than microorganisms, and essentially biological processes for the production of plants or animals other than non-biological and microbiological processes".[710] At the same time, TRIPs further underlines that States are obliged to protect intellectual property rights on any plant variety by using patents or any other legal instruments, or a combination of different legal instruments.[711] Taking into account that genetic material found in forests and other ecosystems and products of metabolism are not living organisms, the products developed on the basis thereof cannot be excluded from patentability based on Article 27.3(b). Such products are often used in pharmacology, cosmetics, and chemical industries. They are rarely used in rearing and breeding living plants and animals. On the other hand, a large part of pharmacological medicines is developed on the basis of genetic material. This means that the Parties to TRIPs can potentially limit the patentability of such products given that they play an important role in the treatment of humans and animals, and that the limitation of access to them can represent a threat for public order and an offence against morality. Nevertheless, States rarely take advantage of this possibility as the patent system delivers a number of practical benefits, which will be further explained below. However, the Doha WTO Ministerial Declaration confirmed that TRIPs should be interpreted and implemented in a way that will allow promoting the protection of public health, particularly by facilitating

707 Article 27.1 of TRIPs.
708 Article 27.2.
709 Article 27.3(a).
710 Article 27.3(b).
711 Ibid.

access to medication for all.[712] Actually, the exclusion from patentability according to Articles 27.2 and 27.3 of TRIPs give States a certain freedom of choice with regard to legislative measures that will permit the population to acquire pharmacological drugs at lower and fairer prices. Most of the States apply a wide range of approaches to the patentability of medication.

One of the strictest patent legislations exists in India.[713] For instance, in 2006, the Indian Patent Office denied the issue of a patent for the cancer drug Glivec, produced by Swiss drug manufacturer Novartis, by arguing that the patent did not satisfy criteria for inventions according to Article 3(d) of the Indian Patents Act.[714] According to this Article, the discovery of a new form of a substance does not constitute an invention if it does not contribute to the efficacy of the substance in question.[715] Curiously, the IPAB referred to possible infringements of public order, and not to the morality or specific provisions related to the protection of human health contained in the same article.[716] In 2009, the Intellectual Property Appellate Body confirmed the denied of approval by the Patent Office.[717] Subsequently, the Indian Supreme Court also supported this decision by interpreting Article 3(d) of the Act, and by arguing that Glivec did not satisfy the general patentability criteria of the Act.[718] The Court refused to address the issue of the compatibility of this decision with TRIPs, arguing that only the WTO dispute settlement body would be authorized to decide upon this issue. The Headquarter of Novartis is located in Switzerland. Thus, only this State is able to exercise diplomatic protection in favor of this company by raising a claim

712 DOHA WTO MINISTERIAL 2001: TRIPS WT/MIN(01)/DEC/2, 20 November 2001, Declaration on the TRIPS agreement and public health, 14 November 2001 – Internet resource – https://www.wto.org/english/thewto_e/minist_e/min01_e/mindecl_trips_e.htm [Last visit: 30.04.2023]

713 *Gabble, Ravinder/Kohler, Jillian Clare*, To Patent or Not to Patent? The Case of Novartis' Cancer Drug Glivec in India, in: Globalization and Health, Volume 10, Article Number 3, 2014, pp. 1 ff.

714 India, Intelectual Property, The Patents Act with Amendments, first adopted in 1970. – Internet resource – https://ipindia.gov.in/writereaddata/Portal/IPOAct/1_113_1_The_Patents_Act_1970 -_Updated_till_23_June_2017.pdf [Last visit: 30.04.2023]; *Gabble/Kohler* [n. 713], pp. 1 ff.

715 *Gabble/Kohler* [n. 713], pp. 1 ff.

716 Ibid.

717 Ibid.

718 Ibid.

before the WTO's dispute settlement bodies. However, Switzerland has never raised such claim.[719]

3. *Patent System and Its Benefits*

Patents grant exclusive rights to produce, utilize, or commercialize a certain innovation or product in the State where the patent is granted. All other actors and interested groups can be excluded from any industrial, commercial, or even scientific utilization of the relevant invention for the whole period of patent validity, which usually does not exceed 20 years in most States.[720] The exclusive rights provided by the patent give a developer or a manufacturer a certain level of confidence that the investments in research and development activities will be rewarded through dividends from the sales of commercialized innovations or products.[721] Such practice is supposed to be an effective way for the promotion of technological and scientific progress, while every innovation becomes a monetary source for further investment in research and development activities. That way, inventors receive additional incentives for the development of new products and innovations.[722] From this point of view, the patent system has the potential to increase scientific and technological progress.

There is also another important feature of the patent system that stimulates invention. This can be seen when an inventor makes an innovation available to the public and does not keep it in secret, as it sometimes happens when the acceptable level of intellectual property protection is not warranted. In addition, from this point of view, the patent system is supposed to promote the invention, while other scholars can use publicly accessible information about most recent innovations.[723] However, in practice, the application of the patent system can have controversial outcomes. Indeed, patents are sometimes likely to generate significant obstacles for scientific and technological progress, given that the utilization of innovations in the new research and development process will require the official permission of a patent holder.[724] The acquisition of such a permission can be challenging and requires a series of negotiations. By setting too high a

---

719 Ibid.

720 *Hall, Bronwyll H.*, Patents and Patent Policy, in: Oxford Review of Economic Policy, Volume 23, Number 4, 2007, pp. 570 ff.

721 Ibid.

722 *Hall* [n. 720], p. 571; *Lochen* [n. 648], pp. 64 f.

723 *Hall* [n. 720], pp. 571 f.

724 Ibid.

price for such a permission, the patent holder can factually impede further utilization of the invention by other inventors, and make the entire research and development process unaffordable.

For this reason, the real impact of patents for innovation is unclear. Theoretical modelling and analysis of empirical data have shown that the positive impact of patents on technological progress can be insignificant, and sometimes patents even result in the decline of research and development activities.[725] However, a number of commentators have demonstrated that the most positive effects of patenting occur in the area of pharmacy and biotechnology.[726]

There are some controversies concerning the impact of patents on development. It can be argued that States with a higher level of socio-economic development are likely to introduce a strong patent system, whereas less developed States are supposed to benefit from a weaker patenting regulation.[727]

Thus, at this point it can be argued that the patenting of products received in the process of bioprospecting is both legally allowed and economically effective. Consequently, the question arises whether patents can really contribute to the protection of forest ecosystems and forest biodiversity. First of all, the role of patenting innovations obtained in the course of bioprospecting should be analyzed in the light of the general practical approach to the conservation of biodiversity, which assumes that the protection of nature runs counter to the general principles of market economy. According to these principles, those taking and implementing decisions that influence nature have almost no incentives to promote the conservation of biodiversity.[728] As is the case with other public goods, biodiversity compounds found in forests and other ecosystems usually cannot be excluded from public access.[729] As a result, although society as a whole can benefit from biodiversity conservation, those stakeholders with the most potential to promote the protection of nature are not motivated to do so. At the same time, the unsustainable management of natural ecosystems and biodiversity is

---

725 *Hall* [n. 720], pp. 571 f.; *Heller, Michael A./Eisenberg, Rebecca S.*, Can Patents Deter Innovation? The Anticommons in Biomedical Research, in: Science, Volume 280, 1998, pp. 698 ff.

726 *Heller* [n. 726], p. 698.

727 Ibid, p. 699.

728 *Lawson, Charles*, The Role of Patents in Biodiversity Conservation, in: Nature Biotechnology, Volume 27, 2009, pp. 994 f.

729 *Humphreys, David*, Logjam: Deforestation and the Crisis of Global Governance, 2006, pp. 5 ff.; *Oldham/Hall/Forero* [n. 700], e78737.

economically beneficial for them.[730] It can be argued that this failure of modern market economy to evaluate the contribution of genetic resources and products of metabolism to research and development activities can be resolved by establishing an effective access and benefit-sharing mechanism supported by the patenting system. In order to achieve fair and equitable sharing of benefits, the patents should grant the legal rights ensuring the participation in the profits from commercialization of products to the developers, manufacturers, and providers of genetic resources.[731] In this way, biodiversity would be associated with certain tangible assets that turn to objects of transaction.[732]

However, the actual impact of such a scheme involving biodiversity protection is also controversial. A major concern in this regard is that it is not easy to determine a fair price for the socio-economic and ecological benefits derived from the conservation and sustainable utilization of forest biodiversity.[733] The benefits arising from research and development activities based on genetic resources cannot encompass the entire variety of multiple functions of forests and forest biodiversity. Usually, only a tiny share of multiple benefits related to the utilization of biodiversity can be considered in one tangible asset. Even if it were possible to take into account a wide range of properties of forest biodiversity in one unique tangible asset, it would hardly be possible to find a customer who would be ready to acquire it, as every buyer is usually only interested in one small part of an entire range of benefits delivered by forests and forest biodiversity.[734] A number of biodiversity compounds' natural properties are so complex that it is impossible to determine a possible customer who would acquire the hypothetical asset encompassing all of them.[735] As will be shown in the next chapter, several initiatives are currently in the process of economically evaluating the ability of forests to sink atmospheric carbon dioxide. Relevant assets can be acquired by major industrial companies emitting carbon dioxide, by States seeking to reduce

730 *Oldham/Hall/Forero* [n. 700], e78737.

731 Ibid.

732 *Lawson* [n. 728], pp. 994 f.

733 Ibid.

734 See *Richards, Michael/Jenkins, Michael*, Potential and Challenges of Payments for Ecosystem Services from Tropical Forests, 2007, pp. 5 f.; *Muradian, Roldan/Pellegrini, Lorenzo/Aguilar, Bernardo/Agarwal, Bina*, Payments for Ecosystem Services and the Fatal Attraction of Win-Win Solutions, in: Conservation Letters, Volume 6, Issue 4, 2013, pp. 275 ff.

735 See *Richards/Jenkins* [n. 734], p. 6; *Muradian/Pellegrini/Aguilar/Agarwal* [n. 734], p. 278.

their carbon dioxide emissions, and by other interested actors. However, sometimes it is difficult to determine the exact consumer of certain ecological services delivered by forests. More accurately speaking, humanity as a whole can be regarded as a consumer of these ecological services of forests. Some values of forests, such as aesthetic ones, are difficult to express in monetary terms.

The patent system cannot consider all these issues. Consequently, its contribution to the protection of forest ecosystems can be enhanced through the promotion of synergies with other conservationist instruments and tools, including non-monetary ones.[736] Otherwise, there is a risk that the revenues delivered by the patent system will not achieve the level upon which the forest owners will be incentivized to promote the protection of forests. Another issue is the conservation of nature being dependent on unreliable fluctuations of the market price for products developed based on genetic material.[737]

It is important to note that patents can have a negative impact on the price of accessed biodiversity compounds and products developed on their basis. The problem is that the legal effects of patents are normally extended not only to the product or innovation itself, but also to the genetic sequence and metabolic processes out of which they were developed.[738] Subsequently, the utilization of the same or a very similar genetic sequence in other innovations will require an official permission of the patent holder who has previously obtained exclusive access to this genetic sequence. The utilization of the same or similar genetic sequence by other scientists and manufacturers can be restricted or even totally forbidden for the entire period of patent validity.[739] That way, the real value of biodiversity compounds will be undermined, and their initial holders will have fewer incentives for biodiversity conservation. This issue can be partially resolved through establishing more accurate and precise standards and criteria for distinguishing between already accessed genetic sequences and those that are available only in-situ. Another option may be the limitation of patentability to products developed based on genetic resources and products of metabolism, as well as the exclusion of genetic sequences from patentability. However, in practice it will be difficult to draw a line between similar genetic sequences as well as to distinguish between the genetic resources and the products developed on their basis.

---

736 *Lawson* [n. 728], pp. 994 f.
737 *Humphreys* [n. 729], 2006, pp. 5 ff.; *Oldham/Hall/Forero* [n. 700], e78737.
738 *Lawson* [n. 728], pp. 994 f.
739 *Hall* [n. 720], pp. 570 f.

Additionally, the promotion of forest conservation through the patent system will depend on the regulatory framework encouraging the redirection of biodiversity-related technology and finance in the conservation of natural ecosystems. In fact, the main objective of the patent system is the protection of intellectual property rights and the enhancement of return on investment. For this reason, providers of genetic resources, developers, and manufacturers are not incentivized to include conservation clauses in patents. By choosing a proper patent design, the relevant stakeholder rarely considers whether the conservation of forests and forest biodiversity should be taken into consideration or not. This circumstance becomes particularly challenging when the return on investments does not cover the full biodiversity conservation costs, which include both the expenditures needed in order to implement the natural conservation measures and the lost economic opportunities related to the unsustainable utilization of biodiversity.

In light of these considerations, it can be argued that the access and benefit-sharing regime with regard to genetic resources based on patents or sui-generis systems can contribute to forest conservation only if robust and effective environmental regulations are adopted at international and national levels. Currently, the application of patents in order to promote access and benefit sharing is a widely accepted practice.

It should be underlined that, if applied to biodiversity resources, the patent system should consider the interests of all the stakeholders involved, including both investors in and providers of genetic resources. At the same time, they should also effectively promote the conservation of forest ecosystems and biodiversity. As it has already been argued above, the primary concern of the relevant stakeholders consists in the protection of intellectual property rights, ensuring the marginal investment return, and warranting fair and equitable sharing of benefits. Astonishingly, the conservation of natural habitats as pools of genetic resources necessary for future research and development activities plays only a secondary role. Providers of access can be interested in promoting the conservation of biodiversity resources that they "own", but only if the costs of conservation are compensated by payments provided in accordance with the patents.

4. *Biopiracy*

The term "biopiracy" can be interpreted as the acquisition of biodiversity resources and their utilization, while violating the legitimate rights of developing States under whose jurisdiction such resources are situated, and while violating the rights of ILCs managing such resources to grant access to them, and to participate in the benefits from their utilization.[740] Biopiracy can have adverse impacts on forest protection, as developing States loose incentives for the conservation of biodiversity. Moreover, they do not receive their fair share of benefits from the utilization of genetic resources, and do not benefit from the transfer of technology and finance that both can serve as a source of funding for the conservation of forests and sustainable development of forest-dependent communities, while combatting driving forces of deforestation. Biopiracy also occurs when genetic material is not accessed in-situ in forest ecosystems but received from ex-situ collections, for example, botanic gardens, where it was introduced decades or centuries before legislation governing access to genetic resources and sharing of benefits arising from their utilization was adopted.[741] In addition, in this case the rights of States where such species are found in-situ can be violated. The problem is further complicated by the fact that once certain useful chemical substances are developed based on genetic sequences and natural metabolism products, initial biodiversity compounds are not further needed in order to reproduce them. As discussed above, the narrow focus of the CBD and the Nagoya Protocol on the utilization of genetic resources leads to the lack of regulation surrounding benefit-sharing regarding utilization of derivatives. This gap also gives rise to biopiracy-related concerns.

Biopiracy can also occur when the intellectual property rights of the biodiversity-based product or innovation's developer, but not of the provider of genetic resources, are protected by the patent. In such a case, the manufacturer can impede the utilization of the relevant innovation in the State of origin of genetic resources. This can become particular challenging in the case of essential products including medicine, methods of food production, or freshwater purification. For this reason, the State of origin usually seeks to protect its right to utilize those innovations covered by the patent.[742]

Indigenous and local communities can be particularly discriminated through biopiracy, as genetic resources under their ownership as well as traditional biodiversity-related knowledge held by them can be acquired without appropriate compensation.[743] Moreover, ILCs are among the most socially unprotected and

---

740 *von Saint André* [n. 653], pp. 24 ff.; *Delgado, Gian Carlo,* Biopiracy and Intellectual Property as the Basis for Biotechnological Development: The Case of Mexico, in: International Journal of Politics, Culture and Society, Volume 16, Number 2, 2002, pp. 297 f.

741 *von Saint André* [n. 653], pp. 24 ff.

742 Ibid.

743 *Lochen* [n. 648], pp. 41 ff.

economically disadvantaged parts of population. Biopiracy can deprive them of a potential source of income. Moreover, the utilization of products and innovations developed based on genetic resources can be denied for them. The problem will be further complicated with regard to the fact that a number of innovations is solely a technological reproduction of methods and practices existing in the traditional knowledge and practices of ILCs.[744] Such traditional knowledge, in the case of forests TFRK, often serves as a basis for the development of certain bioproducts. In a number of cases, the ILCs are ready to grant access to this knowledge without being aware that it will be used for the development of biodiversity-based products and innovations.[745]

In the case of the medication developed from the neem tree, ILCs' representatives have managed to defend their rights, and to convince competent authorities to declare invalid already granted patents.[746] In this case, the medication in question was a simple reproduction of traditional knowledge of Indian folk healers.[747] However, even in such cases it is extremely difficult to assign the benefits from the commercialization of the relevant products to the original holder of TFRK ex post.[748]

In all that, biopiracy is not a legal term, but rather a political one, which serves to consolidate actions for the promotion of rights and interests of the States of origin of genetic resources.[749] Article 15 of the CBD gives a more or less clear idea of the conditions under which access to genetic resources should take place in order to avoid biopiracy. According to this article, access should take place on mutually agreed terms (MAT), and with the Prior Informed Consent (PIC) of the States or communities of origin.

However, it is obvious that international legal provisions for the utilization of genetic resources can only have positive effects on forest conservation when positive incentives are created, and the sufficient funding is available. Moreover, the provisions in question must be sufficiently robust in order to combat biopiracy. On the other hand, taking investors' interests into consideration is also very important, as interpreting the term "biopiracy" too widely can make the process of research and development costly and unaffordable.

---

744 Ibid.

745 Ibid.

746 *Federle, Christine*, Biopiraterie und Patentrecht, 2005, pp. 71 f.

747 Ibid.

748 Ibid.

749 *von Saint André* [n. 653], pp. 26 ff.

Since the meaning of the main terms widely used with regard to bioprospecting has been clarified, the analysis of the ABS-relevant legal provisions can now be undertaken. Regarding the scope of this research, this analysis should be initiated based on briefly considering forest-related provisions of the CBD.

C. Relationship Between the ABS Regime under the CBD and the Conservation of Forests

I. Forest Issues in the CBD

The protection of biodiversity as promoted by the CBD plays an important role in the conservation and sustainable use of forests. With regard to forest protection, it can be appreciated that the Convention distinguishes between the conservation of biodiversity in-situ (in natural habitats) and ex-situ (in botanic gardens, genetic banks etc.), emphasizing that one form cannot replace the other.[750] This means that biodiversity compounds, once removed from their natural habitats, cannot fulfil their role in ecosystems. In this regard, Article 8 of the CBD is particularly important, as it foresees the conservation and restoration of natural habitats, including forests. First of all, a system of protected areas should be established and managed based on special guidelines in order to protect biodiversity.[751] Biological resources should be managed sustainably not only within the territory of protected areas but also outside of them.[752] Notably, the CBD does not only foresee the establishment of these protected areas, but also the protection of ecosystems, natural habitats and the conservation of populations of species living in the wild.[753] Furthermore, degraded ecosystems should be restored and rehabilitated, and the recovery of threatened species should be promoted.[754] All these provisions have a direct impact on forest conservation. However, their legal value is fairly relativized, as States only *shall* implement *them as far as possible and as appropriate*, meaning that the obligations of Article 8 of the CBD are not specified. The CBD delivers no clear guideline concerning the necessary level

750 *Schulte zu Sodingen* [n. 233], pp. 201 f.; *Krohn* [n. 652], pp. 112 f.; *Eikermann, Anja*, Forests in International Law: Is There Really a Need for an International Forest Convention?, 2015, pp. 99 ff.
751 Article 8(a).
752 Article 8(b).
753 Article 8(d).
754 Article 8(e).

of protection that should be ensured in the protected area once such an area has been established.

Thus, the CBD promotes a holistic approach to the conservation of biodiversity resources, which can have a positive impact on forest conservation. Then again, the contribution of the CBD to forest protection can be limited due to the lack of a compliance mechanism, or other enforcement mechanisms. According to the text of the CBD, its compliance mechanism is limited to the monitoring and revision of its implementation process by the COP. Additionally, forest-related issues are not specifically covered by the treaty text.

The protection of forests through the CBD could further be promoted by the Work Programme on Forest Biological Diversity (WPFBD),[755] elaborated on by its COP. When the WPFBD was established by COP 4 in 1998, it was predominantly dedicated to the promotion of forestry research, scientific cooperation, and the development of technologies designed to contribute to the preservation of forest biodiversity.[756] Such activities had to be based on the holistic and inter-sectoral *ecosystem approach*. The approach in question is based on the comprehensive analysis of the human influence on forest biodiversity and of the ways to combat the drivers of deforestation.[757] Additionally, the WPFBD was dedicated to the development and implementation of criteria and indicators for forest biodiversity. The WPFBD is also open to other research and technological goals proposed by the Subsidiary Body of the CBD on Scientific, Technical and Technological Advice, or those identified in the process of its implementation. Notably, during the first phase of its implementation, the WPFBD was highly dependent on the activities promoted by the IFF.

With the WPFBD initially only dedicated to research activities, COP 5 made the decision to extend its scope of application to include practical issues. In the light of this decision, COP 6 adopted an Expanded Programme of Work on Forest Biological Diversity (EPWFBD). In addition to the scientific issues mentioned above, three further program elements were included in this document, namely the promotion of conservation, sustainable use and benefit sharing,

755 CBD COP 4 Decision IV/7 "Forest Biological Diversity", Decision adopted by the Conference of the Parties to the Convention on Biological Diversity at its fourth Session, held at Bratislava from 4 to 15 May 1998, United Nations, 1998, Annex. – Internet resource – https://www.cbd.int/doc/decisions/cop-11/cop-11-dec-06-en.pdf [Last visit: 30.04.2023]

756 *Eikermann* [n. 750], pp. 100 ff.

757 Ibid.

creating institutional and socio-economic enabling environment, and facilitating assessment and monitoring.

However, the EWPFBD also suffers from a lack of precise and detailed provisions, and does not provide an explicit list of measures that should be implemented in order to achieve its goals. Without such measures, the EWPFBD cannot contribute significantly to the protection of forests, yet does provide an extensive practical guideline indicating which actions should be undertaken for this purpose.

Thus, both the CBD and EWPFBD contain important and valuable provisions concerning the promotion of forest protection. However, their general framework for the conservation and sustainable use of biodiversity does not create enough incentives for the protection of forests. For this reason, the potential of the ABS regime as established by the CBD to contribute to forest protection should be further analyzed.

II. ABS under the CBD

In the light of the CBD's third objective, which constitutes the sustainable exploitation of genetic resources and equitable sharing of the benefits arising from their utilization, Article 15 provides a detailed framework for the access and benefit-sharing regime (ABS). During the first negotiation rounds, developed States were particularly advocating the concept of *free access* to genetic resources, which was one of few manifestations of the idea of the common heritage of humanity when applied to biodiversity that could not find its way even to the first Draft of the CBD.[758] It was assumed that access to genetic resources should be fair and equal, free or open; however, it was commonly acknowledged that "free access" did not mean "free of charge".[759] The actual meaning of this concept was interpreted in such a way that developing States were not supposed to restrict access to biodiversity resources to a measure which would make any research and development activities impossible.[760] Yet, this concept was unacceptable for the developing States seeking to protect their sovereign rights. The consensus achieved during the negotiations is reflected in Article 15 of the CBD. In the context of the sovereign rights of States over their biological resources, rights yet limited by the obligation to respect the interests of the international community and of future generations, this provision underlines that States are supposed to

758 *Henne* [n. 648], pp. 117 f.; *Sdunzig* [n. 652], pp. 65 ff.
759 *Henne* [n. 648], pp. 117 f.; *Sdunzig* [n. 652], pp. 65 ff.
760 *Henne* [n. 648], pp. 117 f.

create conditions in order to facilitate the access to genetic resources for environmentally sound use, without imposing any restrictions running counter to the scope of the Convention.[761]

The application of the ABS regime, as well as of the Convention as a whole, to some particular types of genetic resources was a disputable issue from the very beginning.[762] Ever since negotiations began, some parties argued that the CBD should not be applied to migratory species, as it was difficult to determine their State of origin, meaning the State providing access thereto.[763] A separate international agreement was proposed with the intention to regulate the specific issues related to these species. Finally, migratory species were also covered by the Convention thanks to the extensive definition of biodiversity resources in Article 2 of the CBD, with the possibility of resolving problematic issues by adopting the Protocols to the Convention.[764] It is noteworthy that the applicability of the ABS regime to migratory species can also be important in the context of the protection of forests, as many species fulfill an important role in these ecosystems by providing important ecological functions.

Another point of contention was the applicability of the Convention to genetic resources accessed before its adoption and preserved in the genetic banks and other ex-situ collections in the territory of developed States.[765] The initial attempt to define "State providing the genetic resources" as the State that possesses genetic resources both in natural habitats and ex-situ collections led to a serious misunderstanding. Indeed, developed States, which acquired genetic resources according to the historically prevailing concept of free access exempted from charges, would now be able to benefit from ABS by granting access to such biodiversity compounds preserved in their ex-situ collections.[766] However, the attempt of developing States to extend the application of the CBD on endemic plant and animal species found in such ex-situ collections also did not find its way into the final text of the CBD. It was argued that in most cases it was already impossible to determine the State of origin of such resources, with most of them occurring naturally in different regions of the world. Finally, it was decided that

---

761 Ibid.

762 *Henne* [n. 648], pp. 117 f.; *McConnell, Fiona*, The Biodiversity Convention. A Negotiating History, A Personal Account of Negotiating the United Nations Convention on Biological Diversity – and After, 1996, pp. 85 f.

763 *Henne* [n. 648], pp. 124 f.

764 Ibid.

765 *Henne* [n. 648], pp. 117 f.; *McConnell* [n. 762], pp. 85 f.

766 *Henne* [n. 648], pp. 117 f.; *McConnell* [n. 762], pp. 85 f.

genetic resources are those provided by the State of origin or by the State which possesses these resources in accordance with the Convention.[767] Thus, the genetic resources acquired before the adoption of the CBD were exempt from the scope of the Convention, and their regulation was conceded to the already existing rules of international common law. Consequently, ABS cannot be applied to them. It can be argued that this decision deserves approval, as the research and development process would otherwise become unjustifiably challenging.[768] Still, it has negative implications on forest protection, as the benefits arising from the utilization of such genetic resources do not contribute in any way to the conservation of forest ecosystems from which they originate. Such an approach is also counter-productive as these ecosystems can deliver a bulk of genetic resources in the future. Insofar as the State of origin cannot be traced back, it may be proposed to establish a mechanism to channel a part of revenues from commercializing such products into the global forest conservation fund. However, such fund has not been established yet.

Once the genetic resources were accessed, scientific research activities should take place with the full participation of, and, if applicable, in the State providing the access.[769] Although the CBD itself does not explain what full participation means, it was clear from the very beginning of the negotiations that States of origin were expected to participate in research and development activities, and that the majority of such activities should take place in the territory of the providing State.[770] However, the legal value of this provision is relativized by the weak wording "Contracting Parties shall endeavor". It can be seen critically that, instead of establishing a duty to share the relevant benefits, a weaker language was chosen, according to which every Contracting Party should take legislative, administrative and policy measures with the aim of the fair and equitable sharing of research and development results as well as the benefits from commercial and other exploitation of genetic resources with the State of origin.

1. Mutually Agreed Terms (MAT)

Access to genetic resources on Mutually Agreed Terms, as it is stipulated by the Article 15.4 of the CBD, foresees the conclusion of a bilateral agreement between the State of origin and the user State. The first draft of the Convention already

767 Article 15.3 of CBD.

768 von Saint André [n. 653], pp. 39 f.

769 Article 15.3 of CBD.

770 *von Saint André* [n. 653], pp. 39 f.

included the requirement of MAT for any case in which genetic resources were accessed for commercial purposes.[771] Later negotiation rounds extended this to all other cases of access.[772] The concept of MAT is to be applied not only to the access to genetic resources, but also to the transfer of technology.[773] The CBD does not contain detailed provisions concerning the content of MAT. Instead, this concept foresees that the controversial interests of parties are to be reconciled in course of negotiations that take place in every specific case of access. Such an approach permits to take due regard to issues of sovereignty, although it can also be criticized from several points of view. The main problem is that with the parties themselves entirely responsible for determining the conditions of access, the interests of ILCs and environmental issues can be undermined in national legislation once ILCs are not equally represented in the national legislative and executive bodies, and when the environmental issues are not properly promoted on the national level or when they are undermined in the decision-making process.

2. *Prior Informed Consent*

The requirement of the Prior Informed Consent (PIC) regarding ABS appears in the second draft of the CBD.[774] According to the wording of Article 15.5, the requirement of Prior Informed Consent is not obligatory, and the States of origin can waive it in their national legislative acts. However, from a practical point of view, the negotiation of MAT can only take place once States have acquired a Prior Informed Consent.[775] This makes PIC an indispensable part of the ABS process. This assumption cannot be contested by the general systematic of Article 15, which places the condition of acquiring PIC only after prescribing the requirement of MAT. Additionally, the authority to decide on the necessity of PIC in every specific case was only delegated to developing States by the latest draft of the CBD in order to take into regard those states' sovereignty concerns.[776]

771 *von Saint André* [n. 653], pp. 48 f.; *Henne* [n. 648], pp. 71 f.
772 *Henne* [n. 648], p. 124.
773 CBD, Articles 15.7 and 16.2; See also *Henne* [n. 648], p. 124.
774 *Henne* [n. 648], p. 124.
775 *Von Saint André* [n. 653], p. 45.
776 Ibid.

3. Deficiencies of the ABS Regime

Thanks to the adoption of the CBD, the national sovereignty of States over the biodiversity resources found in their territory was reaffirmed, and the ABS regime came into being. This regime delivers a general framework that regulates access to genetic resources and the sharing of benefits arising from their utilization.[777] Such a system may contribute to the protection of forest ecosystems by creating economic incentives for their conservation. However, the treaty text does not contain detailed provisions on how ABS should be implemented, granting the parties to the Convention the authority to determine the content of MAT. In such a situation, a possibility exists that the interests of ILCs and environmental issues will be undermined in the process of access. The lack of detailed provisions generated mistrust on both the users' and owners' side.[778] Users were worried about the possible violation of their intellectual property rights. Due to the absence of detailed international guidelines, mutually agreed terms had to be newly negotiated for every specific case. However, in practice, developing States have fewer possibilities to promote their interests during negotiations with more economically powerful foreign investors. For this reason, only few States adopted national legislation necessary for the application of ABS.[779] Even when such laws regulating access were adopted, they contained usually very strict provisions, like the Executive Order of Philippines, that way contradicting the very

777 *Proelß* [n. 652], pp. 420 f.

778 *Koutouki, Konstantia/Rogalla von Bieberstein, Katharina*, The Nagoya Protocol: Sustainable Access and Benefits-Sharing for Indigenous and Local Communities, in: Vermont Journal of Environmental Law, Volume 13, 2012, p. 519.

779 Republic of the Phillipines, Executive Order N. 247 of the Philippines "Prescribing Guidelines and Establishing a Regulatory Framework for the Prospecting of Biological and Genetic Resources, their By-Products and Derivatives, for Scientific and Commercial Purposes; and for other Purposes", in: Official Gazette, May 18, 1995; Republic of Costa Rica, Biodiversity Law, No. 7788, 1998; Federal Democratic Republic of Ethiopia, Proclamation NO. 482/2006, A Proclamation to Provide for Access to Genetic Resources and Community Knowledge and Community Right, in: Federal Negarit Gazeta 13th Year, No. 13, Addis Ababa, 27 February, 2006; Cf. *Swiderska, Krystyna/Dano, Elenita/Dubois, Olivier,* Developing the Philippines' Executive Order No. 247 on Access to Genetic Resources, International Institute for Environment and Development, 2001, pp. 1 f.; *Smagadi, Aphrodite*, National Measures on Access to Genetic Resources and Benefit Sharing – "The Case of the Philippines", in: Law Environment and Development Journal, Volume 1, 2020, pp. 52 ff.

scope of the CBD, which consists of facilitating the access.[780] For this reason, the ABS-treaties were rarely concluded and implemented.[781]

D. Nagoya Protocol and the Conservation of Forests

I. From the Bonn Guidelines to the Nagoya Protocol

In April 2002, the CBD COP adopted the non-legally binding Bonn Guidelines on Access to Genetic Resources and Fair and Equitable Sharing of the Benefits Arising out of their Utilization[782] that was deemed to further develop the ABS regime, to promote transparency and legal certainty, and to protect the interests both of users and owners of biodiversity resources. By achieving these goals, the Bonn guidelines may contribute to the implementation of ABS, and thereby create incentives for the protection of forests,[783] while delivering practical guidance for the development and implementation of national systems for ABS. Given that a non-legally binding document could not be sufficient in order to provide a remedy for the gaps of the initial convention text, developing States, especially the group of Like-Minded Megadiverse States, have argued that this document represents only the first step on the way to a comprehensive ABS regime that would promote the rights and interests of different relevant groups of stakeholders.[784]

What is important is the fact that the Bonn Guidelines require the PIC of ILCs not only regarding access to genetic resources, but also concerning the related traditional knowledge, although the "approval and involvement" of ILCs as an alternative to their PIC is also foreseen.[785] However, the provision refers to the existing legal rights of ILCs without introducing any innovations specifically related to genetic resources, and emphasizes that its application is subject to the domestic legislation of the State of origin. The introduction of specific ABS-related rights of ILCs could contribute to the legal certainty and promote the consideration of rights of indigenous people during negotiations of mutually

780 *Von Saint André* [n. 653], p. 48.

781 *Proelß* [n. 652], p. 436.

782 Decision VI/24, "Bonn Guidelines on Access to Genetic Resources and Fair and Equitable Sharing of the Benefits Arising out of their Utilization" – Internet resource – https://www.cbd.int/doc/publications/cbd-bonn-gdls-en.pdf [Last visit: 30.04.2023].

783 *Lochen* [n. 648], pp. 148 ff.

784 *Koutouki/Rogalla von Bieberstein* [n. 778], pp. 523 f.

785 Bonn Guidelines, Article 31.

agreed terms in every specific case of access. It is also problematic that the role of ILCs during negotiations of MAT is not specified.[786] Factually, ILCs can still be excluded from negotiations, and their rights can be undermined in the MAT agreement if national legislation does not sufficiently protect their rights.[787]

The Bonn Guidelines can also be criticized for the lack of provisions dealing with the legislative measures to be taken under the national law of user States. In fact, regulations on the user's side are particularly important in order to ensure compliance with the national legislation of the State of origin.[788] Research, development, manufacturing, and commercialization of bioproducts often take place beyond the national jurisdiction of the provider State of genetic resources, which, therefore, is unable to monitor the maintenance of the user's obligation related to benefit-sharing.[789] Another disadvantage of the Bonn Guidelines consists in the document's lack of consideration towards the necessity of international monitoring measures, even though the document promotes the development of national monitoring and reporting systems and voluntary verification mechanisms.[790]

The Bonn Guidelines contain only scarce provisions directly related to the environmental protection, despite one of its objectives being the contribution to the conservation and sustainable use of biological resources.[791] Among other things, the Bonn Guidelines stipulate that the entire access and benefit-sharing process should have as its scope the conservation and sustainable use of biodiversity.[792] It also indicates that the competent national authorities responsible for providing access can also give advice on biodiversity conservation.[793] In this way, the guidelines contribute to the establishment of synergies between the utilization of genetic resources and the conservation of forests. Furthermore, users of genetic resources seeking to obtain PIC should provide information on possible impacts of access on biodiversity conservation.[794] Given that the general scope

---

786 *Koutouki/Rogalla von Bieberstein* [n. 778], p. 523.
787 Ibid, p. 523.
788 *Dross, Miriam/Wolff, Franziska*, New Elements of the International Regime on Access and Benefit-Sharing of Genetic Resources-the Role of Certificates of Origin, Bundesamt für Naturschutz, 2005, pp. 19 ff.
789 *Koutouki/Rogalla von Bieberstein* [n. 778], p. 521.
790 Bonn Guidelines, Articles 55–58.
791 See *Dross/Wolff* [n. 788], p. 18.
792 Bonn Guidelines, Article 11.
793 Bonn Guidelines, Article 14(f).
794 Bonn Guidelines, Article 36.

of ABS also includes the promotion of biodiversity conservation,[795] PIC can be denied if there is a risk that access will have negative impacts on wildlife.[796] However, the final decision on the provision of access is totally left to the discretion of the national authority.[797] The Bonn Guidelines also determine that the benefits arising from utilization of genetic resources should be used in such a way that will help to promote the conservation and sustainable use of biodiversity.[798] What is noteworthy is the fact that the removal of possible obstacles for the conservation and sustainable use of biological resources through access and benefit sharing and the creation of markets are foreseen as incentives for the promotion of biodiversity protection. However, the Bonn Guidelines do not further specify how these measures should be implemented.[799]

It is submitted that the Bonn Guidelines consider that the conservation and sustainable use of biodiversity intrinsically belongs to the scope of the ABS regime. Following this assumption, one can conclude that detailed provisions specifically dealing with biodiversity conservation are not necessary.[800] However, such an approach ignores the fact that owners of genetic resources seek to secure for them a share in the economical, technological, and scientific benefits arising from the utilization of these resources, while users are seeking to strengthen their intellectual property rights and to facilitate access to genetic resources. However, neither side has as its preliminary goal the conservation of forest biodiversity and forest ecosystems. Thus, owner and user States are both likely to undermine the relevant environmental issues if necessary to promote their economic interests. From this point of view, detailed provisions dealing with the conservation of biodiversity are needed on both international and national levels. The Bonn Guidelines do not contain such provisions, and, consequently, do not sufficiently promote the integration of biodiversity conservation in ABS.

In the light of all these insufficiencies and gaps of the Bonn Guidelines, particularly under the consideration of the lack of measures to be implemented on the user's side, the Group of Like-Minded Megadiverse Countries requested the establishment of the International regime for ABS.[801] As a result of numerous complicated negotiations, the Nagoya Protocol came into being.

---

795 Bonn Guidelines, Article 22.
796 Bonn Guidelines, Article 36.
797 Bonn Guidelines, Article 36.2.
798 Bonn Guidelines, Article 48.
799 Bonn Guidelines, Article 51.
800 *Dross/Wolff* [n. 788], p. 18.
801 *Koutouki/Rogalla von Bieberstein* [n. 778], pp. 523 f.

II. Development of the ABS Regime by the Nagoya Protocol

The Nagoya Protocol has introduced a number of important innovations regarding ABS. What is important is the fact that this document can be applied to the genetic resources covered by Article 15 of the CBD, and to the benefits arising from their utilization as well as to the traditional knowledge and the benefits arising from the application of such knowledge.[802] It is noteworthy that Article 5 of the Nagoya Protocol determines that not only States have to ensure the fair and equitable sharing of benefits arising from the utilization of biodiversity resources with the State of origin, but also the State in question has to adopt legal provisions in order to ensure that the interests of ILCs, including their right to participate in the benefit-sharing, will be protected.[803]. Additionally, the wording of Article 5 is far stricter than the more broadly framed provisions of the CBD. It determines explicitly that the sharing of benefits shall take place, whereas Article 15.7 of the CBD stipulates that States only have to take "legislative, administrative, and political measures with the scope of sharing" these benefits.[804] Such a vague obligation in the CBD is explained by the fact that most benefits arise within the private sector, and that, therefore, sharing them will be regulated by the contracts negotiated and concluded on a case-by-case basis.[805]

However, it can be argued that the existence of the obligation to share benefits can foster States' efforts related to the development and implementation of a legal framework for ABS. If a duty to ensure the sharing of benefits is not established, the State is less likely to prioritize benefit sharing while developing the pertinent national legislation. Notably, the Nagoya Protocol underlines that the sharing of benefits is linked directly to the utilization of genetic resources and to the process of access.[806] The importance of such regulation consists of

---

802 Article 5 of NP; *Morgera/Tsioumani/Buck* [n. 650], pp. 110, 114, 117 f.

803 Article 5.2 of NP; *Morgera/Tsioumani/Buck* [n. 650], p. 114; *Greiber, Thomas/Moreno, Sonia Peña/Mattias Åhrén/Carrasco, Jimena Nieto/Kamau, Evanson Chege/Medaglia, Jorge Cabrera/Oliva, Maria Julia/Perron-Welch, Frederic/Ali, Natasha/Williams, China*, An Explanatory Guide to the Nagoya Protocol on Access and Benefit-sharing, IUCN, 2012, pp. 83 ff.; *Fodella, Allessandro*, "Recent Developments on Access and Benefit Sharing Relating to Genetic Resources (ABS) in International Law", in: *Casonato/Busatta/Penasa/Piciocchi/Tomasi* (a cura di), Il biodiritto e i suoi confini: Definizioni, dialoghi, interazioni, 2014, p. 99.

804 *Greiber/Moreno/Mattias/Carrasco/et. al.* [n. 803], p. 84.

805 Ibid.

806 Ibid.

the circumstance that at the very moment of access, only scarce information is available regarding the benefits arising from the utilization of these resources, as the commercialization may take place with a great spatial and temporal distance and without a clear correlation between the moment of access and actual utilization.[807] In other words, the Nagoya Protocol seeks to make available the widest range of benefits arising from the exploitation of genetic resources, and to share them with the States of origin.[808] The innovations that were introduced by the Nagoya Protocol allow for increasing the confidence of the States of origin in the ABS regime. In this way, it encourages them to participate in ABS-related projects and to protect biodiversity resources. Although Article 5 does not explicitly mention forests, strengthening obligations related to benefit sharing can deliver further incentives for forest protection.

1. Development of PIC and MAT by the Nagoya Protocol

The Nagoya Protocol reaffirms that access to genetic resources occurs upon mutually agreed terms and is subject to the PIC of the State of origin.[809] The regulations of the Nagoya Protocol dealing with MAT are far more detailed and specific than the relevant regulations of the CBD, even though the great freedom of choice regarding MAT design is still left to the Parties.[810] Article 6.3.g has crucial importance as it stipulates a requirement for the establishment of precise rules and procedures in order to achieve MAT. Moreover, it prescribes an obligatory written form for MAT and includes a minimum list of clauses that have to be regulated therewith.[811] They include a dispute settlement clause; terms on benefit sharing, including when related to intellectual property rights; terms on subsequent third-party use, if any; and terms on changes of intent where applicable.[812] Although this list is non-exhaustive and non-prescriptive, its inclusion in the Nagoya Protocol represents an important breakthrough in the development of the ABS regime, as both the CBD and the Bonn Guidelines did not provide any guidance for the development of MAT.

---

807 *Morgera/Tsioumani/Buck* [n. 650], p. 115.

808 Ibid.

809 Article 5(1–5), Article 6(1).

810 *Fodella* [n. 803], p. 100; *Morgera/Tsioumani/Buck* [n. 650], pp. 167 ff.; *Greiber/Moreno/Mattias/Carrasco/et. al.* [n. 803], pp. 93 ff.

811 *Morgera/Tsioumani/Buck* [n. 650], p. 168.

812 *Fodella* [n. 803], p. 100; *Morgera/Tsioumani/Buck* [n. 650], pp. 167 ff.; *Greiber/Moreno/Mattias/Carrasco/et. al.* [n. 803], pp. 93 ff.

The terms on subsequent third-party use as well as the terms on changes of intent were included in this list of minimal requirements to MAT in order to tackle those cases where a chain of providers and users are involved in the utilization process of genetic resources. This occurs when genetic resources are initially acquired for purely scientific purposes, and subsequently transferred to a third party, which seeks to commercialize the results of research.[813] Another case involves the scenario when the Party seeking access changes its initial intent and uses the acquired genetic resources for commercial purposes. In all these cases, the right of owners to participate in the fair and equal benefit sharing must be ensured.[814] Therefore, these provisions contribute to owners' awareness that if they provide access to genetic resources, they will be able to pursue their rights related to benefit sharing. The reference to intellectual property rights can contribute to the promotion of rights of the owners of genetic resources by affirming that the sharing of benefits arising from intellectual property rights shall be addressed by MAT. With the inclusion of a unique provision dealing with the sharing of benefits related to patents in the Nagoya Protocol, this important issue has appeared for the first time on the agenda of the ABS regime.[815]

From their side, any State requiring PIC "has to take the necessary legislative, administrative or policy measures" with the aim to ensure transparency, clarity and legal certainty of their national legislation systems regarding ABS.[816] They have to ensure "fair and non-arbitrary" access-related rules and procedures, and to provide the information concerning the application of PIC. The granting of PIC has to be confirmed by the respective national authority issuing a clear and transparent written decision in a cost-effective and timely manner.[817]

It should be noted that the Nagoya Protocol contains a list of provisions on dispute settlement to be included in MAT in order to promote compliance. This list is neither exhaustive nor prescriptive, whereas States are supposed to encourage the users and providers of genetic resources to include them in MAT. Thereby, relevant stakeholders should determine the jurisdiction to which any

813 *Fodella* [n. 803], p. 100; *Morgera/Tsioumani/Buck* [n. 650], pp. 167 ff.; *Greiber/Moreno/Mattias/Carrasco/et. al.* [n. 803], pp. 93 ff.

814 *Greiber/Moreno/Mattias/Carrasco/et. al.* [n. 803], pp. 93 ff.; See also *Glowka, Lyle/Burhenne-Guilmin Françoise/Synge, Hugh*, A Guide to the Convention on Biological Diversity, IUCN, 1994, p. 30.

815 See *Morgera/Tsioumani/Buck* [n. 650], p. 169.

816 Article 6.3(a) of the NP.

817 Article 6.3(b) of the NP; *Fodella* [n. 803], p. 100; *Morgera/Tsioumani/Buck* [n. 650], pp. 167 ff.; *Greiber/Moreno/Mattias/Carrasco/et. al.* [n. 803], pp. 183 ff.

eventual dispute will be subjected, the applicable law as well as the alternative procedures for dispute settlement, such as arbitration or mediation.[818] In the absence of such clauses, legal uncertainty may arise concerning the court in charge, before which an eventual dispute should be brought, and concerning the law to be applied.[819] In this case, the general rules of international private law may help to resolve uncertainty, but such an approach represents a more cost- and time-consuming way, and often leads to controversial conclusions.[820] What is noteworthy is that forest-rich States providing genetic resources may prefer to seek recourse under the jurisdiction of a user State, as the judgement can be directly enforced in the user State's national jurisdiction.[821] Provisions on arbitration and mediation can also be important, as these alternative procedures of dispute settlement can be preferred by the owners of genetic resources as less time- and cost-consuming.[822]

Moreover, the Nagoya Protocol foresees that every party shall ensure the access to legal recourse under its jurisdiction in cases of MAT-related disputes.[823] Thus, the Nagoya Protocol calls upon its Parties to remove procedural barriers and to create special opportunities in order to facilitate the legal recourse to be available in case of MAT infringement in each State regardless of the claimant's nationality.[824] Despite the fact that access to domestic courts is usually granted to foreign plaintiffs in the majority of States, in practice foreign citizens, such as the representatives of ILCs, are likely to face serious obstacles such as lacking knowledge of a pertinent legal system or a national language.[825] At the same time, the provision of Article 18.2 is not per se sufficient to ensure access to the legal recourse in such cases.[826] Arguably, it can only be operational in combination with

818 *Fodella* [n. 803], p. 101; *Morgera/Tsioumani/Buck* [n. 650], pp. 283 ff.; *Greiber/Moreno/Mattias/Carrasco/et. al.* [n. 803], pp. 183 ff.

819 *Morgera/Tsioumani/Buck* [n. 650], pp. 286 f.

820 Ibid.

821 Ibid.

822 *Isozaki, Hiroji*, Enforcement of ABS Agreements in User States, in: *Kamau, Evanson C./Winter, Gerd* (eds.), Genetic Resources, Traditional Knowledge, and the Law: Solutions for Access and Benefit Sharing, 2009, p. 446.

823 Article 18.2; *Fodella* [n. 803], p. 101; *Morgera/Tsioumani/Buck* [n. 650], pp. 283 ff.; *Greiber/Moreno/Mattias/Carrasco/et. al.* [n. 803], pp. 183 ff.

824 *Morgera/Tsioumani/Buck* [n. 650], p. 289.

825 *Isozaki* [n. 822], p. 442.

826 See *Greiber/Moreno/Mattias/Carrasco/et. al.* [n. 803], pp. 186 f.

Article 18.1 that promotes the inclusion of special dispute settlement provisions in the MAT. In absence of such provisions, the pertinent domestic provisions as well as international private law will have to intervene in case the MAT obligations are breached.[827] To support this assumption, it should be noted that the Nagoya Protocol does not specify upon which criteria the national court will decide whether the MAT-related dispute falls under its jurisdiction or not.[828] For this reason, the issue of jurisdiction eventually depends on the national law of the State in question, and on the relevant clauses included in the MAT. Hence, these factors will vary from case to case.[829]

Despite certain progress in facilitating access to legal recourse, not all relevant issues have been resolved by the Nagoya Protocol. It can be argued that special international provisions are needed in order to define the possible types of infringements of MAT and of the entire access procedure, in case of which access to justice must be ensured. Moreover, it is necessary to establish international standards of procedure as well as standards of interpreting pertinent domestic and international law and contractual clauses.[830]

Furthermore, the Nagoya Protocol stipulates that States must take measures in order to ensure access to justice and the availability of mechanisms for mutual recognition and enforcement of judgements of foreign courts and arbitral awards.[831] Thereby, the Nagoya Protocol seeks to support the providers of genetic resources that usually do not have easy access to justice in user States.[832] It should be noted that this provision does not promote the harmonization of judicial procedures amongst the Parties to the Nagoya Protocol that could be achieved only based on a multilateral agreement.[833] Instead, the provision obliges both user and provider States to adopt such measures unilaterally.[834] The obstacles to the recognition of foreign judgements still represent a serious challenge in the promotion of compliance with MAT, whereas the recognition of foreign arbitral awards can be facilitated taking into account that most States are also parties to the New York Convention on the Recognition and Enforcement of

827 Ibid.

828 *Morgera/Tsioumani/Buck* [n. 650], p. 289.

829 Ibid.

830 *Isozaki* [n. 822], pp. 441 f.

831 Article 18.3(a-b).

832 *Morgera/Tsioumani/Buck* [n. 650], p. 289.

833 *Morgera/Tsioumani/Buck* [n. 650], p. 289; *Greiber/Moreno/Mattias/Carrasco/et. al.* [n. 803], p. 187.

834 *Morgera/Tsioumani/Buck* [n. 650], p. 291.

Foreign Arbitral Awards from 1958,[835] and taking into account that there is the possibility to negotiate an ex-post arbitration agreement.[836] The Nagoya Protocol does not specify which measures the Parties should take in order to facilitate the access to justice and the mutual recognition of court decisions, leaving the freedom of choice to them. In this regard, it is important to mention that there is a special need to provide the representatives of ILCs with the access to justice as their interests can be particularly influenced by ABS, also considering that the Nagoya Protocol foresees that ILCs can be a part of MAT.

2. *The Establishment of the ABS Clearing-House Mechanism*

Another important innovation introduced by the Nagoya Protocol is the establishment of the ABS Clearing-House Mechanism, designed to be a part of the CBD Clearing-House Mechanism, established in accordance with CBD Article 18(3).[837] The CBD Clearing-House Mechanism was created in order to promote international cooperation in relation to the conservation and sustainable use of biodiversity.[838] It is noteworthy that the ABS Clearing-House Mechanism is supposed to be need-driven, decentralized, and compatible with national capacities.[839] Additionally, it is intended to support the decision-making process and ensure access to metadata. Finally, it is envisioned to foster the involvement of the private sector.[840]

Thus, the ABS Clearing-House Mechanism is developed in order to facilitate the exchange of ABS-relevant information.[841] Particularly, it should promote the exchange of information provided directly by the parties and playing a significant role in the implementation of the Nagoya Protocol.[842] Despite the

835 Convention on the Recognition and Enforcement of Foreign Arbitral Awards, adopted on 10 June 1958 and entered into force on 7 June 1959, 330 UNTS 3.

836 *Morgera/Tsioumani/Buck* [n. 650], p. 291; *Chiarolla, Claudio*, The Role of Private International Law under the Nagoya Protocol, in: *Morgera/Buck/Tsioumani*, (eds.), The 2010 Nagoya Protocol on Access and Benefit-Sharing in Perspective: Implications for International Law and Implementation Challenges, 2010, p. 446.

837 Article 14.1.

838 *Glowka/Burhenne-Guilmin/Synge* [n. 814], p. 91.

839 CBD Expert Meeting on the Modalities of Operation of the Access and Benefit-Sharing Clearing-house, "Issues for consideration in the Establishment of the Access and Benefit Sharing Clearing-house" (22 March 2011), UN Doc UNEP/CBD/ABS/EM-CH/1/2, Paragraph 10; *Morgera/Tsioumani/Buck* [n. 650], p. 238.

840 *Morgera/Tsioumani/Buck* [n. 650], p. 238.

841 Article 14.1; *Morgera/Tsioumani/Buck* [n. 650], p. 240.

842 *Morgera/Tsioumani/Buck* [n. 650], p. 240.

Nagoya Protocol not clarifying whether or not access to information will also be provided for the general public, the current practice under the CBD Clearing-House Mechanism permits to conclude that this is likely to be the case.[843] This is particularly important, considering that the information provided by the ABS Clearing-House Mechanism can be relevant for indigenous and local communities as well as other forest-dependent people whose rights can be particularly affected by ABS.[844] Consequently, the ABS Clearing-House Mechanism can improve the position of those forest-dependent people who are the owners of genetic resources and of associated traditional knowledge.

The Nagoya Protocol stipulates that with due regard to the domestic laws dealing with the protection of confidential information, parties shall provide any information to the ABS Clearing-House Mechanism as determined by the Protocol and/or required by the Conference of the Parties to the Nagoya Protocol.[845] It is further intended that such information should include legislative, administrative and policy measures related to ABS and information on competent national authorities. It should also include a written certificate or some other official document serving as evidence that PIC was granted and MAT was established.[846] Notably, the Nagoya Protocol stipulates that, in addition to all this information, a State may also provide information concerning relevant competent authorities of ILCs, model contractual clauses, methods and tools applied to monitor genetic resources, codes of conduct and best practices.[847] The provision for voluntary inclusion of information related to ILCs is particularly important in the light of promoting the rights of forest-dependent people. It is to be expected that information about access and benefit-sharing regarding traditional knowledge will be shared.[848]

The ABS Clearing-House Mechanism can be a particularly important innovation, considering that national access certificates, including MAT submitted to the Clearing-House Mechanism, can be seen as "internationally recognized certificates of compliance".[849] That way, the ABS Clearing-House Mechanism can contribute to the legal certainty and transparency of the ABS regime. However, it

---

843 See also *Morgera/Tsioumani/Buck* [n. 650], pp. 238, 240, 241.

844 See also *Morgera/Tsioumani/Buck* [n. 650], p. 240.

845 Article 14.2; See also *Morgera/Tsioumani/Buck* [n. 650], p. 238.

846 Article 14.2(a–c).

847 Article 14.3.

848 *Morgera/Tsioumani/Buck* [n. 650], p. 238; *Greiber/Moreno/Mattias/Carrasco/et. al.* [n. 803], pp. 149 ff.

849 Article 6.3(e), 17.2.

is not completely clear to which extent this process can contribute to the development of the compliance system for ABS. Indeed, the Clearing-House Mechanism is not competent to undertake any steps in case a breach of MAT takes place or if the access in some particular case results in a serious threat or damage to biodiversity. In this regard, the Nagoya Protocol can also be criticized for the absent obligation to submit the information relevant to the protection of biodiversity and natural habitats, including forest ecosystems, as pools of genetic resources.

3. *Compliance Provisions of the Nagoya Protocol*

Some elements of the compliance system established by the Nagoya Protocol have already been mentioned above. They include the introduction of clauses on dispute resolution, cross-border access to justice, jurisdiction applicable in every case of eventual breaches of contractual clauses in MAT, mechanisms for mutual recognition and enforcement of foreign judgements and arbitral awards, and the establishment of the ABS Clearing-House Mechanism. Justice-related provisions of the Protocol seek to provide the owners of genetic resources with an instrument to protect their rights in case of a violation of MAT. Acquiring the status of internationally recognized certificates by national access permits after their submission to the ABS Clearing-House Mechanism can help to ensure legal certainty for users. However, the Nagoya Protocol does not include enough provisions on compliance. One of the great shortcomings of the Nagoya Protocol consists in the lack of an effective international mechanism, which should ensure compliance with international legal provisions on ABS, especially in relation to combatting biopiracy and ensuring benefit-sharing.[850] The Protocol itself promotes the inclusion of compliance-related clauses in MAT, but does not provide any effective remedy for cases of misappropriation of genetic resources when they are accessed without establishing MAT or applying for PIC. The absence of a compliance mechanism that can effectively protect the rights and interests of providers, particularly of ILCs, sparked a wave of disapproval among developing States. However, they finally agreed to sign the document as it provided them with at least facilitated access to justice in user States.[851] At any time, the Nagoya Protocol authorized the first meeting of Parties serving as the COP (MOP-COP) to consider and approve cooperative measures and procedures to promote

850 *Fodella* [n. 803], p. 103.

851 *Nijar, Gurdial S.*, The Nagoya Protocol on Access and Benefit-Sharing of Genetic Resources: An Analysis, Centre of Excellence for Biodiversity Law, University of Malay, 2011, p. 18.

compliance, and to tackle the cases of non-compliance. It was advised to add special provisions on assisting Parties struggling to achieve compliance.[852]

In order to fulfill this task, the Nagoya Protocol MOP-COP I held in PyeongChang, Republic of Korea, on 13–17 October 2014 adopted the "Cooperative procedures and institutional mechanisms to promote compliance with the Nagoya Protocol and to address cases of non-compliance".[853] The compliance procedures and institutional mechanisms under the Nagoya Protocol are intended to be completely independent from the dispute settlement procedures and mechanisms under Article 27 of the CBD.[854] At the same time, they must be flexible, cost-effective, non-adversarial, advisory, and facilitative.[855]

To begin with, it was decided to establish a Compliance Committee.[856] The Committee includes 15 members nominated by the Parties, meaning that there will be three members nominated for each of the five regional groups.[857] What is important is the fact that representatives of ILCs can also be nominated for the Compliance Committee, but only on a voluntary basis.[858] However, the Decision in question stipulates that at least two ILCs' representatives will be nominated as observers, participating in almost any activity of the Committee. It can be evaluated positively that two observers should be nominated by the ILCs themselves, and that at least one of them should come from a developing State.[859] Nevertheless, while ILCs representatives are entitled to participate in the deliberations of the committee, they cannot do so when final decisions are adopted.[860] Additionally, in certain cases, when a deliberation concerns issues wherein only Parties are involved, ILCs' interests are not affected, and the Party concerned prefers to keep the meeting closed, ILCs' observers can be excluded from the deliberation. Even though the general excludability of the public from deliberations is

852 Article 30.

853 UNEP/CBD/NP/COP-MOP/DEC/*1/4*, COP-MOP Decision NP-1/4, Cooperative procedures and institutional mechanisms to promote compliance with the Nagoya Protocol and to address cases of non-compliance, 20 October 2014. – Internet resource – https://www.cbd.int/doc/decisions/np-mop-01/np-mop-01-dec-04-en.pdf [Last visit: 30.04.2023].

854 Annex to COP-MOP Decision NP-1/4, Section A, Paragraph 1.

855 Annex to COP-MOP Decision NP-1/4, Section A, Paragraph 2.

856 Annex to COP-MOP Decision NP-1/4, Section B, Paragraph 1.

857 Annex to COP-MOP Decision NP-1/4, Section B, Paragraph 2.

858 Annex to COP-MOP Decision NP-1/4, Section B, Paragraph 2.

859 Annex to COP-MOP Decision NP-1/4, Section B, Paragraph 2.

860 Annex to COP-MOP Decision NP-1/4, Section B, Paragraph 2.

foreseen for cases when the individual Party's case is discussed,[861] the provision does not clarify how it can be ensured that the deliberation in question really would not have any impacts on ILCs. It can thus be argued that the interests and rights of ILCs are not sufficiently pursued by the legal framework guiding the activities of the Compliance Committee.

The Committee receives submissions related to the issues of compliance and non-compliance with the Protocol from the Parties regarding themselves, regarding other Parties, and regarding the submissions from the MOP-COP of the Nagoya Protocol.[862] Submissions must be filed with the Committee through the Secretariat.[863] Submissions by a Party with respect to another Party as well as submissions by the MOP-COP of the Nagoya Protocol will be redirected by the Secretariat to the Party in question.[864] Once the Secretary of the Nagoya Protocol receives the response from the Party concerned, the submission along with the response and other relevant information has to be redirected to the Compliance Committee.[865] The Committee elaborates on and adopts recommendations for the Party concerned on how it can achieve compliance with the international legal regime for ABS.[866] In order to ensure the transparency of deliberations, both the Party concerned and the Party that filed the submission to the Committee are allowed to take part only in the consideration of the case, but not in the development and adoption of recommendations.[867] At the same time, the Party concerned is given an opportunity to examine and to comment the draft recommendations.[868] Consequently, any such comment is submitted to the COP-MOP together with a special report.[869]

Moreover, the Committee on Compliance has to examine each case where the Party fails to submit its national report according to Article 29 of the Nagoya Protocol, and where information contained in the national report or provided by the ABS Clearing-House and Secretariat indicates that the Party needs assistance in complying with the Protocol.[870] The committee is also authorized to

861 Annex to COP-MOP Decision NP-1/4, Section B, Paragraph 12.
862 Annex to COP-MOP Decision NP-1/4, Section D, Paragraph 1.
863 Annex to COP-MOP Decision NP-1/4, Section D, Paragraph 2.
864 Annex to COP-MOP Decision NP-1/4, Section D, Paragraph 5.
865 Annex to COP-MOP Decision NP-1/4, Section D, Paragraph 6.
866 Annex to COP-MOP Decision NP-1/4, Section E, Paragraph 2(a-c).
867 Annex to COP-MOP Decision NP-1/4, Section B, Paragraph 8.
868 Ibid.
869 Ibid.
870 Annex to COP-MOP Decision NP-1/4, Section D, Paragraph 9.

undertake a number of measures in order to promote compliance and to address cases of non-compliance, including offering advice or assistance to the Party concerned. It may also request the Party in question to develop an action plan in order to ensure compliance with the protocol and provide a Party with necessary assistance in developing and implementing such a plan.[871]

However, the Committee is not authorized to undertake any other measures in order to incite the Party, which constantly breaches its obligations, to ensure compliance. No sanctions and no direct incentives are foreseen for such cases. Despite the Compliance Committee's existence, assuring and promoting compliance predominantly remains the task of the Parties themselves, whereby provisions on promoting access to justice in user States play a great role. However, in reality there are serious doubts about the ability of owner States to effectively defend their rights during negotiations with economically and politically more influential developed nations.[872] The compliance mechanism can be criticized for a lack of provisions dealing with the promotion of the conservation of biodiversity and of natural habitats as pools of genetic resources. The promotion of rights of ILCs remains highly dependent on the recognition of their rights in the national legislation of the State in question.

The promotion of compliance is further complicated by omissions and gaps in the treaty text. Inter alia, the States failed to include in the Nagoya Protocol a legally binding list of activities that fall under the category of "utilization of genetic resources", failed to create a multilateral instrument for verifying whether benefit-sharing had taken place and whether it was fair and equitable, and failed to establish legally binding criteria and standards of fairness and equity when applied to such benefit-sharing, meaning the level of benefit-sharing acceptable under the ABS regime.[873] Moreover, no guidance is provided on the interpretation of MAT.

---

871 Annex to COP-MOP Decision NP-1/4, Section E, Paragraph 2(a–c).

872 Cf. *Krohn* [n. 652], p. 111.

873 *Tvedt, Morten Valløe*, Beyond Nagoya: Towards a Legally Functional System of Access and Benefit-Sharing, in: Oberthür/Rosendal (eds.), Global Governance of Genetic resources: Access and Benefit Sharing after the Nagoya Protocol, 2010, pp. 158 ff.; See also *Tvedt, Morten Valløe/Thomme, Young*, Beyond Access: Exploring Implementation of the Fair and Equitable Sharing Commitment in the CBD, IUCN Environmental Policy and Law Paper No. 67/2, 2007, pp. 1 ff.; *Tvedt, Morten Valløe/Fauchald, Ole Kristian*, Implementing the Nagoya Protocol on ABS: A Hypothetical Case Study on Enforcing Benefit Sharing in Norway, in: The Journal of World Intellectual Property, 2011, pp. 383 ff.

The promotion of compliance can be achieved, among other things, through introducing effective sanction mechanisms against those Parties constantly breaching the provisions of the Nagoya Protocol or failing to force the private users of genetic resources to comply therewith. However, this way can be particularly challenging as the infringement of international rules related to access and benefit-sharing can hardly be attributed to an entire State. The option of introducing sanctions against private users of genetic resources can be considered, yet it would require full recognition of private companies' legal subjectivity under international law. Such legal subjectivity is not fully supported by the current practice of States, even though the duty of States to act with due diligence can arguably be extended also on multinational companies and other private entities, as the relations between them and developing States are often governed by the provisions of international public law.[874] It is submitted that multinational companies are still able to acquire rights and undergo obligations under international public law.[875] It should be noted, however, that these rights and obligations are usually established through the provisions of treaties that are adopted by States.[876] In case of "internationalized contracts" concluded between developing States and multinational corporations, such rights and obligations can also be generated directly by these contracts.[877] Based on these considerations, it can be suggested that the obligation to exercise due diligence currently also applies to multinational companies. Despite this, it is submitted that developing States can only pursue their interests that can be affected by the activities of foreign private entities through their domestic legislation.[878] Thus, even if the obligation of multinational companies to exercise due diligence was recognized, it would not practically help to promote the conservation of forests unless these companies were subjugated to the jurisdiction of international courts within respective treaties. In any case, international arbitrations involving a State from one side

874 *Wouters, Jan/Chané, Anne-Louise*, Multinational Corporations in International Law, KU Leuven Working Paper No. 129, 2013, p. 20; *Chetail, Vincent*, the Legal Personality of Multinational Corporations, State Responsibility and Due Diligence: The Way Forward, in: *Chetail/de Frouville/Vinuales*, Unity and Diversity of International Law. Essays in Honour of Prof. Pierre-Marie Dupuy, 2014, p. 113.

875 *Chetail* [n. 874], p. 113.

876 Ibid.

877 Ibid.

878 Ibid.

and a multinational company from the other side have already taken place, as it happened in the Libyan nationalization case.[879]

Ideally, the trade sanction mechanism developed under CITES could serve as a good example for the development of a compliance mechanism under the Nagoya Protocol. Access to genetic resources and the ability to commercialize products developed on the basis thereof can be restricted for users steadily breaching the Protocol. As it has already been pointed out, such sanction mechanism can be more effective than other tools for the promotion of compliance. The compatibility of such a regime with international law can be resolved as it is reconciled in the case of trade sanctions under the CITES, if equal rules are applied to domestic and foreign entities engaged in the access and benefit-sharing concerning genetic resources.[880]

For this reason, priority should be given to promoting transparency and assisting States in improving compliance with the Nagoya Protocol. For instance, the nomination of ILCs representatives as active members of the Compliance Committee should be obligatory and their participation in the decision-making process should always be ensured. As ILCs are particularly interested in the conservation of forests and gentle use of their resources, such rules would contribute to the promotion of forest protection through the Nagoya Protocol.

4. Environmentally Provisions of the Nagoya Protocol

Despite conservation and sustainable use of biodiversity being among the main goals of the CBD, both the Bonn Guidelines and Nagoya Protocol contain only scarce environmentally relevant provisions. The Nagoya Protocol underlines that its objective consists in the establishment of fair and equitable access and a benefit-sharing regime regarding genetic resources by contributing to the conservation and sustainable use of biodiversity at the same time.[881] Among other things, the Protocol calls its Parties to create conditions, which facilitate and encourage research contributing to the conservation and sustainable use of biodiversity, while developing and implementing national legislation relevant to ABS.[882] One of the measures proposed by the Protocol is facilitating access to biological resources for non-commercial scientific purposes.[883] In such cases, the respective

879 See Texaco Calasiatic v. Libyan Arab Republic (Merits), in: International Legal Materials, Volume 17, 1978, p. 17, Paragraph 47.

880 See Subsection B.IV.2 of Chapter 2.

881 Article 1.

882 Article 8(a).

883 Ibid.

legislative provisions should consider the possible change of intent. Additionally, the Parties should encourage users and providers to redirect benefits arising from the exploitation of genetic resources into the conservation of biological diversity and the sustainable use of its components.[884] Moreover, measures that are undertaken by the Parties in order to promote capacity building may also include promoting the contribution of ABS-related activities to the conservation and sustainable use of biodiversity.[885] As capacity building precludes the transfer of financial and technological resources to the developing States, this provision promotes wider engagement of developed States in global initiatives aiming at the conservation of nature. However, all these provisions are relativized by vague and declaratory wording such as "Parties are encouraged" and "as appropriate".

The Preamble of the Nagoya Protocol underlines that its implementation should be pursued in accordance with the provisions of the CBD.[886] However, the relevant provisions dealing with promoting the conservation of nature are vague and deprived of any binding targets, so that they are unlikely to be sufficient in order to stop and reverse the loss of forest ecosystems. The Nagoya Protocol does not remedy this general vagueness of pertinent CBD provisions. The access and benefit-sharing regime was also developed, among other things, with the intention to create incentives for the implementation of such ecologically relevant provisions.[887] However, States were seeking to promote their economic interests while negotiating the Nagoya Protocol. Therefore, it is not fully reasonable to expect that the conservation of biodiversity will be promoted by the ABS regime alone. For this purpose, it would be necessary to establish a legal duty to redirect the benefits arising from the utilization of genetic resources in the conservation of biodiversity and the promotion of local livelihoods. More detailed and precise provisions, which would foresee the effective implementation measures, are needed in order to ensure that ABS will really contribute to the biodiversity conservation. For this purpose, either the Nagoya Protocol should be modified, or a new protocol aimed at promoting the conservation of biodiversity under the CBD should be adopted. However, no such negotiations are taking place at the moment, and are unlikely to take place in the near future. For

884 Article 9.

885 Article 22.5(h).

886 The Preamble of the Nagoya Protocol: "*Reaffirming* the sovereign rights of States over their natural resources and *according* to the provisions of the Convention".

887 *Morgera/Tsioumani/Buck* [n. 650], pp. 48 ff.

this reason, it is reasonable to promote the implementation of the existing legal framework for the ABS regime, as provided by the Nagoya Protocol.

In this regard, it should be mentioned that the Nagoya Protocol stipulates the possibility to establish a global multilateral benefit-sharing mechanism,[888] which should be used in order to promote the fair and equitable sharing of benefits arising from the utilization of genetic resources and traditional knowledge that are accessed in transboundary or other situations, when it is impossible to apply for PIC. In case of its creation, the benefits from the utilization of genetic resources and associated traditional knowledge shared through this mechanism should be used in order to promote the conservation and sustainable use of biological diversity worldwide. However, the Nagoya Protocol only calls the Parties to consider the need for such a mechanism without stipulating any relevant obligations. The two first MOP-COPs sessions did not manage to make any substantial decision on the issue, but only general recommendations were adopted.[889] Currently, it is difficult to say to which extent the global multilateral benefit-sharing mechanism could promote forest conservation. It is submitted that the main challenge, i.e., the possible undermining of environmentally relevant issues in the ABS regime, would hardly be resolved by its establishment.

Additionally, the Nagoya Protocol can be criticized for failing to consider multiple ecological functions of biodiversity and natural habitats, functions that go far beyond their ability to serve as natural pools of genetic resources. In this situation, the possibility exists that forests not involved in the ABS regime will still be exposed to unsustainable forest exploitation. This implies the possibility that those forests, which are used for this goal, are not managed in an environmental-friendly manner, unless it is required for the sake of the ABS regime. In other

888 Article 10.

889 NP COP-MOP Decision 1/10, The need for and modalities of a global multilateral benefit-sharing mechanism (Article 10), First meeting of the COP-MOP of the Nagoya Protocol, PyeongChang, Republic of Korea, 13–17 October 2014. – Internet resource – https://www.cbd.int/doc/decisions/np-mop-01/np-mop-01-dec-10-en.pdf [Last visit: 30.04.2023]; NP COP-MOP Decision 2/10, The need for and modalities of a global multilateral benefit-sharing mechanism (Article 10), Second meeting of the COP-MOP of Nagoya Protocol in Cancun, Mexico, 4–17 December 2016. – Internet resource – https://www.cbd.int/doc/decisions/np-mop-02/np-mop-02-dec-10-en.pdf [Last visit: 30.04.2023]; NP COP-MOP Decision 3/10, Global multilateral benefit-sharing mechanism (Article 10), Third meeting of the COP-MOP of Nagoya Protocol in Sharm El-Sheikh, Egypt, 17–29 November 2018. – Internet resource – https://www.cbd.int/doc/decisions/np-mop-02/np-mop-02-dec-10-en.pdf [Last visit: 30.04.2023].

words, although the CBD and the EPWFBD promote a holistic approach to the conservation of forest biodiversity and of biodiversity in general, this approach is not anchored in the ABS regime, with economic issues prevailing over environmental ones. This is particularly challenging considering that ABS is the crucial element of the entire international regime for biodiversity, and plays an important role in the CBD.

5. *Nagoya Protocol and the Rights of Forest-Dependent People*

Forest-dependent people are particularly influenced by the provisions of the Nagoya Protocol dealing with the rights of ILCs. The pertinent Nagoya Protocol provisions are more detailed in comparison to the respective Articles of the CBD. One important innovation is that the Nagoya Protocol seeks to promote the protection of ILCs' rights through the domestic legislation of provider States.[890] Based on the logic of Article 5.2 and the general systematic of the Nagoya Protocol, it can be argued that the Party in question has to determine ILCs that possess genetic resources. They also have to adopt legislative and executive measures aimed to ensure the fair and equitable sharing of benefits arising from the utilization of genetic resources with them.[891] Article 5.2 represents an important threshold in the development not only of the ABS framework under the CBD, but also of international environmental law in general. For the first time, an international legally binding document has established an obligation for the parties to introduce measures to reward ILCs for their contribution to research and development activities on the basis of genetic resources. Such an innovation is very important not only from the point of view of justice, but also for practical reasons.

First of all, ILCs, including forest-dependent people, are inherently entrusted with the stewardship of genetic resources. At the same time, the communities in question highly depend upon biodiversity and natural habitats for their livelihoods and spiritual needs. Secondly, the obligation to ensure their participation in benefit-sharing owed directly to ILCs can encourage them to facilitate access to genetic resources and traditional knowledge. Consequently, the contribution of ILCs to scientific progress can be promoted. The inclusion of the provision on ILCs can be seen as a major achievement, as it goes beyond the initial Protocol's

---

890 Article 5.2.

891 Cf. *Morgera/Tsioumani/Buck* [n. 650], pp. 117, 120 ff.

negotiating mandate.[892] Arguably, Article 5.2 of the Nagoya Protocol is based on indigenous peoples' rights as recognized by the United Nations Declaration on the Rights of Indigenous Peoples (UNDRIP).[893] At the same time, the Nagoya Protocol goes beyond the declaratory listing of rights of ILCs by recognizing specific rights related to genetic resources. Moreover, the treaty further substantiates these rights by establishing a specific obligation to ensure the fair and equitable sharing of relevant benefits owed directly to ILCs. Additionally, it should be mentioned that the provisions of the Nagoya Protocol are characterized by a higher normative value in comparison to a non-binding UNDRIP.

All in all, the Nagoya Protocol must be appreciated for explicitly obliging its parties to respect specific rights of ILCs related to genetic resources. It does not limit itself to encouraging them to respect the internationally recognized rights of ILCs, as, e.g., the Bonn Guidelines do. Moreover, the precise and direct obligation does not mean that the Nagoya Protocol ignores those rights of ILCs that are already recognized in other documents. On the contrary, it explicitly mentions them when referring to the national legislation of States of origin.[894] This means that other ILCs' rights that are not directly linked to ABS are also respected and promoted by the Protocol, such as the right to use genetic resources after access takes place as well as cultural rights.[895]

Problems arise, however, as domestic legislation regulating ownership is missing in many biodiversity-rich States. If interpreted broadly, Article 5.2 of the Nagoya Protocol will ensure the sharing of benefits also in cases where relevant national legislation is absent. However, in such cases, promoting ILCs' rights would depend on the interpretation of the Protocol's provisions in a court and by the parties to MAT. Consequently, the absence of a duty to recognize ILCs' rights in national legislation can be seen as a weakness of the Nagoya Protocol.

Recognized ILCs' rights are further administered by Article 12.4 of the Nagoya Protocol. This article determines that the Parties should avoid activities able to restrict the customary use and exchange of genetic resources and traditional knowledge by ILCs. However, the provision further refers to the main objectives of the CBD, meaning that such a restriction can still be legitimized by a necessity

892 Ibid.

893 Ibid, p. 118.

894 *Morgera/Tsioumani/Buck* [n. 650], p. 118; *Greiber/Moreno/Mattias/Carrasco/et. al.* [n. 803], pp. 83 ff.

895 *Morgera/Tsioumani/Buck* [n. 650], p. 118; *Greiber/Moreno/Mattias/Carrasco/et. al.* [n. 803], pp. 83 ff.

to avoid extensive harm to biodiversity, and to promote its conservation as well as the sustainable use of its components.[896] Consequently, the Nagoya Protocol does not clarify further how the conflict of interests should be resolved in a case of eventual incompatibility of biodiversity conservation and the promotion of rights of indigenous people. Moreover, it is unclear how to resolve possible tensions between the right of ILCs to use and exchange the genetic resources and associated traditional knowledge, and facilitating of access to genetic resources. Such situations can occur, for example, when the gathering of biodiversity samples in a certain forest ecosystem will require the limitation of the right of ILCs to access a certain forest ecosystem. However, the Nagoya Protocol does not provide any guidance on how such situations can be resolved.

Moreover, the Nagoya Protocol stipulates an obligation for the States to take measures in order to ensure that the PIC of ILCs or their approval and involvement is ensured for the cases of access to genetic resources. This should always be the case when the ownership rights of ILCs in relation to these resources are recognized in the domestic legislation. Notably, this provision is relativized as it foresees the possibility to obtain the approval and involvement of ILCs instead of their PIC, whereas the legal meaning of the term "approval and involvement" is vague and unclear.[897] Despite providing the Parties with greater flexibility while implementing the Nagoya Protocol, the approval and involvement sets a lower standard of ensuring ILCs participation than the PIC, which is likely to be a more robust and effective tool for the promotion of ILCs' rights.[898] For this reason, some scholars advise the Nagoya Protocol Parties to consider both expressions as identical.[899] However, such an advice or interpretation is not supported by the fact that the parties deliberately introduced both terms in the treaty text. The internal contradiction of Article 6.2 also manifests itself in the discrepancy between its last part, indicating that the provision as a whole should be applied to situations where ILCs are entitled to grant access, and its first part,

---

896 *Morgera/Tsioumani/Buck* [n. 650], p. 224; *Greiber/Moreno/Mattias/Carrasco/et al.* [n. 803], pp. 140 f.; *Curci* [n. 651], p. 59.

897 Article 6.2.

898 See *Morgera/Tsioumani/Buck* [n. 650], p. 154.

899 *Nijar, Guardial Singh*, The Nagoya Protocol on Access And Benefit Sharing of Genetic Resources: Analysis And Implementation Options for Developing Countries, Research Paper 36, South Centre, 2011, pp. 25 f.; *Buck, Matthias/Hamilton, Clare*, The Nagoya Protocol on Access to Genetic Resources and the Fair and Equitable Sharing of Benefits Arising from their Utilization to the Convention on Biological Diversity, in: RECIEL, Volume 20, Issue 1, 2011, p. 55; *Morgera/Tsioumani/Buck* [n. 650], p. 154.

which stipulates that the participation of ILCs can be restricted to "approval and involvement". Similar rules are foreseen for granting access to the traditional knowledge associated with genetic resources, the main difference being that in this case the MAT should also be established.[900]

The Nagoya Protocol's scheme for the promotion of rights of forest-dependent people is more robust and effective in comparison to other pertinent international agreements dealing with forest protection and sustainable forest management. The document in question entitles them to grant PIC or approval, and the access should occur only after MAT is established. These provisions deliver an example for other international legal regimes related to forest protection to ensure the participation of forest-dependent people in forest-related projects at all stages of their development and implementation. At the same time, the promotion of the rights of ILCs under the Nagoya Protocol still suffers from serious gaps. The requirement of negotiating MAT is established only in relation to the access to traditional knowledge, and not to those genetic resources to which ILCs enjoy the right to grant access. Moreover, the promotion of the ILC's rights remain highly dependent on their legal status in the domestic legislation of provider States. Consequently, further regulations should be adopted in order to promote the rights of ILCs, inter alia, through a broader involvement of ILCs' representatives in the activities of the Committee of Compliance, especially at the stage of adopting recommendations.

6. *Combating Biopiracy in the Nagoya Protocol*

At the negotiation stage, developing States were particularly keen to foster international measures deemed fit to combat biopiracy. For this sake, they required the user States to commit themselves to more strict internal compliance measures.[901] Consequently, the Nagoya Protocol establishes compliance obligations on the user side. Accordingly, every Contract Party "shall take appropriate, effective and proportionate legislative, administrative or policy measures to provide that" the genetic resources utilized by public and private entities under its jurisdiction have been accessed after the PIC was granted and MAT were established in accordance with the national legislation of "other Parties".[902]

It can be seen negatively that the Nagoya Protocol does not contain any indicative list of measures to be adopted by user States in order to comply with

900 Article 7.
901 *Fodella* [n. 803], pp. 103 ff.
902 Article 15.1.

the provision in question. Despite the Bonn Guidelines containing such a list, the measures in question did not find their way into the Nagoya Protocol. This omission to provide guidance for the development of the compliance measures to be implemented on the user side can partially be attributed to the fact that the measures proposed by the Bonn Guidelines have not been applied in any State. Additionally, States were granted by the Nagoya Protocol with the flexibility to identify measures to be adopted.[903] However, it can be argued that this explanation is unlikely to justify the omission in question, as the absent implementation of the Bonn Guidelines' list of measures stems from this document's non-legally binding nature. The inclusion of such an indicative list of measures in the legally binding Nagoya Protocol could foster their implementation and promote the compliance with the domestic legislation of provider States by the users of genetic resources. At any time, it is affirmed that the development of a national framework for ABS, including user compliance measures, is likely to require the collaboration of different Parties for practical reasons in order to create a coherent and operational international framework for ABS. Yet, such an assumption can be contested on the grounds that States are likely to prioritize the promotion of their national interests instead of promoting the robust international ABS regime. No guidance for the development of user compliance measures is provided by the Nagoya Protocol, and no operational criteria are established upon which the coherency and effectiveness of such measures could be insured. Therefore, it is likely that compliance measures on the side of users will be insufficient in order to combat biopiracy.

The fact that the provision in question refers to "other Parties" instead of the "Party of origin of genetic resources or the Party that acquired genetic resources in accordance with the Convention" deserves particular attention. It results in the assumption that user States must accept also those internationally recognized certificates of compliance and decisions to grant PIC that were issued by any intermediary State from which genetic resources have been acquired. It would be the case even when the State in question did not acquire genetic resources from the State of origin in accordance with the Convention.[904] From another point of view, the wording was chosen in order to consider the possibility that different States can raise sovereignty claims over the same genetic resources.

---

903 *Morgera/Tsioumani/Buck* [n. 650], p. 254; *Greiber/Moreno/Mattias/Carrasco/et al.* [n. 803], pp. 162 f.

904 *Morgera/Tsioumani/Buck* [n. 650], pp. 249, 251, 254; *Greiber/Moreno/Mattias/Carrasco/et. al.* [n. 803], pp. 162 f.

In this case, using the words "other Parties" in the treaty text can be justified, as the users of genetic resources, which acquired genetic resources in accordance with the Convention, should not bear negative consequences resulting from such sovereignty disputes, which – from this point of view – should be resolved through negotiations between the owners of genetic resources, on one side, and intermediary States, on the other. However, it can be seen as one of the Protocol's weaknesses that this document fails to pursue the rights of States of origin in cases where an intermediary State is involved. It can hence be proposed that, in such cases, the ex-post negotiation of mutually agreed terms with the initial State of origin should be undertaken. Such a possibility is generally admissible under the Nagoya Protocol's framework. However, the final user State is able to reject any negotiations by legitimately arguing that the utilization of genetic resources occurs in accordance with the internationally recognized certificates of compliance issued by the intermediary State.

Another important issue arises in relation to the temporal scope of the Protocol. In short, general principles of international law foresee that an international treaty cannot be applied to any scenario that occurred before its entry into force.[905] This means that the provisions of the Nagoya Protocol are not applicable to genetic resources acquired before its entry into force. However, there is nothing withholding the Nagoya Protocol's application to the sharing of benefits arising from continuous or new utilization of these resources.[906] Additionally, this holds true for such cases where the ex-post establishment of MAT can represent a realistic option, if the State of origin can be identified.[907] For other cases, when such a State cannot be identified, or when the application for PIC is not possible, the global multilateral benefit-sharing mechanism should be established in accordance with Article 10 of the Convention. However, the Parties have not yet considered this possibility.

Another important provision of the Nagoya Protocol deemed to promote compliance and to combat the misappropriation as well as the misuse of genetic resources foresees the designation of checkpoints that will collect the relevant information concerning PIC, MAT, and the origin of genetic resources used under its jurisdiction.[908] Private and public entities utilizing genetic resources are

905 Vienna Convention on the Law of Treaties (with annex), adopted on 23 May 1969, in: 1155 UNTS 331, Article 28.

906 *Morgera/Tsioumani/Buck* [n. 650], p. 79.

907 Ibid, p. 80.

908 Article 17.1(a.i).

supposed to be obligated to submit the relevant information to the designated checkpoints.[909] Subsequently, the information collected will be submitted to a relevant national authority, to the Party granting PIC, and to the ABS Clearing-House.[910] It can be seen positively that the ABS Clearing-House serves as an international checkpoint, and that internationally recognized certificates are introduced.[911] However, the provider States' initial demand to establish a list of mandatory checkpoints was not fulfilled.[912]

Thus, the Nagoya Protocol has introduced a number of important provisions serving for the promotion of legality and transparency as well as combating biopiracy. However, the enforcement of relevant provisions and the ability of developing States to pursue their rights remain highly dependent on the user States' internal ABS legislation. Robust international measures for combating biopiracy were not established, and some important provisions, such as the prohibition to use genetic resources acquired in violation of the provider states' legal framework for ABS, were not included in the final treaty text. It can be also criticized that the above-mentioned checkpoints are not empowered to promote the compliance. Consequently, the Nagoya Protocol's provisions are not robust enough in order to ensure compliance and combat biopiracy. This situation may have negative impacts on forest conservation, as forest-rich States are not endowed with enough incentives to protect their forest ecosystems. For this reason, more detailed and robust documents on the promotion of compliance with the protocol and the combating of biopiracy should be elaborated on by the NP COP-MOP.

Moreover, for the purpose of forest conservation, it should be proposed to draft a separate document on the utilization of genetic resources of forests. As the most important terrestrial biodiversity pools, forests deserve the Governing Bodies' specific attention. These documents could consider the specific interests of forest-dependent people, which are not yet covered by the Nagoya Protocol. For example, they could consider the role of forests in supporting indigenous livelihoods. It could also promote the redirection of benefits arising from the utilization of genetic resources of forests into the conservation of forest ecosystems. It can be also proposed to elaborate on provisions that would establish a duty to preserve forest ecosystems where genetic resources are gathered.

909 Article 17.1(a.ii).

910 Article 17.1(a.iii).

911 Article 17.2.

912 *Koutouki/Rogalla von Bieberstein* [n. 778], pp. 530 f.

7. Financial Mechanism of the Nagoya Protocol

Taking into account that the legal framework and infrastructure needed for the implementation of the ABS regime is not available in many developing States, the transfer of financial and technological resources, inter alia, with the goal of promoting capacity building is required for the implementation of the Nagoya Protocol.[913] For this purpose, the financial mechanism of the CBD also serves as the financial mechanism of the Nagoya Protocol.[914] Currently, the Global Environmental Facility (GEF) acts as the financial mechanism of both instruments. Financial assistance should be designed and directed in such a way that it considers the States' development needs as well as the interests and rights of ILCs.[915] Moreover, it is possible that developed States will provide financial resources through other mechanisms, including bilateral, regional, and multilateral channels. However, no obligation is established to provide financial resources through the GEF or other mechanisms, and no targets for fundraising are set. Consequently, both the predictability and availability of financial resources are not ensured. The financial mechanisms of the Nagoya Protocol can potentially contribute to forest conservation as States are supposed to promote the conservation of biodiversity and the sustainable use of its compounds while implementing the Nagoya Protocol. At the same time, the Protocol does not stipulate the obligation to redirect a certain part of funds into forest protection. Consequently, the contribution of the financial mechanism of the Nagoya Protocol to forest conservation will depend on the internal legislation of forest-rich States. Thus, more detailed provisions stipulating specific obligations to provide financial resource and to redirect them partially into the conservation of biodiversity are needed.

E. Conclusions

With the adoption of the CBD, the international legal regime for the conservation and sustainable use of biodiversity has emerged. With the Convention's entry into force, many controversial issues related to the ownership over genetic resources and the modalities of their utilization have been resolved. It is possible to reconcile controversial approaches, one of which advocating unlimited national sovereignty over genetic resources, and another promoting unlimited

913 *Morgera/Tsioumani/Buck* [n. 650], pp. 325 f.

914 Article 25.2.

915 Article 25.3.; *Morgera/Tsioumani/Buck* [n. 650], pp. 327 ff.; *Greiber/Moreno/Mattias/Carrasco/et al.* [n. 803], pp. 225 f.

access to them for all stakeholders, while recognizing the relevant resources as a common heritage of humanity. Consensus was achieved that biodiversity resources are subject to the national sovereignty of States within the State territory on which they are found, whereas the respective States are obliged to manage these resources sustainably, and to facilitate access to them. From the point of view that forests represent the most important terrestrial biodiversity hotspots, the CBD can have an extensive impact on their protection.

The Convention foresees an extensive list of measures dedicated to the protection of biodiversity both ex-situ and in-situ. Moreover, the treaty deals with promoting access to genetic resources and ensuring fair and equal sharing of benefits arising from their utilization. The ABS regime can promote research and development activities based on genetic resources and create incentives for forest protection. However, there is a number of controversial issues related to the utilization of genetic resources, beginning with the scope of the ABS regime all the way through biopiracy, including the particular risk of violating ILCs' rights, as well as potentially adverse impacts on the environment. For these reasons, the implementation of ABS proceeded slowly. With time, the notion of genetic resources was interpreted more broadly, so that natural products of metabolism that do not contain functional units of heredity were also covered by the ABS regime. This made way to finally include the utilization of "derivatives" in the Nagoya Protocol's scope of application. This document also delivered detailed rules regarding the procedure of applying for PIC and negotiating the MAT, and introduced a kind of "internationally recognized certificate of compliance", while promoting transparency and legal certainty for users.

The Nagoya Protocol has also promoted the interests of indigenous people, by recognizing their rights to participate in access and benefit sharing, and has strengthened the position of provider States. It has supported their right to have access to justice in user States, and promoted the internal compliance measures in the user States with the aim of avoiding the utilization of genetic resources acquired against a State of origin's national law. However, the international compliance mechanism of the Nagoya Protocol is weak, and an effective international system for monitoring ABS transactions in order to combat biopiracy has not been established. Protecting the rights of indigenous people in the Nagoya Protocol goes far further than the relevant provisions of any other international agreement, but still depends too strongly on the national legislation of States concerned. Additionally, a number of issues have to be clarified. Environmental issues are among the most undermined in the Nagoya Protocol. The treaty failed to introduce strong environmental commitments. The assumption that ABS itself, without introducing further provisions specifically dedicated to the protection of nature, indirectly strengthens biodiversity conservation, is highly debatable. Both providers and users of environmental resources are more concerned with the promotion of their economic interests than with the conservation of the environment. Moreover, the benefits shared with the States of origin do not always cover the full costs of environmental conservation, and, even if they do, there is no duty to redirect such resources into the funding of the conservation or sustainable use of biodiversity.

In addition to these issues, despite the generally holistic approach to biodiversity conservation promoted by the CBD, the ABS regime – at least when applied to forests – is likely to underestimate the multiple ecological and socio-economic functions of forests unrelated to biodiversity conservation and utilization. Forests with a naturally low level of biodiversity and forests not used for gathering biodiversity samples will still be exposed to unsustainable forest management.

It is also questionable whether the general approach to forest conservation based on the incentives emerged from the access and benefit-sharing regime is the best way to promote forest conservation. In this regard, it is particularly problematic that the benefits from the utilization of genetic resources do not cover all the costs of the conservation of forest biodiversity.

However, the implementation of ABS has the potential to contribute to the conservation of forest ecosystems. As the Nagoya Protocol cannot consider non-biodiversity-related functions of forests, it is necessary to analyze other mechanisms serving for the evaluation of ecological services provided by forests.

Chapter Four Forest Protection, Climate Change, and International Law

A. Introduction

As the analysis of the international legal regime for forests has shown, the problem of deforestation has not been resolved by the pertinent international treaties. The international community has constantly failed to elaborate and implement effective and robust international regulations and mechanisms aimed to halt and reverse deforestation and forest degradation. International documents specifically dealing with forests are non-legally binding and do not establish enough incentives for their implementation. The forest-related provisions of other international environmental agreements have only had scarce positive impacts on the promotion of forest conservation. Astonishingly, despite the fact that the conservation of forest ecosystems as major biodiversity pools is likely to play a major role in the achievement of the goals of the CBD, the non-legally binding Expanded Working Programme on Forest Biodiversity (EWPFB) is the only forest-specific document elaborated under the umbrella of the Convention. Just like other forest-specific documents, the EWPFB does not include incentives for its implementation. Against this backdrop, the existing international consensus on forests and forest-related issues is insufficient in order to pursue a comprehensive, effective, and robust strategy for forest protection.

However, the history of forest-related negotiations has shown that the particular interest of States in the regulation of certain forest-related environmental issues can potentially serve as a "least common denominator", upon which a wide-reaching consensus on forest issues can be established. As demonstrated in Chapter 3, the conservation and sustainable use of biodiversity and its compounds represents one of these issues. However, the incentivizing effect of the CBD on the protection of forests was insignificant until today.

The natural ability of forests to act as sinks and reservoirs of carbon dioxide has the potential to promote their conservation. It is already known that forests are able to contribute to the stabilization and reduction of CO_2-levels in the atmosphere. This important ecological function of forests can also deliver the "least common denominator" that can facilitate the development of international legal regime for forests. Along with the genetic resources suitable for bioprospecting, the ability of forests to sink and store carbon dioxide is a measurable and tradable tangible asset. It is a major asset to the industries and States that are struggling to achieve the requisite reduction of CO_2 emissions. Moreover, this means that forest-rich States are able to contribute to climate change mitigation by conserving and promoting their forest ecosystems. These forest-rich States and private forest owners argue that they have the right to receive a fair compensation, as they abandon unsustainable forest management practices.[916] Certain forms of forest use, such as deforestation and forestland conversion, bring higher profits in the short-term, but result in the depletion of climate-relevant forest functions. Currently, an approach is being developed, by which forest-rich developing States are supposed to receive fair compensation that will be estimated based on the contribution of forests, under their jurisdiction, to the reduction of net CO_2-emissions.[917] However, this approach is questionable from the ethical point of view. Firstly, it promotes a purely utilitarian anthropocentric approach to nature that neglects the intrinsic value of forests ecosystems. Secondly, whereas the right to development and the need to protect natural ecosystems for their intrinsic value are undisputed issues from the ethical point of view, the existence of the right of forest owners to receive financial compensation for not logging their forests is dubious. It gives rise to concerns that the owners will cut their forests intentionally in order to exert pressure on developed States interested in the conservation thereof. Despite all these issues, it is submitted that the approach in question can be perceived as a realistic avenue to the development of effective mechanisms for the promotion of forest conservation.[918]

It is noteworthy that the development and implementation of these mechanisms is not stimulated by the understanding of the ecological significance of forest ecosystems on their own, but by the concerns in relation to climate change. This last issue has received much attention in international negotiations and certain progress in the promotion of climate change mitigation has been made after the UNFCCC[919] and the Kyoto Protocol (KP)[920] were adopted. Despite certain

916 See *Schulte zu Sodingen, Beate,* Der völkerrechtliche Schutz der Wälder, 2002, pp. 189 f.

917 See Section D.I of this Chapter.

918 *Beyerlin, Ulrich,* Internationaler Waldschutz am Scheideweg? Von den "Rio-Waldprinzipien" zu "REDD plus", in: *Hecker, Bernd/Henlder, Reinhard/Proelß, Alexander/Reiff, Peter* (eds.), Jahrbuch des Umwelt- und Technikrecht, Schriftenreihe des Instituts für Umwelt- und Technikrecht der Universität Trier, 2012, pp. 194 ff.

919 United Nations Framework Convention on Climate Change, adopted on 9 May 1992 and entered into force on 21 March 1994 in: 1771 UNTS 107.

920 Kyoto Protocol to the United Nations Framework Convention on Climate Change, adopted on 11 December 1997 in Kyoto, Japan, in: 2303 UNTS 162.

critical claims that are raised by a number of so-called "climate sceptics",[921] anthropogenic climate change is widely recognized by the international organizations and by most governments.[922] Notwithstanding this, the promotion of the conservation of forests through climate change mitigation and adaptation can have negative impacts on forests.

Firstly, the promotion of the protection of forests becomes highly dependent on the perception of climate change problem, which can vary in different States. Favorable conditions for the conservation of forests could be created if the carbon dioxide emissions were recognized as the main cause of global climate change, and the conservation of forests was acknowledged as the most effective and robust way to curb these emissions.[923] At the same time, if the problem of CO_2 emissions and their role in climate change do not receive great attention in a certain State, or if technical solutions are recognized as the best way to reduce these emissions, forest ecosystems will not benefit from climate change mitigation and adaptation activities. Secondly, the narrow focus on the climate-related functions of forests can be counterproductive for their conservation and sustainable use. For instance, the estimation of the amount of CO_2 sunk and stored by forests can be particularly challenging from the technical point of view. The forests, whose ability to sink and store CO_2 is low, can be overlooked by climate change mitigation activities. Moreover, approaches to the conservation of forests that are based on their ability to sink and store CO_2 neglect the fact that forest ecosystems are very complex, and every component thereof must be protected in order to ensure their survival, enhance their vitality and promote their ecological functions. Consequently, non-carbon functions and natural characteristics of forest ecosystems do not receive sufficient consideration in relevant activities. It is also problematic that these policies tend to consider forests as mere tree stands and forest carbon stocks without distinguishing between natural forests, near-natural forests, and monoculture plantations. The race for promoting forest

921 *Capstick, Stuart Bryce/Pidgeon, Nicholas Frank*, What Is Climate Change Scepticism? Examination of the Concept Using a Mixed Methods Study of the UK Public, in: Global Environmental Change, Volume 24, 2014, pp. 389–401.

922 See *Gupta, Joyeeta,* The Climate Change Convention and Developing Countries – From Conflict to Consensus? 1997, pp. 7 f.

923 Cf. *Matsumoto, Ken'ichi/Hasegawa, Tomoko/Morita, Kanako/Fujimori, Shinichiro*, Synergy Potential between Climate Change Mitigation and Forest Conservation Policies in the Indonesian Forest Sector: Implications for Achieving Multiple Sustainable Development Objectives, in: Sustainability Science, Volume 14, 2018, p. 1660.

carbon stocks can accelerate the conversion of highly biodiverse and resilient natural forests into carbon plantations.

The approach to the mitigation of climate change that foresees the conservation and enhancement of forest carbon stocks and the reduction of emissions from deforestation is currently developed under the umbrella of the UNFCCC. The relevant mechanism is described via the acronym "REDD+". This new mechanism delivers an opportunity for the enhanced cooperation between forest-rich developing States and developed States. Their effective implementation will require the promotion of transparency, accountability, and legal certainty in the forest sectors of developing States. Moreover, it will require strong commitments from the side of developed States to provide the financial resources set aside to reward the achieved reduction of deforestation and the enhancement of forest carbon stocks. While implementing REDD+, it is important to ensure that the implementation of these mechanisms contributes to the conservation of forest ecosystems and of forest biodiversity, and to the promotion of the rights of ILCs, and in no way promotes the conversion of natural forests into monoculture plantations. Finally, the new mechanism should be designed in a way that will make it attractive for developed States and foreign investors in order to ensure the reliability and availability of funding.

With this in mind, it is necessary to analyze the international legal provisions that serve as a basis for the international forest-related cooperation in relation to the role of forests in climate change mitigation. It will be particularly examined whether and to what extent these provisions promote the conservation of forest ecosystems and support their multiple functions. Another important issue is whether the existing decisions respect the rights of ILCs and forest-rich States.

B. Role of Forests under the UNFCCC

I. Preparatory Activities that Precluded the Adoption of the UNFCCC

The first major international conferences, dedicated to the problem of climate change, were held in Geneva in 1979,[924] and in Villach, Austria, in October of

924 *Jahrmarkt, Lena*, Internationales Klimaschutzrecht, Der Weg zu einem Weltklimavertrag im Sinne gemeinsamer, aber differenzierter Verantwortlichkeit, 2016, pp. 54 f.; *Bolle, Ulrike*, Das Intergovernmental Panel on Climate Change (IPCC): eine völkerrechtliche Untersuchung, 2011, p. 10; *Brunnengraeber, Achim/Dietz, Kristina/Hirschl, Bernd/Walk, Heike/Weber, Melanie*, Das Klima neu denken: Eine sozial-ökologische Perspektive auf die lokale, nationale und internationale Klimapolitik, 2008, p. 88;

1985.[925] In the aftermath of the second conference, two workshops, held in Bellagio and Villach in 1987, stressed for the first time the necessity of climate mitigation activities in the forest sector.[926]

Subsequently, the "World Conference on the Changing Atmosphere", held in Toronto on 27–30 of June 1988, adopted the final statement, which included several specific recommendations,[927] inter alia, on the reduction of deforestation and promotion of afforestation. For this purpose, it was proposed to create a global climate fund to provide developing States with the adequate funding needed for the promotion of sustainable forest management in the tropics.[928]

The next major conference dedicated to the problem of climate change took place in Noordwijk on 6th July 1989. The conference was well attended with more than 69 States taking part in the conference. Despite the expectations of some participants, the envisaged development of a legally binding agreement on CO_2-reduction was not achieved.[929] Based on the Discussion Paper prepared by the former Dutch Minister of Environment Pieter Wisenius, the idea of an integrated approach was developed. It deserves particular attention because it permits the inclusion of both reduction of emissions and promotion of natural sinks of CO_2.[930] The advocates of this approach insisted on the necessity of

---

Bodansky, Daniel/Brunnee, Jutta/ Rajamani, Lavanya, International Climate Change Law, 2017, p. 98; *Steffan, Martin*, die Bemühungen um eine Internationale Klimakonvention, Verhandlungen, Interessen, Akteure, 1994, p. 60; *Kaiser, Karl/Weizsäcker, Ernst Ulrich von/Raimund, Bleischwitz/Comes, Stefan,* Internationale Klimapolitik. Eine Zwischenbilanz und ein Vorschlag um Abschluss einer Klimakonvention, (Arbeitspapiere zur internationalen Politik 65), 1991, pp. 13 f.

925 *Bodansky/Brunnee/Rajamani* [n. 924], p. 98; *Steffan* [n. 924], p. 60; *Kaiser/Weizsäcker/Raimund/Comes* [n. 924], pp. 13 f.

926 *Steffan* [n. 924], p. 60.

927 *Steffan* [n. 924], p. 62; *Enquete-Kommission „Vorsorge zum Schutz der Erdatmosphäre,* Schutz der Erde – Eine Bestandsaufnahme mit Vorschlägen zu einer neuen Energiepolitik. Dritter Bericht der Enquete-Kommission des 11. Deutschen Bundestages, Deutscher Bundestag, 1990, pp. 10 ff.

928 *WMO, UNEP*, L'Atmosphère en évolution, implications pour la sécurité du globe, 27–30 juin 1988, Toronto, Canada, documentations et rapports de la conférence, 1989, p. 298; *Steffan* [n. 924], p. 62.

929 *Steffan* [n. 924], p. 64

930 Atmospheric pollution and climate change Ministerial Declaration held at Noordwijk, the Netherlands, on 6 and 7 Novermber 1989, The Noordwijk Declaration on Climate Change, November 1989, Paragraphs 9 and 20. – Internet resource – https://ntrl.ntis.gov/NTRL/dashboard/searchResults/titleDetail/PB90210196.xhtml [Last visit: 30.04.2023]

strengthening forest conservation efforts and of promoting afforestation and reforestation.[931] The measurable and verifiable targets were proposed including the restoration of 12 Mio Hectares of deforested areas.[932] It was proposed to establish a number of financial facilities with the aim to support different climate-related measures in developing States, including afforestation and forest conservation. However, these and other interesting propositions did not find their way into the final declaration, which is a vague and declaratory document.[933]

The "White House Conference", held in 1990, was also dedicated to climate change issues, but it could not contribute significantly to the protection of forests, because the whole meeting was overshadowed by the controversies between the European Community and the USA.[934] This is why the participants of the Conference failed to develop any document. However, it is noteworthy that some issues, including "scientific uncertainty" and "economic practicability" acquired particular importance during these negotiations.[935] It is obvious that these issues play an important role in any international and national forest-related policies, including policies beyond those that are relevant to climate change.

Subsequently, the second Climate Conference held in The Hague from 29 October until 7 November 1990 did not contribute to the development of international forest-related policies.[936] During the conference, the research results presented in the reports of IPCC[937] were reconfirmed. Its main outcome

931 Ibid, Paragraph 20.

932 Ibid, Paragraph 21.

933 See also *Steffan* [n. 924], p. 62.

934 *Topping, John C.*, U. S. Law and Public Policy Concerning Global Warming, in: *Iwama, Turu* (ed.), Politics and Laws in Global Warming: International and Comparative Analysis, 1991, p. 127; *Steffan* [n. 924], p. 65.

935 *Steffan* [n. 924], p. 65.

936 *Steffan* [n. 924], pp. 65–66.

937 See *IPCC*, Climate Change: The 1990 and 1992 IPCC Assessments, Intergovernmental Panel on Climate Change, 1992. In general, forests play a secondary role in the IPCC reports. However, these documents confirm that deforestation is one of the major sources of CO_2-emissions and state further that afforestation and reforestation can contribute substantially to the reduction of CO_2-emissions. (p. 82) Noteworthy, the reports argue that deforestation can also have other direct effects on climate, by changing the albedo index of the surface and by decreasing evapotranspiration. (p. 82) In general, the consequences of such changes are not precisely described in these documents. Nevertheless, it should be noted that the authors of IPCC reports estimate that the replacement of tropical rainforests in the Amazon Basin by grassland "would reduce the annual rainfall in the region by some 20% and increase medium temperatures

consisted in the confirmation of necessity to initiate the negotiations of a climate convention.[938]

II. Negotiations of the UNFCCC

On the eve of the 1992 Rio Summit, one of the two Intergovernmental Negotiation Committees (INC) was entrusted with the development of the UNFCCC draft text.[939]. Before the final draft text of the UNFCCC was completed, five meetings of the INC had taken place. At the beginning of negotiations, different States and groups of States presented "informal papers" containing their propositions for the future text of the Convention.[940] Although the representatives of the European Community followed the same approach to climate change mitigation at all previous conferences and meetings, they did not manage to develop a draft common project.[941] Instead, France, Germany, United Kingdom, and the Netherlands presented their own "informal papers" respectively.[942] All of them acknowledged that climate change mitigation should be based on respect for national sovereignty and guided by the precautionary principle. They particularly underlined that developed States bear special responsibility and must play a leading role in combating climate change, *inter alia*, by providing financial and

---

by several degrees". (p. 82) Notably, the newest research about the forest's influence on the atmospheric precipitations permits to conclude that these calculations are underestimating the effects of deforestation on temperature and precipitations (See Subsection A.II.3 of Chapter 1 of this dissertation).

938 *Steffan* [n. 924], p. 66.

939 *Bodansky, Daniel*, The History and Legal Structure of the Global Climate Change Regime, in: *Sprinz, Detlef/Luterbacher, Urs* (eds.), International Relations and Global Climate Change, 2001, pp. 17 ff.; *Ramakrishna, Kilaparti/Young, Oran R.*, International organizations in a Warming World: Building a global Climate Regime, in: *Mintzer, Irving M.* (ed.), Confronting Climate Change: Risks, Implications, and Responses, 1992, pp. 253 ff.; *Zaelke, Durwood/Cameroon, James*, Global Warming and Climate Change – An Overview of the International Legal Process, in: American University International Law Review, Volume 5, Issue 2, 1990, pp. 250 ff.

940 *Bodansky* [n. 939], pp. 17 f.; *Ramakrishna/Young* [n. 939], pp. 254 f.; *Zaelke/Cameroon* [n. 939], pp. 251 f.

941 *Steffan* [n. 924], p. 86.

942 United Nations A/AC.237/Misc.1/Add.1, Preparation of a Framework Convention on Climate Change: Set of Informal Papers Provided by Delegations, related to the Preparation of a Framework Convention on Climate Change, Note by the Secretariat, Addendum 1. Geneva, 19–28 June 1991. – Internet resource – https://undocs.org/en/a/ac.237/misc.1/add.1 [Last visit: 30.04.2021]

technological support to developing States.[943] Thus, the principal of common but differentiated responsibility has also found its way into the draft documents developed by the European States. The inclusion of this principle paved the way for the establishment of forest-related climate change mitigation mechanisms. However, in general, forestry issues did not play a major role in the informal papers in question.[944]

In this regard, the Dutch "informal paper" deserves particular attention. It called for the adoption of a special protocol on "sustainable forest management/ reforestation/afforestation, including funding mechanisms".[945] In this regard, the representatives of the Netherlands argued that the Contracting Parties had to commit themselves to ensure the most extensive conservation of forests in "their best possible structure", particularly by combating the drivers of deforestation, such as the unsustainable utilization of forest resources.[946] Moreover, they requested the Contracting Parties to set precise goals for the reduction of deforestation rates with the aim to end deforestation completely.[947] In addition, according to the Dutch "informal papers", the afforestation and reforestation measures had to be undertaken in order to increase the capacity of existing forests to bound CO_2 and in order to create new forests serving as sinks and reservoirs of this greenhouse gas.[948] The document called for the promotion of afforestation not only in developing States, but also in developed European States.[949] Moreover, the Netherlands proposed the creation of special funds to be used for the promotion of "sustainable forest management practices", deemed, inter alia, to promote the reduction of greenhouse gas emissions.[950]

The preamble of the draft convention text prepared by the UK also recognized the necessity to conserve and promote natural sinks and reservoirs of greenhouse gases, including forests.[951] The UK delegation developed further

943 *Steffan* [n. 924], p. 86.
944 See *Steffan* [n. 924], p. 86.
945 United Nations, A/AC.237/Misc.1/Add.1 [n. 942], Annex 1 to the Proposal of Netherlands, p. 41.
946 Ibid.
947 United Nations, A/AC.237/Misc.1/Add.1 [n. 942], Proposal of Netherlands, paragraph 2.8, p. 91.
948 Ibid.
949 Ibid.
950 Ibid.
951 United Nations, A/AC.237/Misc.1/Add.1 [n. 942], Paper 9: Proposal of United Kingdom, Preamble, p. 54.

recommendations for the future treaty text, which were not included in the draft version, because it was probable that the parties would oppose the inclusion of these elements in the treaty.[952] Inter alia, these recommendations included the promotion of a holistic approach, which would have permitted the integration of both the reduction of greenhouse gas emissions and the promotion of sinks.[953] The delegation from the UK urged States to commit themselves to the development and exchange of technologies, practices, and training both with the aim to reduce emissions and to conserve and enhance natural carbon sinks.[954] Additionally, the UK delegation argued that the final text of the Convention had to stipulate a specific obligation for States to develop their national strategies for the conservation and enhancement of sinks.[955] It was intended that these strategies would describe the current forest areas, the projected future forest area, currently implemented and projected measures serving to conserve and enhance sinks, and the possible impact of these measures on forests.[956] States would be obliged to determine their national forest-related targets including the targets for the preservation of current forest area and re- and afforestation targets. The UK paper further underlined that developing States should be assisted as they seek to achieve all these targets.[957] Like the "informal paper" prepared by the Netherlands, the recommendations proposed by the United Kingdom encouraged States to adopt a protocol on conservation and enhancement of sinks, particularly forests.[958]

The text proposed by the USA also encouraged States to adopt a comprehensive approach that would take into consideration both the necessity to reduce emissions of greenhouse gases by source and the need to promote carbon sinks.[959] However, the propositions of the USA were more general and did not contain any precise goals and targets in relation to the reduction and a/reforestation activities. Importantly, the USA representatives had underlined that any climate change mitigation activities should be based on the principle of

952 United Nations, A/AC.237/Misc.1/Add.1 [n. 942], Paper 10: Draft UK Paper on Possible Elements for Inclusion in a Framework Convention on Climate Changes, p. 83.
953 Ibid, p. 84.
954 Ibid.
955 Ibid.
956 Ibid, p. 85.
957 Ibid, pp. 86 f.
958 Ibid, pp. 86 f.
959 United Nations, A/AC.237/Misc.1/Add.1 [n. 942], Paper 11: Proposition by the USA, p. 89.

economic rationality and the national circumstances of each State should be considered.[960] General obligations for the Parties to the UNFCCC, as proposed by the USA, would not mention explicitly the conservation and enhancement of sinks. However, it referred to the "net" reduction of all emissions based on the comprehensive approach, which arguably presumed the promotion of natural sinks and reservoirs of greenhouse gases.[961] However, the document did not specify further, which role sinks were intended to play in combating climate change. In contrast to the draft papers submitted by the European delegations, the American document foresees the possibility to omit mitigation actions in the forest sector while achieving the goals of the convention, as there was no duty to include sinks in the national strategies for reducing emissions. At the same time, States were indirectly encouraged to include forest-related measures in their strategies, as the conservation and enhancement of sinks could be a cost-effective way to achieve the convention's goal. In spite of this fact, the vague and unspecified obligations contained in the draft text prepared by the USA would not provide for the promotion of multiple functions of forests. At least, it contained a vague obligation for States to avoid other types of environmental harm while implementing their national strategies.[962] Notwithstanding this, no mechanisms were proposed in order to monitor the compliance with this obligation. Thus, the draft text as proposed by the USA would not contribute to the combating of drivers of deforestation.

The draft text of the convention prepared and presented by India, which played a prominent role in the Rio Summit and its preparatory activities,[963] deserves mentioning in the context of this research.[964] Accordingly, developed States were supposed to bear the main costs of climate change mitigation, whereas developing States would only have to consider the reduction of emissions in their

960 Ibid, p. 91.

961 Ibid.

962 Ibid.

963 *Fischer, Wolfgang*, Die Klimapolitik der Staaten zwischen globaler Verantwortung und nationalen Interessen, in: *Borsch, Peter/Wiedemann, Peter* (eds.), Was wird aus unserem Klima? Fakten, Analysen und Perspektiven, 1992, p. 211; *Steffan* [n. 924], p. 88.

964 United Nations A/AC.237/Misc.1/Add.3, Preparation of a Framework Convention on Climate Change: Set of Informal Papers Provided by Delegations, related to the Preparation of a Framework Convention on Climate Change, Note by the Secretariat, Addendum 1. Geneva, 19–28 June 1991, Paper No. 15: India, p. 3. – Internet resource – https://undocs.org/en/a/ac.237/misc.1/add.1 [Last visit: 30.04.2023]

development programmes.[965] Furthermore, India emphasized the necessity of international financial support and technological transfer for the benefits of developing States.[966] The emission reduction goals were formulated in a vague and superfluous manner.

Thus, the Dutch proposition can be evaluated positively from the perspective of forest conservation. It is important that this document addressed the problem of deforestation and forest degradation in all States and was not limited to the deforestation in tropics. Moreover, it underlined the necessity to conserve natural forests and to promote afforestation. At the same time, the document could be criticized for the lack of attention to the promotion of non-carbon related multiple forest functions. Nevertheless, the very fact that the forest-relevant provisions proposed by the Netherlands did not find their way into the final text of the UNFCCC can be seen as one of the major failures of the 1992 Rio summit.

At the same time, the propositions of the United Kingdom were more restrained in comparison to the forest-related recommendations made by the delegation from the Netherlands. The Dutch "informal paper" tended to treat forests as ecosystems and particularly underlined the necessity to preserve them "in their best possible structure". In such a way, the numerous non-carbon-related functions and natural properties of forests could be considered.[967] The text of the "informal paper" submitted by the UK was only focused on the forests' ability to sink and store CO_2, while it referred to "forest areas" and not to "forests". Even in relation to the conservation of sinks, it did not mention the necessity to preserve and promote multiple forest functions.[968] Consequently, the UK's proposal failed to consider the interests of forest-dependent people and the need to conserve and promote natural forests and their biodiversity. Had the UK's document been adopted, the relevant issues could have been partially tackled by subsequent decisions adopted by the UNFCCC COP and through the cooperation with other international environmental treaties. However, the subsequent experience with REDD+ demonstrates the weakness of this approach.[969] Consequently, the

965 Ibid, pp. 4 f.
966 Ibid.
967 United Nations, A/AC.237/Misc.1/Add.1 [n. 942], Annex 1 to the Proposal of Netherlands, p. 41.
968 United Nations, A/AC.237/Misc.1/Add.1 [n. 942], Paper 10: Draft UK Paper on Possible Elements for Inclusion in a Framework Convention on Climate Changes, p. 85.
969 See Section D of this Chapter.

Dutch proposals had higher potential to contribute to the development of the international legal regime for forests.

The American text also encouraged States to create national inventories of emissions and sinks and to update them regularly.[970] However, these measures could only constitute one of the elements of a deforestation monitoring system but did not promote the conservation of forests. In general, the American text was too vague to settle any more or less verifiable and measurable forest targets even in the future. The American delegation accepted the UK's proposition to adopt a specific protocol on natural sinks of greenhouse gases, particularly forests, but was opposed to any obligation to negotiate, adopt, and implement this protocol in the treaty text.[971]

What concerns the Indian draft version, it can be noted that the notion of "sinks" was not mentioned among the main definitions that are clarified in its Article 2.[972] This can be explained by the fact that the coalition of developing States was very reluctant to include forests and other natural sinks and reservoirs of greenhouse gases in the Convention because they feared the erosion of emission reduction activities in the industrial sectors of developed States.[973] According to the Indian draft version, financial resources provided by developed States under the framework of the Convention were deemed necessary to contribute to the achievement of the development goals of developing States, including climate change adaptation. At the same time, developing States were unwilling to undergo any obligations to redirect the funding provided to them into the conservation of forests. Therefore, it is unclear if these financial transfers could contribute to the conservation of forests. To sum up, if adopted, the Indian version of the UNFCCC would hardly contribute to the development of the international legal regime for forests.

III. Role of Forests in the UNFCCC

The UNFCCC was adopted on 9 May 1992 and was opened for signature during the Rio Conference held from 3 to 14 June 1992. The Convention entered into force on 21 March 1994. Currently, the UNFCCC is ratified by 196 States and by the European Union. Consequently, it enjoys almost universal membership.

---

970 United Nations, A/AC.237/Misc.1/Add.1 [n. 942], Paper 10: Draft UK Paper on Possible Elements for Inclusion in a Framework Convention on Climate Changes, p. 85.

971 Ibid, p. 93.

972 Ibid.

973 *Steffan* [n. 924], p. 88; *Schulte zu Sodingen* [n. 916], pp. 190 ff.

The preamble of the UNFCCC refers implicitly to forests, indicating that States are "aware" of the significance of natural ecosystems acting as sinks and reservoirs of greenhouse gases.[974] However, this final wording is more restrained in comparison to the version initially proposed by the United Kingdom, according to which States would have recognized the necessity to "develop strategies for the absorption of greenhouse gases", *inter alia*, through the conservation and promotion of sinks, "including the Earth's forests".[975] As Preamble plays an important role in the implementation of a treaty as one of the means of its interpretation,[976] once adopted, the British draft version could have promoted the development of forest-related mechanisms for climate change mitigation.

Moreover, the Preamble underlines that climate change can have adverse impacts on natural ecosystems.[977] It can be argued that this definition is not comprehensive, as it does not consider the possibility of certain positive effects of global warming on natural ecosystems. Possibly, a more adequate wording could be "recognizing the unpredictable impacts of the emission of greenhouse gases on natural ecosystems that are likely to have adverse effects on them". At any time, the wording of the preamble promotes the understanding of the fact that forest ecosystems may act either as sources or as sinks of greenhouse gases under different circumstances.[978]

In its main part, the UNFCCC stipulates the following: "the Parties shall promote sustainable management, and promote and cooperate in the conservation and enhancement of sinks and reservoirs of all greenhouse gases, including biomass, *forests* and oceans".[979] Although the stronger modal verb "shall" is used, the legal value of this article is weakened by the reference to "national and regional development priorities, objectives, and circumstances".

According to the Convention, only developed States listed in its Annexes I and II are obliged to reduce their emissions of greenhouse gases.[980] Moreover, developed States are obliged to provide financial and technological assistance in helping developing States to achieve their development goals in a climate-friendly and sustainable manner.[981] According to the principle of common but

974 UNFCCC, Preamble.
975 See Subsection B.II of this Chapter.
976 Vienna Convention on the Law of Treaties (with annex), adopted on 23 May 1969, in: 1155 UNTS 331, Article 31.2.
977 UNFCCC, Preamble.
978 *Schulte zu Sodingen* [n. 916], pp. 206 f.
979 UNFCCC Article 4.1(d).
980 *Bodansky/Brunnee/Rajamani* [n. 924], pp. 131 ff.
981 UNFCCC, Article 4.3.

differentiated responsibility, developed States are perceived as primarily responsible for the historical emissions of greenhouse gases. Therefore, they are supposed to bear the main burden of climate change mitigation and adaptation costs.[982]

It was argued that the reduction of emissions should be achieved through mitigation measures in the industrial sectors of developed States. Therefore, the Convention does not oblige the States in question to conserve and enhance their forest carbon sinks. Although the protection and enhancement of sinks is foreseen as a national climate mitigation measure, it is not clear which role the conservation of forests should play in the process.[983] Moreover, the UNFCCC does not oblige or urge developing States to redirect the international funding provided under the Convention's framework into forest conservation and afforestation.[984] Overall, the UNFCCC provides a foundation for the wider inclusion of forest-related activities in climate change mitigation policies. However, as no relevant obligation was established, the activities in transport, energy, and industry sectors were favored by the national and international climate-related policies.[985] Importantly, forest management in developed States was widely perceived as sustainable, meaning that no additional actions were needed.[986] Thus, the inclusion of forest-related activities in their national climate change mitigation strategies was left to the discretion of developed States.

Notably, forests can be included in climate-related activities through the Joint Implementation (JI) procedure that is established by the UNFCCC.[987] According to Articles 4.2.(d) and 4.2.(a) of the UNFCCC, States are allowed to achieve the reduction of CO_2-emissions jointly, by assisting other parties in achieving the

982 *Bodansky/Brunnee/Rajamani* [n. 924], pp. 131 ff.

983 Article 4.2(a).

984 See *Schulte zu Sodingen* [n. 916], pp. 206 f.; *Krohn, Susan Nicole*, Die Bewahrung der Tropischen Regenwälder durch völkerrechtliche Kooperationsmechanismen, 2002, pp. 123 ff.; *Eikermann, Anja*, Forests in International Law: Is There Really a Need for an International Forest Convention?, 2015, pp. 107 ff., 126.

985 *Boyd, William*, Ways of Seeing in Environmental Law: How Deforestation Became an Object of Climate Governance, in: Ecology Law Quarterly, Volume 37, Issue 3, 2010, p. 869.

986 *Schulte zu Sodingen* [n. 916], pp. 206 f.; *Krohn* [n. 984], p. 126.

987 *Schulte zu Sodingen* [n. 916], pp. 207 f.; *Krohn* [n. 984], p. 127; *Eikermann* [n. 984], p. 112.

objective of the Convention.[988] These mechanisms can potentially make the reduction of emissions more cost-effective and promote the transfer of financial resources and technologies. Due to the reluctance of developing States to undergo emission reduction commitments, the applicability of JI in relation to them was not clear at first. Nevertheless, a number of projects were implemented in tropical rainforests situated in developing States, particularly in Costa Rica, Ecuador, and Paraguay.[989] However, the main beneficiaries of JI are the economies in transition. The relevant activities in these States included re- and afforestation. In this manner, JI has contributed to the conservation of some temperate and boreal forests. Under these projects, a number of afforestation activities have taken place in Russia, Poland, and Ukraine.[990] One of the main downsides of the projects in question is the cultivation of monoculture plantations that were chosen as a standard afforestation policy, whereas forest biodiversity and multiple forest functions are widely neglected.[991] Admittedly, some projects, particularly those implemented in tropical rainforests in developing States, have also involved the conservation of natural forests.[992] Nevertheless, the scale of implementation of the JI procedure was not sufficient in order to achieve any significant improvement in the conditions of forest ecosystems. Overall, only a small number of projects included forest-related activities. For this reason, the contribution of JI to the conservation of forests was not too significant.

At the same time, from a practical point of view, developing States could contribute to the mitigation of climate change by conserving natural rainforests and promoting reforestation. However, the option of including the reduction of emissions from developing States' forest sectors did not find its way into the Convention's text.[993] In fact, from the point of view of the principle of common

988 *Michaelowa, Alex/Dutschke, Michael*, Climate Policy and Development, 2000, pp. 10 ff.; *Freestone, David/Streck, Charlotte* (eds.), Legal Aspects of Implementing the Kyoto Protocol Mechanisms: Making Kyoto Work, 2009, pp. 107 f.

989 *Cullet, Philippe/Kameri-Mbote, Patricia*, Joint Implementation and Forestry Projects, Conceptual and Operational Fallacies, in: International Affairs, Volume 74, Issue 2, 1998, pp. 393 ff.; *Dutschke, Michael/ Michaelow, Axel*, Joint Implementation as Development Policy – The Case of Costa Rica, 1997, pp. 1 ff.

990 *Evans Meredydd/Legro, Susan/Popov, Ilya*, The Climate for Joint Implementation: Case Studies from Russia, Ukraine, and Poland, in: Mitigation and Adaptation Strategies for Global Change, Volume 5, Issue 4, 2000, pp. 322 ff.

991 Ibid.

992 *Cullet/Kameri-Mbote* [n. 989], p. 393.

993 *Schulte zu Sodingen* [n. 916], pp. 207 f.; *Krohn* [n. 984], pp. 123 ff.; *Eikermann* [n. 984], p. 126.

but differentiated responsibility, developing States were reluctant to accept any specific legally binding emission reduction obligations. The general obligation to promote the conservation of forests as carbon sinks according to Article 4.1(d) is not precise enough to promote the conservation of forests in developing States. The obligation contained in Article 4.2(a) to adopt policies and to take measures on climate change mitigation, inter alia, by conserving and promoting forests as greenhouse gas sinks, applies only to the Annex I Parties. Moreover, the obligations under Article 4.2(a) are vague and unspecified, meaning that they cannot strengthen the obligation to promote natural sinks and reservoirs as established under Article 4.1(d).[994] Among other things, it can be criticized that the provision in question does not specify further the role of the conservation of forests in climate change mitigation policies. Annex-I States are left free to decide in which sectors climate change mitigation activities should be adopted. Therefore, Article 4.2(a) would not have effectively promoted the conservation of forests even if it had also referred to developing States. Of course, developing States could voluntarily apply climate change mitigation policies in forests, but they were not motivated to do so by the UNFCCC. Consequently, the UNFCCC underlines the role of forests in climate change mitigation but falls short of establishing a robust framework for the promotion of forest protection. Moreover, the treaty fails to consider the impacts of deforestation on global and regional precipitation patterns.

Importantly, any measures in relation to the combating of deforestation, the conservation and sustainable management of forests, and a/reforestation that are developed under international climate change law have to be subordinated to the main objective of the UNFCCC as determined in its Article 2. It consists of the stabilization of greenhouse gases in the atmosphere at a level at which dangerous anthropogenic climate change would not take place. Thus, the ultimate scope of any relevant activities is not to protect forests, but to promote the reduction of emissions, and the removal of greenhouse gases from the atmosphere.[995] The UNFCCC was initially developed as a non-conservationist treaty. However, the inclusion of the conservation of natural forests and their multiple climate-related functions, which naturally go beyond the absorption of CO_2, in

---

994 *Birnie, Patricia/Boyle, Alan/Redgewell, Catherine*, International Law and the Environment, 3rd ed. 2009, p. 360.; *Stoll, Peter-Tobias/Krüger, Hagen, Peter-Tobias/Krüger, Hagen*, Klimawandel, in: *Proelß, Alexander* (ed.), Internationales Umweltrecht, 2017, p. 297.

995 *Schulte zu Sodingen* [n. 916], pp. 209 f.

its main objective, could provide for a more comprehensive implementation of the Convention in question. Arguably, the long-term stabilization of CO_2 levels presupposes the conservation and promotion of healthy and resilient natural ecosystems.[996]

On the other hand, the promotion of the conservation of forests under the umbrella of the UNFCCC can have negative impacts on forest ecosystems. Indeed, the narrow focus on the forest's ability to absorb and store greenhouse gases does not favor a holistic approach to their conservation. It is conceivable that climate change mitigation activities can be implemented even if they represent a damage or threat to forest ecosystems. As the convention does not contain any provision on the relationship with other international environmental agreements and does not establish robust and effective environmental safeguards, it does not provide sufficient guidance on how such a situation should be avoided. Supposedly, climate change mitigation activities are likely to have priority over the conservation of forests.[997] For example, at the time of the UNFCCC negotiations it was erroneously presumed that young planted forests, including monoculture plantations, absorb greater amounts of CO_2 than mature natural forests.[998] It was also erroneously argued that mature forests cannot absorb more carbon dioxide because of the lack of visible tree growth in these ecosystems and, therefore, they can act only as carbon stocks.[999] From this point of view, mature natural forests can be cleared in order to plant monoculture plantations. More recent research has shown that natural forests are still able to absorb and store carbon dioxide, whereas this greenhouse gas is stored more efficiently in these forest ecosystems, whose vitality is much higher than in the case of monoculture plantations.[1000] Relevant observations have shown that despite the apparent stability of forest biomass in mature natural or near-natural forests, their carbon stocks are constantly increasing. These forests are far less prone to adverse natural phenomena, and thus, they can sustainably store carbon dioxide.[1001] In addition to the inability of monoculture plantations, which should not be treated in a same

996 *Stoll/Krüger* [n. 994], pp. 287 f.

997 See Section B of Chapter 5.

998 *Luyssaert, Sebastiaan /Schulze, E Detlef/Börner, Annett/Knohl, Alexander/Hessenmöller, Dominik/Law, Beverly E./Ciais, Philipp/Grace, John*, Old-Growth Forests as Global Carbon Sinks, in: Nature, Volume 455, Number 7210, 2008, pp. 213 f.

999 Ibid.

1000 Ibid.

1001 Ibid.

way as natural forests,[1002] to fulfil multiple forest functions, their carbon stocks are anything but stable. They are more prone to forest fires and pest outbreaks, which both result in the release of significant amounts of carbon dioxide into the atmosphere. Furthermore, monoculture plantations cannot absorb the same amount of carbon dioxide, as was emitted during the logging of mature natural forests. However, the UNFCCC does not consider all these issues and does not draw a clear line between natural forests and monoculture plantations.[1003] The protection of rights of forest-dependent people by the UNFCCC is another point of concern. Notably, unlike the CBD, the UNFCCC does not establish any obligation for States to protect the rights of ILCs. The lack of attention to the relevant issues in the treaty text can be seen as another oversight of the Convention, as ILCs are not only particularly vulnerable to the adverse impacts of climate change, but also their rights can be violated by climate change mitigation activities.

C. Role of Forests in Climate Change Mitigation under the KP

I. Need for the KP

As already addressed above, the obligations of the UNFCCC are vague and unspecified. As a framework convention, the UNFCCC does not stipulate any measurable and verifiable reduction commitments and does not provide detailed strategies and measures for its implementation. The lack of strong and enforceable obligations impedes the progress in implementing the UNFCCC. However, general obligations contained in a framework convention can be strengthened and specified via the adoption of legally binding protocols. From this background, the UNFCCC COP 1 held in Berlin from 28 March until 7 April 1995 adopted the so-called Berlin Mandate, according to which Parties

---

1002 A comprehensive notion of forests is absent in international law. If the qualitative criteria and indicators of forest viability, biodiversity, and the ability to fulfill multiple forest functions are taken into account, the monoculture plantations should not be regarded on a par with natural forests. See also: *Hance, Jeremy*, Monoculture tree plantations are "green deserts" not forests, say activists, in: Mongabay, 19 September, 2008. – Internet resource – https://news.mongabay.com/2008/09/monoculture-tree-plantations-are-green-deserts-not-forests-say-activists/ [Last visit: 30.04.2023]

1003 *Schulte zu Sodingen* [n. 916], pp. 207 f.; *Krohn* [n. 984], pp. 123 ff.; *Eikermann* [n. 984], p. 126.

agreed to launch the negotiation of a protocol to the UNFCCC or another instrument deemed to strengthen the general framework of the international legal regime for climate changes.[1004] Thereby, the Parties aimed, inter alia, to elaborate precise reduction policies and establish measurable and verifiable reduction targets.[1005] Notably, the Berlin mandate foresaw, inter alia, the enhancement of the role of greenhouse gases by sinks.[1006] Just like the initial treaty text, the Berlin Mandate envisaged the establishment of relevant obligations for developed States only, meaning that the tropical forests of developing States were automatically excluded from the scope of the future protocol. After 2 years of negotiations, the Kyoto Protocol was adopted on 11 December 1997. It was decided that the Protocol could enter into force only after the 55th ratification, including, inter alia, ratifications of a number of Annex I States that are responsible altogether for at least 55% of emissions. This complex ratification procedure resulted in the unusually late entrance into force of the Protocol, which happened only on 16 February 2005.[1007] The KP further specifies the provisions and principles of the UNFCCC. Its measurable emission reduction targets are applicable uniquely to Annex I States.[1008]

II. Role of Sinks in the KP

The inclusion of sinks became one of the most contentious issues during the negotiations of the KP.[1009] It was pointed out that the term "sink" is sometimes used synonymously with removals in the documents handling the so-called

---

1004 *Grubb, Michael/Vrolijk, Christiaan/Brack, Duncan,* The Kyoto protocol: a guide and assessment, 1999, p. 62 ff.; *Schütz, Sebastian,* The Kyoto Protocol with an emphasis on its flexible instrument: the Clean Development Mechanism, 2011, p. 24; *Carraro, Carlo,* International Environmental Agreements on Climate Change, 1999, pp. 1–2, 210 f.; *Bodansky/Brunnee/Rajamani* [n. 924], pp. 105 ff.

1005 *Carraro* [n. 1004], pp. 210 f.; *Bodansky/Brunnee/Rajamani* [n. 924], pp. 105 ff.

1006 *Bodansky/Brunnee/Rajamani* [n. 924], pp. 105 ff.

1007 Ibid.

1008 Article 3(1).

1009 United Nations FCCC/AGBM/1997/INF.2, Compilation of Responses from Parties on Issues Related to Sinks, Ad Hoc Group on the Berlin Mandate, Eighth Session, Agenda item 3, Kyoto, 30 November 1997. – Internet resource – https://unfccc.int/resource/docs/1997/agbm/inf02.pdf [Last visit: 30.04.2023]; See also: *Abidin, Handa,* The Protection of Indigenous Peoples and Reduction of Forest Carbon Emissions: The REDD-Plus Regime and International Law, 2015, pp. 16 ff.; *Eikermann* [n. 984], pp. 113 f.

"land use, land-use change and forestry" (LUCF) issues.[1010] Whereas in the sense of the UNFCCC "sink" means any process or activity which removes greenhouse gases from the atmosphere and a "source" means any process or activity that releases these gases, the LUCF category operates with the terms "sinks" and "emissions", including 4 subcategories: land conversion, abandoned land, forests, forest biomass and others.[1011] All of them can act as sinks or emissions. Some Parties advocated an approach according to which only "net removals by sinks" should be taken into account in the international legal regime for climate change.[1012] Finally, the Parties were not able to agree on a comprehensive notion of "sinks". Unlike the UNFCCC, which generally considers all possible activities in the LULUCF category, the scope of application of the Kyoto Protocol was restricted in order to make it operational.[1013] If the comprehensive notion of sinks as adopted under the UNFCCC had been included in the KP, its text would have been too vague and would not specify the provisions of the Convention. For this reason, Annex B States are allowed to include in their estimations of national emissions only the emissions of greenhouse gases resulting from deforestation and their removals resulting from afforestation and reforestation.[1014] Consequently, a bulk of forest-related activities can be considered in the national climate-related strategies and policies. However, these restrictions do not permit to include the emission reductions resulting from sustainable forest use and do not consider the removals of greenhouse gases by natural forests.[1015] It was decided that the forthcoming COPs will take a decision on whether and how emissions and removals resulting from other activities in the land-use and forest category, including from the improvement of forest management, should be calculated in the emissions of Annex-B States.[1016] A number of participants advocated for emissions and removals from forest management and revegetation also being accounted for, but the issue remained highly controversial.[1017] Finally, it was

1010 United Nations FCCC/AGBM/1997/INF.2 [n. 1009], Paragraph 6; see also *Eikermann* [n. 984], pp. 112 f.

1011 United Nations FCCC/AGBM/1997/INF.2 [n. 1009], Paragraph 6; see *Eikermann* [n. 984], pp. 117 f.

1012 United Nations FCCC/AGBM/1997/INF.2 [n. 1009], Summary table of responses by Parties to the questionnaire regarding sinks: part A.

1013 See also *Eikermann* [n. 984], pp. 117 f.

1014 Ibid.

1015 Kyoto Protocol, Article 3.3.

1016 Kyoto Protocol, Article 3.4; see *Eikermann* [n. 984], pp. 117 f.

1017 *Abidin* [n. 1009], p. 17.

decided that the enhancement of forest carbon stocks achieved by a gentle forest use could be included only in the cases where the net changes of forest carbon stocks compensate entirely for the emissions resulting from deforestation.[1018]

The scheme for accounting emissions and removals within the LULUCF category as it emerged under the KP incorporates one fundamental error. According to it, the harvesting of biomass for bioenergy is reported as land-use associated emissions. However, carbon dioxide emissions from biomass combustion are not included in the energy sector emission reporting. This means that the emissions from the combustion of imported wood and wood pallets are not included in the overall emission of States.[1019] This fact can have enormous adverse impacts on the conservation of forests. As the result, deforestation and forest degradation can significantly increase in many forest-rich States.

It is also problematic that the provisions of the KP do not consider the impact of climate change mitigation activities on the conditions of forest biodiversity and multiple ecological functions of forests. The management of forests is intended widely for the scope of Article 3.4. of the KP. This means that the enhancement of forest carbon stocks can be included as removals by sinks eligible under the KP even in the cases where the removals of CO_2 resulting from establishing monoculture plantations surpasses the emissions from the conversion of natural forests.[1020] Just like the UNFCCC, the KP does not establish any effective environmental safeguards for the conservation of natural forests. Thus, the KP does not promote a holistic approach to the protection of forests, which would include the conservation of natural forests; the enhancement of forest areas not used for industrial timber and biomass harvesting; and the promotion of the sustainable use of managed forests.

1018 KP CMP, Decision 16/CMP.1 "Land use, land-use change and forestry", Paragraph 1(d), in: United Nations FCCC/KP/CMP/2005/8/Add.2, Report of the Conference of the Parties serving as the meeting of the Parties to the Kyoto Protocol on its first session, held at Montreal from 28 November to 10 December 2005, 2006. – Internet resource – https://unfccc.int/resource/docs/2005/cmp1/eng/08a03.pdf [Last visit: 30.04.2023]

1019 *Haberl, Helmut/Sprinz, Detlef/Bonazountas, Marc/Cocco, Pierluigi/Desaubies, Yves/Henze, Mogens et al.*, Correcting a fundamental error in greenhouse gas accounting related to bioenergy, in: Energy Policy, Volume 45, 2012, pp. 18 ff.

1020 *Eikermann* [n. 984], p. 119; Fry, Ian, Twists and Turns in the Jungle: Exploring the Evolution of Land Use, Land–Use Change and Forestry Decisions within the Kyoto Protocol, in: RECIEL, Volume 11, Issue 2, 2002, p. 163.

III. Contribution of Flexible Mechanisms under the Kyoto Protocol to the Protection of Forests

The Kyoto Protocol has established and further promoted flexible mechanisms for its implementation, which were already introduced by the UNFCCC, particularly in its Article 4.2(a). The idea behind them is that developed States listed in Annex B of the Protocol commit themselves to not exceed their "assigned amounts" of greenhouse gases listed in Annex A.[1021] However, these States are allowed to maintain emissions on prescribed levels or reduce them jointly. This means that any State that is able to reduce its emissions of greenhouse gases beyond the level negotiated in the Protocol can sell its unused emission units to other developed States.[1022] These emissions units are referred to as "assigned amount units" or "carbon credits".

Flexible mechanisms under the Kyoto Protocol include Joint Implementation, the Clean Development Mechanisms (CDM), and Emissions Trading. As already noted, JI originates directly from the UNFCCC, whereas the CDM is an innovation introduced by the KP. However, the Kyoto Protocol further develops and specifies the concept of JI. Accordingly, JI is to be used as a cooperation mechanism only between the Annex-I States. However, non-Annex I States are also able to join such schemes voluntarily.[1023] At the same time, the CDM is a flexible mechanism that is specifically developed for the promotion of cooperation between developed and developing States. For this reason, the group of developing States were initially opposed to its inclusion in the Protocol because they were concerned about the emission reduction efforts in developed States being undermined.[1024]

It was agreed that the emission reductions and the removals by sinks achieved during the pilot phase of implementing JI in the period 1995–2002 would not be converted into carbon credits in accordance with the protocol.[1025] Only schemes initiated after 1 January 2002 can be considered under international JI projects, but only if all the requirements are met. These decisions can be criticized insofar as the restrictions concerning the conversion of emission reductions into carbon credits can keep investors away from forest-related JI projects. At the

1021 *Grubb/Vrolijk/Brack* [n. 1004], pp. 87 ff.; *Bodansky/Brunnee/Rajamani* [n. 924], p. 180.

1022 *Bodansky/Brunnee/Rajamani* [n. 924], p. 181.

1023 Ibid.

1024 *Grubb/Vrolijk/Brack* [n. 1004], pp. 87 ff; *Bodansky/Brunnee/Rajamani* [n. 924], p. 180.

1025 *Grubb/Vrolijk/Brack* [n. 1004], pp. 97 ff.

very end, most JI projects were implemented in so-called transition States. States are permitted to include all the activities mentioned in Articles 3.3 and 3.4 of Kyoto Protocol in JI projects, meaning that emissions and removals originating from deforestation and a/reforestation can be included in the JI-reduction schemes.[1026] JI projects have rarely included forest-related activities, and, once included, these activities were mostly limited to the re/afforestation with monoculture tree stands.[1027]

Despite the initial opposition of developing States, the flexible mechanisms were finally included in the KP after they were linked to the promotion of sustainable development in non-Annex I States. By participating in the CDM, these States are able to contribute voluntarily to the achievement of the UNFCCC's objective. Unlike the JI, which includes emission reductions resulting from the prevented deforestation, the CDM is limited to the reforestation and afforestation activities.[1028] The issue of inclusion of emission reductions from avoided deforestation was not regulated in the KP itself. Subsequently, the UNFCCC COP 7 included the detailed provisions on the CDM in the Marrakesh Accords, adopted in November 2001 in Morocco. The Marrakesh Accords serve as a guideline for the CDM implementation. The document stipulates explicitly that avoided deforestation will not be eligible under the CDM.[1029]

States opposed the inclusion of the emission reductions resulting from avoided deforestation in the CDM for a number of technical and ethical issues.[1030] Major concerns arose in relation to the so-called additionality issue, which means that it is difficult to attribute the achieved reductions of deforestation rates exactly to the CDM projects, while excluding all other factors, such as economic ones,

---

1026 Ibid.

1027 *Schulte zu Sodingen* [n. 916], pp. 207 f.; *Krohn* [n. 984], pp. 130 f; *Eikermann* [n. 984], p. 115.

1028 *Schulte zu Sodingen* [n. 916], pp. 207 f.; *Krohn* [n. 984], pp. 130 f; *Eikermann* [n. 984], p. 115.

1029 UNFCCC COP Decision 17/CP.7, "Modalities and procedures for a clean development mechanism as defined in Article 12 of the Kyoto Protocol", in: United Nations FCCC/CP/2001/13/Add.2, Report of the Conference of the Parties on its Seventh Session, held at Marrakesh from 29 October to 10 November 2001, United Nations, 2002. – Internet resource – https://unfccc.int/resource/docs/cop7/13a01.pdf [Last visit: 30.04.2023]

1030 *Eikermann* [n. 984], p. 115; *Streck, Charlotte/Scholz, Sebastian M.*, The Role of forests in Global Climate Change: Whence We Come and Where We Go, in: International Affairs, Volume 82, Issue 5, 2006, pp. 867 ff.

was a challenging process.[1031] For example, a forest-rich State can claim that the avoided deforestation was achieved through the CDM, whereas the exploitation of forest ecosystems was economically unaffordable. This practice would constitute a violation of Article 12.5(c) of the KP, according to which the certification of emission reductions can take place only if these emissions could not be avoided without a CDM project. These technical issues played a significant role in the omission of avoided emissions from deforestation in the CDM. However, it is possible to argue that prevented deforestation achieved through the establishment of new natural parks and designated protected areas, which was possible thanks to the financial and technical assistance provided in the course of a CDM project, would fit perfectly to the purpose of this flexible instrument. In such a manner, the incentives for forest conservation could be delivered.

Secondly, the avoided deforestation was not included due to concerns in relation to leakage issues. The leakage of emissions takes place when the avoided deforestation in certain area leads to the shift of industrial logging and forestland clearance into other areas, meaning that the overall emissions from the forest sector remains unchanged or even increases.[1032] Despite this, a developing State hosting a CDM project would receive certified emission reductions. Such a constellation could have at least two negative consequences. Firstly, the financial resources invested in a CDM project would not contribute to the resolution of environmental problems. Secondly, the anthropogenic pressure on forests outside of the project area would increase leading to the further depletion of environmental functions of forests.

In the situation, when eligible activities under the CDM include only re- and afforestation, the contribution of this flexible mechanism to the conservation of forests can be negligible. The protection of forests in developing States can benefit from the CDM projects only insofar as the project area is a/reforested with native and climate change resistant species in their natural composition. The re/afforestation activities must also contribute to the reduction of pressure on natural forests.

1. Forest-Related Activities under the CDM

As stated above, detailed provisions on the development and implementation of CDM projects are contained in the Marrakesh Accords. The implementation

1031 *Streck/Scholz* [n. 1030], p. 868.
1032 Ibid.

of the CDM is subject to the KP CMP's authority and is overseen by the CDM Executive Board.[1033] The Members of the Executive Board are nominated by the Parties to the Convention and are elected by the KP CMP, while the UNFCCC COP was temporarily authorized to elect the members before the entry of the KP into force.[1034] The Executive Board oversees the CDM and is accountable to the KP CMP. It is noteworthy that the Executive Board is entrusted with the accreditation of designated operational entities (DOE). The DOE is another important element of the international infrastructure for the implementation of the CDM. They are entrusted with the validation of proposed CDM project activities, including forest-related activities.[1035] It is important that they also verify and certify the removals of greenhouse gases by sinks, achieved through the re/afforestation.[1036] Transparency and legality are promoted as the DOEs have to ensure their impartiality and objectivity of the project's implementation.[1037] In order to enhance transparency, it is foreseen that generally one single DOE can fulfill only one major function related to the implementation of projects: either the validation of the project itself or the verification and certification of emissions.[1038] However, in some cases and only upon special request, the Executive Board may allow one single DOE to perform different functions within a single CDM project.[1039] For the sake of transparency, the Executive Board has to make the project-related information publicly available, except for the data, which the project participants consider as confidential or proprietary.[1040] It is noteworthy that the relevant provisions specify that the information related to additionality

1033 KP Article 12.4.

1034 UNFCCC COP Decision 17/CP.7, Paragraphs 2 and 3(a).

1035 Ibid, Annex, Paragraph 27(a).

1036 KP CMP Decision 5/CMP.1, "Modalities and procedures for afforestation and reforestation project activities under the clean development mechanism in the first commitment period of the Kyoto Protocol", Annex, Paragraph 6, in: United Nations FCCC/KP/CMP/2005/8/Add.2, Report of the Conference of the Parties serving as the meeting of the Parties to the Kyoto Protocol on its first session, held at Montreal from 28 November to 10 December 2005, 2006. – Internet resource – https://unfccc.int/resource/docs/2005/cmp1/eng/08a03.pdf [Last visit: 30.04.2023]

1037 Decision 17/CP.7, Paragraph 27(d).

1038 Decision 17/CP.7, Paragraph 27(e).

1039 Ibid.

1040 Decision 17/CP.7, Paragraph 27(i, k, m).

issues, to baseline methodology issues and environmental impact assessment, cannot be considered as confidential or proprietary.[1041]

Each Party to the Kyoto Protocol, which is not included in Annex I, can participate in the CDM on a voluntary basis after having designated a national authority for the CDM.[1042] For this purpose, the State also has to consider that re/afforestation projects have to fulfill special requirements. Accordingly, a minimum tree crown cover should be between 10% and 30%, a minimum land area between 0,05 and 1 hectare, a minimum tree height between 2 and 5 meters.[1043] The last provision can be criticized for the lack of attention to the promotion of multiple forest functions. It could be proposed to oblige designers of re/afforestation projects under the CDM to promote forest biodiversity and to avoid the introduction of alien species in a forest ecosystem.

After Parties participating in a re/afforestation project under the CDM satisfy all the requirements, they can proceed with the validation and registration of the project activities. In this regard, "validation" means an independent evaluation of the project by the DOE.[1044] The validated project has to be registered by the Executive Board.[1045] For re/afforestation project activities a restriction is established, which foresees that only temporary and long-term CERs (tCERs and lCER respectively) can be issued.[1046] Whereas tCERs expire at the end of the next commitment period,[1047] lCERs expire at the end of a crediting period of a forest-related project activity, for which they were issued.[1048] By applying tCERs and lCERs, the COP seeks to address the issue of non-permanence.[1049] It can be criticized that the provision does not consider that the stability of forest carbon

---

1041 KP CMP Decision 3/CMP.1, "Modalities and procedures for a clean development mechanism as defined in Article 12 of the Kyoto Protocol", Annex, Paragraph 6, in: United Nations FCCC/KP/CMP/2005/8/Add.2, Report of the Conference of the Parties serving as the meeting of the Parties to the Kyoto Protocol on its first session, held at Montreal from 28 November to 10 December 2005, 2006. – Internet resource – https://unfccc.int/resource/docs/2005/cmp1/eng/08a03.pdf [Last visit: 30.04.2023]

1042 KP CMP Decision 3/CMP.1, Annex Paragraphs 28, 29, 30.

1043 KP CMP Decision 5/CMP.1, Annex, Paragraph 8.

1044 Ibid, Annex, Paragraph 10.

1045 Ibid, Annex, Para 11.

1046 *Eikermann* [n. 984], p. 114.

1047 KP CMP Decision 5/CMP.1, Annex, Paragraph 1(g).

1048 KP CMP Decision 5/CMP.1, Annex, Paragraph 1(h).

1049 Ibid.

stocks can be substantially increased by promoting biodiverse and resilient forests. Moreover, the concept of tCERs/lCERs can keep investors from re/afforestation project activities under the CDM. It can be proposed to introduce the timely unlimited CERs for the forest-related activities under the CDM.

Over the course of the project validation, a DOE selected by participants of a forest-related project must review the project-related documentation and confirm that a number of requirements have been met.[1050] It is noteworthy that requirements refer implicitly also to the ILCs. However, the provision in question does not mention them directly, while referring instead to "local stakeholders". Moreover, instead of ensuring the full participation of the representatives of ILCs at all stages of the project implementation, including the acquiring of their PIC and benefit-sharing, parties are only expected to invite them to "comment" on the project in question and to make a summary of comments. Parties are supposed to consider these comments over the course of project implementation. Consequently, the level of protection of the rights of ILCs under the CDM is lower than in the case with the Nagoya Protocol.[1051]

The selected DOE has also to verify if the non-permanence and leakage issues were addressed by the project participants.[1052] It is noteworthy that it has to check if the socio-economic and environmental issues are considered by the project design.[1053] It is problematic that the DOE verifies the provided documentation, but does not check how the information about the socio-economic and environmental impacts of re/afforestation project activities was gathered. It is not entitled to monitor the conditions of biodiversity and ecosystems on the ground or to get the relevant information from other international agencies.

It is also important to note that prior to submitting of a validated project to the Executive Board, the DOE must check if participants of the project have submitted a confirmation of voluntary participation.[1054] Furthermore, the State, on whose territory the project is implemented, has to submit a confirmation that the project activity assists it in achieving sustainable development.[1055] At the same time, taking into account a lack of a unique internationally recognized definition of sustainable development, it is unclear upon which criteria the contribution of

1050 KP CMP Decision 5/CMP.1, Annex, Paragraph 12.
1051 See Subsection D.II.5 of Chapter 3.
1052 KP CMP Decision 5/CMP.1, Annex, Paragraph 12(c–f).
1053 KP CMP Decision 5/CMP.1, Annex, Paragraph 12(c).
1054 KP CMP Decision 5/CMP.1, Annex, Paragraph 15(a–c).
1055 Ibid.

forest-related projects under the CDM to the sustainable development should be evaluated. In general, the principle of sustainable development foresees that the socio-economic development is achieved by the means, which permit States to safeguard the environment and to avoid the depletion of natural resources.[1056] While promoting the development in underdeveloped regions and considering the needs of present and future generations, the principle seeks to promote both inter- and intra-generational equity. However, the principle should be further specified in order to make it operational. For this reason, it could be proposed that the DOE must be authorized to require the documentation that would specify in which areas the improvement has taken place.

It can be seen negatively that an effective international system for monitoring the implementation of re/afforestation projects is not established. Instead, the DOE has to verify if the baseline and monitoring methodologies adopted by project participants comply with the requirements as approved by the Executive Board.[1057] Before the participants seek to adopt a new methodology, it must initially be approved by the Executive Board.[1058] However, once the project is validated by the DOE, the pertinent provisions entrust the fulfillment of monitoring activities to the project participants. The requirements developed by the Executive Board deal exclusively with the estimation of greenhouse gas removals, while addressing additionality issues.[1059] At the same time, robust provisions for the monitoring of conditions of forest ecosystems are absent. The participation of the representative of ILCs in the monitoring activities is not ensured.

2. *Consideration of Environmental Issues by the Flexible Mechanisms*

To promote consideration of environmental issues in the process of the implementation of LULUCF activities, including forest-related projects under the

---

1056 *United Nations*, Report of the World Commission on Environment and Development: Our Common Future, United Nations, 1987. – Internet resource – https://sustainabledevelopment.un.org/content/documents/5987our-common-future.pdf [Last visit: 30.04.2023]; see also: *Epiney, Astrid/Scheylie, Martin*, Strukturprinzipien des Umweltvölkerrechts, 1998, pp. 35 ff.; *Schröter, Jessica*, Strukturprinzipien des Umweltvölkerrechts und ihr Beitrag zur Eindämmung des Klimawandels, Schriftenreihe des Instituts für Umwelt- und Technikrecht der Universität Trier, Band 130, 2015, p. 241.

1057 KP CMP Decision 5/CMP.1, Annex, Paragraph 12(g).

1058 KP CMP Decision 5/CMP.1, Annex, Paragraph 13.

1059 KP CMP Decision 5/CMP.1, Annex, Paragraph 25–30; KP CMP Decision 3/CMP.1, Appendix C.

CDM, a number of documents were developed by the Governing Bodies of the UNFCCC and the IPCC. Inter alia, the Good Practice Guidance for Land Use, Land-Use Change and Forestry was adopted by the 21st Session of the IPCC, in November 2003.[1060]

The COP encouraged parties to apply the Good Practice Guidance while implementing afforestation and reforestation projects under the CDM.[1061] The document deals predominantly with technical issues in relation to the determination of project boundaries and estimation of changes in forest carbon stocks. However, the IPCC did not address a number of other important issues including "non-permanence, additionality, leakage, uncertainties, and socio-economic and environmental impacts, including the impacts on biodiversity and natural ecosystems", because the UNFCCC COP entrusted the SBSTA with the task to consider all these aspects by elaborating "definitions and modalities for including afforestation and reforestation project activities under the CDM".[1062]

Therefore, environmental issues played only a secondary role in the process of elaboration of rules and guidelines for the LULUCF activities. The environmental issues were considered only in several COP Decisions, such as Decision 19/CP.9[1063] and in Decision 14/CP.10.[1064]

1060 *Penman, Jim/Gytarsky, Michael/Hiraishi, Taka/Krug, Thelma/Kruger, Dina/Pipatti, Riitta/Buendia, Leandro/Miwa, Kyoko/Ngara, Todd/Tanabe, Kiyoto/Wagner, Fabian*, Good Practice Guidance for Land Use, Land-Use Change and Forestry, IPCC National Greenhouse Gas Inventories Programme, Intergovernmental Panel on Climate Change, Good Practice Guidance for Land Use, Land-Use Change and Forestry, IPCC National Greenhouse Gas Inventories Programme, 2003.

1061 UNFCCC COP Decision 19/CP.9 "Modalities and procedures for afforestation and reforestation project activities under the clean development mechanism in the first commitment period of the Kyoto Protocol" in: United Nations FCCC/CP/2003/6/Add.2, Report of the Conference of the Parties on its ninth Session, held at Milan from 1 to 12 December 2003, United Nations, 2004, Paragraph 2. – Internet resource –https://cdm.unfccc.int/Reference/Documents/dec19_CP9/English/decisions_18_19_CP.9.pdf [Last visit: 30.04.2023]

1062 UNFCCC COP Decision 11/CP.7, "Land use, land-use change and forestry", in: United Nations FCCC/CP/2001/13/Add.1, Report of the Conference of the Parties on its seventh Session, held at Marrakesh from 29 October to 10 November 2001, United Nations, 2002, Paragraph 2(e). – Internet resource – https://unfccc.int/sites/default/files/resource/docs/cop7/13a01.pdf?download [Last visit: 30.04.2023]

1063 Decision 19/CP.9.

1064 UNFCCC COP Decision 14/CP.10, "Simplified modalities and procedures for small-scale afforestation and reforestation project activities under the clean development mechanism in the first commitment period of the Kyoto Protocol and measures to

Nevertheless, it can be still evaluated positively that the participants of the CDM projects must provide documentation that contains an analysis of impacts on biodiversity, natural ecosystems, and on environment beyond the project area.[1065] The requirement of provision of documentation concerning the environmental impacts outside the project area can contribute to the mitigation of the risk of leakage. The DOE has to verify if this documentation has been provided before issuing the validation report. Subject to confidentiality and property issues, the DOE is obliged to make the information in question and the whole project design document publicly available.[1066] This provision helps to facilitate public control over the promotion of environmental issues in the course of project activities under the CDM. Importantly, if any of negative impacts determined in the course of primary evaluation is considered significant, the project participants will have to undertake a socio-economic and/or an environmental impact assessment.[1067] Considering that this assessment is to be accomplished in accordance with the requirement of the hosting party,[1068] the national legislation of the party concerned is likely to play a crucial role. This fact makes the promotion of environmental issues in the course of project implementation dependent on the environmental standards adopted by the host party. As environmental protection is just now becoming a priority in many developing States,[1069] it can be advised to establish international environmental and social standards and safeguards to be followed by project participants. However, the decisions guiding the implementation of re/afforestation projects under the CDM do not contain such environmental and social safeguards. Moreover, the DOE is not competent to evaluate the transparency and adequacy of the environmental and/

facilitate their implementation", in: United Nations FCCC/CP/2004/10/Add.2, Report of the Conference of the Parties on its tenth session, held at Buenos Aires from 6 to 18 December 2004, United Nations, 2005. – Internet resource – https://unfccc.int/resource/docs/cop10/10a02.pdf#page=26 [Last visit: 30.04.2023]

1065 Decision 5/CMP.1, Annex, Paragraph 12(c).

1066 Decision 5/CMP.1, Annex, Paragraph 15(b).

1067 Decision 5/CMP.1, Annex, Paragraph 12(c).

1068 Ibid.

1069 Decision 5/CMP.1, Annex, Paragraph 12(c); See *Roque, Patrícia Gonçalves/Severo, Eliana Andréa/Dorion, Eric Charles Henri/Roque, Eliel de Souza/de Guimarães, Julio Cesar Ferro*, The Dilemma of Environmental Sustainability in a Developing Country: Environmental Crimes in Southern Brazil, in: Business Strategy and Development, Volume 1, Issue 1, 2018, pp. 43 ff.

or socio-economic impacts assessment, and to monitor the implementation of subsequent remedial actions.

The provisions that guide the implementation of LULUCF activities, which primarily refer to Annex I States, foresee, among other things, that such activities will contribute to the conservation and sustainable use of biodiversity.[1070] However, this statement is a declaratory one, as no specified obligations and due diligence duties are established, and no mechanisms for the promotion of biodiversity issues are created. The general logic of Decision 16/CMP.1 concludes that the removal of sinks can be accounted for in the national emission reductions independently of the contribution of the LULUCF activities to the conservation of natural forests.[1071]

At the negotiations of the KP, concerns were raised that the inclusion of sinks under its scope can create new hazards for forest ecosystems. The industrial logging in natural forests and the establishment of monoculture plantations could be promoted.[1072] It is also problematic that the quantitative indicators used to define forests for the purpose of the KP do not include the level of biodiversity and the medium age of trees.[1073] Indeed, the medium age of trees in a natural forest significantly exceeds the same value for monoculture plantations.[1074] In general, the protection of natural forests is not promoted effectively by the activities under the Kyoto Protocol. Annex I States are not obliged to conserve the remaining natural forests and other high-value forest ecosystems, to extend the area of non-managed forests, and to promote the forest biodiversity in the managed ones is not established. It is also problematic that the KP does not consider natural forests as carbon sinks. Moreover, avoided deforestation is not included as an eligible activity under the scope of the CDM that is to be applied in forest-rich tropical States.

1070 Decision 16/CMP.1, Paragraph 1(e).

1071 *Maguire, Rowena*, Global Forest Governance, Legal Concepts and Policy trends, 2013, p. 154; *Sagemueller, Imke*, Forest sinks under the United Nations Framework Convention on Climate Change and the Kyoto Protocol: Opportunity or Risk for Biodiversity?, in: Columbia Journal of Environmental Law, Volume 31, Issue 2, 2006, pp. 189 ff.; *Jacquemont, Frédéric, Caparros Alejandro*, The Convention on Biological Diversity and the Climate Change Convention 10 Years After Rio: Towards a Synergy of the Two Regimes?, in: RECIEL, Volume 11, Issue 2, 2002, pp. 170 ff.

1072 See also *Eikermann* [n. 984], p. 116.

1073 Decision 16/CMP.1, Annex, Paragraph 10.

1074 Decision 11/CP.7, Annex, Paragraphs 2 ff.

Additionally, the legal framework for the implementation of the LULUCF activities and for the application of forest-related projects under the CDM does not provide for an effective system of safeguards that would ensure the conservation of forest biodiversity and of natural forest ecosystems.

3. *Rights of ILCs and the Flexible Mechanisms*

The re/afforestation project activities under the CDM can potentially have adverse effects on ILCs. However, the KP and the relevant COP/CMP decisions do not contain effective and robust social safeguards.[1075] The pertinent provisions do not contain any direct references to them. At least, the KP indicates that both private and public entities can participate in the activities under the CDM.[1076] However, the actual ability of ILCs to participate in the re/afforestation projects activities under the CDM activities will depend on the internal legislation of the host party, which means that the rights of ILCs can be violated. The donor States are not obliged to consider the ILC-related issues and to monitor their conditions while participating in the projects.

As mentioned above, the Marrakesh Accords and the pertinent CMP decisions do not mention ILCs. Instead, they use a broad term "stakeholder", including public, groups and communities affected or potentially affected by the CDM projects.[1077] The "Stakeholders" are permitted to attend the meetings of the CDM Executive Board, but only if they are accredited at the UNFCCC. As ILCs are not directly mentioned, and as they rarely have an access to the UNFCCC bodies, this provision does not ensure their participation in the activities of the CDM Executive Board. Additionally, "the right to attend" does not automatically preclude "the right to vote".[1078] As it was already mentioned, the local stakeholders are invited to make comments on the proposed re/afforestation CDM project. However, the participants of the project only have to take "due account" of any comment received, and the DOE is not obliged to deny the validation of a project based on the summary of comments submitted by project participants.[1079] Moreover, it is not clear whether and how ILCs may be invited to make comments on the proposed CDM project. It will also depend on their legal status

1075 See *Carbon Market Watch*, The Clean Development Mechanism: Local Impacts of a Global System, October 2018, pp. 1 ff.

1076 KP, Article 12.9 in combination with Article 12.3(a).

1077 KP CMP Decision 5/CMP.1, Paragraph 12.

1078 Ibid.

1079 KP CMP Decision 5/CMP.1, Annex, Paragraph 15(d).

under the host party's legislation and on their ability to pursue their rights in the domestic legal system.[1080] It is also problematic that none of the provisions oblige the DOEs to deny the validation of the project in the case when the rights of ILCs are violated. Transparency can be further affected as project participants can select the DOE that will validate the proposed project activity. The existing CDM projects have already had negative impacts on ILCs.[1081] Thus, it is recommended that proposed projects should be validated both by independent experts and by the representatives of ILCs.

4. *Issue of Non-Permanence of Forest Carbon Stocks and Other Concerns*

During the negotiations of the KP, States also expressed their concerns in relation to the non-permanence of forest carbon stocks.[1082] It was argued that forests can bind carbon dioxide only for a relatively short period, which corresponds to the tree life span. After the tree is logged or when it falls victim to a forest fire, pest outbreak, or another natural hazard, it will likely release the stored carbon dioxide into the atmosphere in the process of decomposition. For this reason, some NGOs and developing States opposed the accounting of removals by sinks in national emission reductions of Annex I States. They argued that climate change mitigation should be achieved only by reducing emissions from industries and transports.[1083] This opinion was also supported by the EU-Members, although some observers argued that the position of the European States was motivated by their unwillingness to create favorable conditions for the USA, rather than by environmental concerns.[1084] Actually, the USA would particularly benefit from the accountability of LULUCF activities in their national emission reductions.[1085]

1080 Ibid.

1081 *Carbon Market Watch* [n. 1075].

1082 *Eikermann* [n. 984], p. 116.

1083 *Streck, Charlotte/Scholz, Sebastian M.*, The Role of Forests in Global Climate Change: Whence We Come and Where We Go, in: International Affairs, Volume 82, Issue 5, 2006, pp. 867 ff.; *Fearnside, Philip M.*, Saving Tropical Forests as a Global warming Countermeasure: An Issue that Divides the Environmental Movement, in: Ecological Economics, Volume 39, Issue 3, 2001, pp. 167 ff.

1084 *Fearnside* [n. 1083], pp. 171 ff.

1085 *Fearnside* [n. 1083], pp. 172 ff.

It can be argued that the concerns of the NGOs were predominantly unfounded and exaggerated, as the forest stocks in healthy forest ecosystems constantly enhance.[1086] As previously mentioned, carbon stocks in these forests are far more stable than in monoculture plantations that are prone to forest fires and pest outbreaks. In other words, the decomposition of the dead plant biomass constitutes an indispensable part of the natural forest ecosystem functioning, and it should be accepted that some amount of carbon dioxide constantly returns into the atmosphere only to be bounded again by growing and maturing trees together with the new amounts of CO_2. As long as a forest ecosystem remains stable, forest carbon stocks are also stable.[1087] The perception of carbon stocks in natural forest ecosystems as unstable ones is partially based on a lack of understanding of the natural dynamics in these forests. This means that the conservation of natural forests must be promoted simultaneously with the reduction of emissions from industry and transportation.

Another major concern is that if planted forests will be harvested in the future, the CO_2 stored in their biomass and forest soil can be released back into the atmosphere as soon as trees are logged.[1088] However, these estimations do not consider that the harvested timber will be processed and that wooden products can preserve the stored carbon for an indefinitely long time. Moreover, decision-makers should consider the non-carbon related functions of forest ecosystems. Planted forests need a long time to acquire the ability to fulfil multiple forest functions as effectively as natural forests. If diverse and local tree species are used in the course of a/reforestation, the planted forests will be able to fulfill multiple forest functions, but not as efficiently as natural ones. Thus, re/afforestation activities and forest conservation must be promoted for the purpose of climate change mitigation/adaptation, but only if the conservation of natural forests and forest biodiversity and near-natural re/afforestation is ensured.

Developing States were opposed to the CDM projects due to their sovereignty concerns and development issues. It was not clear if relevant activities would contribute to national socio-economic development or if the right of States to exploit their national resources will be violated. For example, Brazil expressed its concerns about the possible loss of its sovereign right to exploit the resources of the Amazon rainforests.[1089] However, the preservation of the Amazon rainforests

---

1086 See Subsection A.II.3 of Chapter 1.
1087 See Subsection A.II.3 of Chapter 1.
1088 *Eikermann* [n. 984], p. 116.
1089 *Fearnside* [n. 1083], pp. 167 ff.

and the promotion of their ecological functions can bring numerous benefits to Brazil and its population. The inclusion of the avoided deforestation as an eligible activity under the CDM might deliver additional revenues that might be invested in the conservation of Amazon Rainforests and in the development of local livelihoods. Thus, those projects that foresee the conservation of natural forests should not be seen as a disguise attempt to violate the sovereign rights of developing States.

Finally, the reduction of emissions from deforestation and the promotion of sinks achieved by the activities under the KP was merely insignificant.[1090] The biggest part of the Certified Emission Reductions from the CDM projects were achieved by reducing emissions of hydrofluorocarbon-23 (HFC-23) or by destroying this greenhouse gas, which has a particularly powerful global warming effect.[1091] Projects involving HFC-23 are far more profitable than activities in the forest sector.[1092] Thus, Kyoto Protocol with its flexible mechanisms could not significantly contribute to forest protection.

D. Emergence of REDD+

I. Initial Proposal of Papua New Guinea

On 17 May 2005, the representative of Papua New Guinea, H.E. Robert G. Aisi, presented his report at the UNFCCC Seminar of Governmental Experts held in Bonn, Germany.[1093] In his speech, he complained that the interests of Papua New Guinea and of other forest-rich nations were not considered by the KP. The treaty would fail to address the emissions from deforestation in tropics, and to provide developing States with the financial and technological resources needed for sustainable development. Even if a developing State would take efforts in order to reduce, halt, and reverse deforestation, it would not receive credits under the

1090 *Schulte zu Sodingen* [n. 973], pp. 207 f.; *Krohn* [n. 984], p. 135; Eikermann [n. 984], p. 126.

1091 See *Montzka, Stephen A./Kuijpers, Lambert/Battle, Mark O./Aydin, Murat/Verhulst, Kristal/Salzman, Eric S./Fahey, David W.*, Recent Increases in Global HFC-23 emissions, in: Geophysical Research Letters, Volume 37, 2010, p. 1.

1092 Ibid.

1093 UNFCCC Seminar of Governmental Experts, Statement by H. E. Robert G. Aisi, Ambassador of Papua New Guinea to the United Nations, Bonn, 17 May 2005. – Internet resource –https://unfccc.int/sites/default/files/sem_sup_papua_new_guinea.pdf [Last visit: 30.04.2023]; See also *Abidin* [n. 1009], pp. 20 f.

Kyoto Protocol.[1094] The main message was that forest-rich States deserved to be treated equally and to have access to the international carbon markets on a "fair and equitable basis". Such access would permit them to receive financial compensation for reducing emissions from tropical deforestation, which could contribute to the socio-economic development and eliminate poverty.[1095] In this context, Aisi referred explicitly to the most vulnerable groups of ILCs living in the rural areas of Papua New Guinea who legally own their forest areas.[1096]

In order to promote the role of reducing deforestation emissions in forest-rich States, Papua New Guinea proposed different approaches to the further development of the international legal regime for combating climate change.[1097] In accordance with one approach, the role of forest-rich tropical States could be promoted through the reform of the Marrakesh Accords.[1098] One disadvantage of this approach consists in that the Marrakesh Accords were adopted as a COP decision. As such, they can be regarded as a manifestation of a subsequent practice or a subsequent agreement in relation to the interpretation and the application of the UNFCCC according to Article 33.3(a–b) of the VCLT.[1099] However, the initial consensus that made the conclusion of the UNFCCC possible cannot be further developed by decisions of the COP to the same extent as by the adoption of a new Protocol to the Convention. The main difference is that a Protocol has to be signed and ratified by parties. This procedure provides for a more wide-going modification of the obligations established by the UNFCCC in comparison to the adoption of a decision by the COP. Consequently, the reform of the Marakkesh Accords would not remedy the lack of attention to forest issues in the UNFCCC and the Kyoto Protocol. At least, it would permit avoided deforestation as an eligible activity under the CDM. However, it would hardly resolve

1094 *Abidin* [n. 1009], pp. 20 f.

1095 Ibid.

1096 See also *Abidin* [n. 1009], p. 21.

1097 UNFCCC "Statement by H. E. Robert G. Aisi" [n. 1093], p. 2.

1098 Ibid.

1099 United Nations A/CN.4/671, Second report on subsequent agreements and subsequent practice in relation to the interpretation of treaties by Georg Nolte, Special Rapporteur, International Law Commission, Sixty-sixth session, Geneva, 5 May–6 June and 7 July–8 August 2014, 2014, pp. 43 f. – Internet resource – https://undocs.org/en/A/CN.4/671 [Last visit: 30.04.2023]; *Churchill, Robin R./Ulfstein, Geir*, Autonomous Institutional Arrangements in Multilateral Environmental Agreements: A Little-Noticed Phenomenon in International Law, in: The American Journal of International Law, Volume 94, Number 4, 2000, p. 626.

the sovereignty, legality, and transparency concerns of different potential project participants and would not deliver incentives for the implementation of forest projects under the CDM.

Another way to promote the deforestation emission reductions proposed by Papua New Guinea consisted in the amendment of the KP itself.[1100] This approach would have provided for the development of a legal framework permitting to account deforestation emission reductions in the context of the UNFCCC's regime. However, forest-rich tropical States were excluded from the Annexes to the KP. Thus, the inclusion of forest conservation activities in developing States would have required a profound reform of the KP. For this purpose, the incentives for the implementation of forest-related activities would have to be created, and developing States would have to be obliged to contribute to climate change mitigation. This obligation would have had to be reconciled with the principle of common but differentiated responsibilities, which would have essentially required a reform of the UNFCCC itself. Moreover, the Kyoto Protocol failed to establish a robust, transparent, and effective legal system for combating climate change in relation to not only the net forest emission reductions, but also the industrial and transport-related emissions in developed States.[1101] For this reason, the adoption of a new Optional Protocol on forests would, according to Aisi, be a better way to include the reduction of emissions from deforestation in tropics in the UNFCCC.[1102]

II. Propositions of Papua New Guinea to Resolve Technical Issues

The statement of Papua New Guinea included propositions addressing the concerns in relation to technical aspects of the forest-related activities, including additionality and leakage issues and non-permanence.[1103] According to these propositions, additionality could be addressed by establishing national deforestation baseline rates. These baselines would have to be precise and clear and follow precise targets. The baseline approach was subsequently incorporated in the REDD+-mechanism.[1104] National deforestation baseline rates can serve to

1100 UNFCCC "Statement by H. E. Robert G. Aisi" [n. 1093], p. 3.

1101 *Victor, David G.*, The Collapse of the Kyoto Protocol and the Struggle to Slow Global Warming, 2001, pp. 25 ff.

1102 UNFCCC, Statement by H. E. Robert G. Aisi [n. 1093], p. 2.

1103 Ibid.

1104 See Subsection D.IV.2.

verify whether the reduction of deforestation rates was really achieved in comparison to the business-as-usual scenario. Notwithstanding this, it will hardly permit to verify if the reduction of deforestation rates was really achieved by climate-related projects or occurred for other reasons. Indeed, the socio-economic and political factors that influence deforestation rates are very complex, and their contribution to the change of deforestation rates is difficult to analyze and calculate precisely. However, considering the crucial importance of forests for the achievement of the main objective of the UNFCCC, their conservation should be promoted despite these issues.

Concerning the problem of leakage, Papua New Guinea proposed to address emissions from deforestation on the national level.[1105] This idea assumed that the protection of forests on the national level would permit to avoid leakage. However, this approach disregarded the fact that leakage could occur not only on a national level but also on the international level. The reduction of deforestation rates in one State or region can cause an undesirable shift of deforestation into other States and regions. From this point of view, emissions from deforestation should be addressed on the global level. To address the issue of leakage, it can be proposed to establish an international monitoring board to ensure that the reduction of deforestation and the enhancement of re/afforestation rates occur on the global level. To mitigate the risk of leakage, forest projects for climate change mitigation/adaptation should be attractive for most forest-rich States. For this purpose, the incentives should be created for them to participate in these projects, and payments for the achieved emission reductions must be sufficient in order to promote the sustainable development in these States. Importantly, those policies that promote the conservation of forests and improved forest management, should also be adopted in forest-rich developed States. Otherwise, tropical developing States may perceive the new forest-related initiative under the UNFCCC as a disguised trade restriction.[1106] Thus, the risk of leakage can be tackled by promoting the conservation and sustainable use of forests all around the world.

1105 UNFCCC "Statement by H. E. Robert G. Aisi [n. 1093], p. 2.

1106 *Wong, Anny*, The Roots of Japan's International Environmental Policies, 2001, pp. 150 ff.

III. Development of REDD+

1. UNFCCC COP 13

The year 2006 did not bring substantial progress regarding the development of REDD+. Despite this, UNFCCC COP 13, held in Bali in 2007, managed to adopt the Bali Action Plan and other decisions dealing with the promotion of reducing emissions from deforestation in developing States.[1107] Despite the title of Decision 2/CP.13 referring only to reducing emissions from deforestation (RED), the document itself mentions emissions from forest degradation and the enhancement of forest carbon stocks through sustainable forest management. In addition, the Bali Action Plan also added "the role of conservation" of forests to the concept of RED.[1108] Thus, the new REDD+-mechanism was finally included in the agenda of international climate talks, with the second "D" in "REDD+" standing for forest degradation, and the "+" referring to "the role of conservation, sustainable management of forests and enhancement of forest carbon stocks in developing countries".[1109] Therefore, with the adoption of the Bali Action Plan and Decision 2/13, the concept of RED was transformed into REDD+, which represents a more comprehensive approach.[1110]

Decision 2/13 underlined the voluntary character of REDD+, meaning that developing States participate voluntarily in activities under the scope of the new mechanism, whereas the developed States provide the financial, technological, and scientific support and advice on a voluntary basis.[1111] Consequently, no measurable and legally binding targets concerning the implementation of REDD+ were established, and parties and stakeholders "in position to do so" did

1107 UNFCCC COP Decision 1/CP.13, "Bali Action Plan", in: United Nations FCCC/CP/2007/6/Add., in: Report of the Conference of the Parties on its thirteenth session, held in Bali from 3 to 15 December 2007, United Nations, 2008. – Internet resource – https://unfccc.int/resource/docs/2007/cop13/eng/06a01.pdf [Last visit: 30.04.2023]

1108 Decision 1/CP.13, Paragraph 1(b.iii).

1109 UNFCCC COP 2/CP.13, "Reducing emissions from deforestation in developing countries: approaches to stimulate action", in: United Nations FCCC/CP/2007/6/Add.1, Report of the Conference of the Parties on its thirteenth session, held in Bali from 3 to 15 December 2007, United Nations, 2008. – Internet resource – https://unfccc.int/resource/docs/2007/cop13/eng/06a01.pdf [Last visit: 30.04.2023]

1110 *La Viña, Antonio G. M./de Leon, Alaya/Barrer, Reginald Rex*, History and future of REDD+ in the UNFCCC: issues and challenges, in: *Voigt, Christina*, Research handbook on REDD+ and International Law, 2016, p. 16; *Maguire* [n. 1071], p. 160.

1111 Decision 2/CP.13, Paragraphs 1, 2, 5.

not assume obligation to financially support the new mechanism. The COP also encouraged all relevant stakeholders to promote the application of the IPCC's Good Practice Guidance for LULUCF and the indicative guidance contained in the Annex to its decision.[1112] Like the Good Practice Guidance, this last indicative guidance deals with the accounting of emission reductions.[1113] It can be evaluated positively that the document in question addresses the issue of leakage by indicating that, while assessing the projects implemented on subnational level, the associated displacement of emissions should be taken into consideration.[1114] The text does not contain any further environmental and social safeguards.

2. Copenhagen Accords

The subsequent COP UNFCCC 15, which took place in 2009 in Copenhagen, adopted a number of important decisions. Decision 4/CP.15 is dedicated to REDD+. This decision contains a methodological guidance for the implementation of REDD+-related activities in developing States.[1115]

Decision 4/CP.15 is characterized by comparatively strong wording, however, it did not change the voluntary character of REDD+. It requests developing States to identify the drivers of deforestation and forest degradation, the means to combat them, and the activities that promote the reduction of deforestation emissions and the enhancement and stabilization of forest carbon stocks.[1116] Developing States are also requested to apply the guidance and guidelines as developed by the IPCC.[1117] They also have to establish effective and transparent national forest monitoring systems and regional subsystems.[1118] However, these requirements are relativized by the reference to "national circumstances and

---

1112 Ibid, Paragraph 6.

1113 Ibid.

1114 Decision 2/CP.13, Annex, Paragraph 5.

1115 UNFCCC COP Decision 4/CP.15 "Methodological guidance for activities relating to reducing emissions from deforestation and forest degradation and the role of conservation, sustainable management of forests and enhancement of forest carbon stocks in developing countries", in: United Nations FCCC/CP/2009/11/Add.1, Report of the Conference of the Parties on its fifteenth session, held in Copenhagen from 7 to 19 December 2009, United Nations, 2010. – Internet resource – https://unfccc.int/resource/docs/2009/cop15/eng/11a01.pdf [Last visit: 30.04.2023]

1116 Ibid, paragraph 1(a).

1117 Ibid, paragraph 1(d).

1118 Ibid, Paragraph 1(d).

capabilities".[1119] It should be appreciated that the results of monitoring should be publicly available and appropriate for review.[1120] This provision permits to increase transparency and legal certainty, and to facilitate public control over the implementation of REDD+. Both the national forest monitoring systems and the forest reference emission levels and/or forest reference levels (FREL/FRL) must be transparent. However, the Decision does not clarify further, how transparency shall be ensured.[1121] Furthermore, the need for promoting SFM, the multiple functions of forests, and the conservation of biodiversity is recognized only in the preamble, but no specific provisions were included in the document.

The Copenhagen Accord contains further provisions for the implementation of REDD+. The text of the Accord characterizes the role of reducing emissions from deforestation and forest degradation as crucial and underlines the need to immediately establish the REDD+-mechanism and to promote the mobilization of resources from developed States.[1122] Importantly, the accord has further stipulated that "new and additional, predictable and adequate resources" must be provided for various climate-related purposes, including the implementation of REDD+. The Green Climate Fund (GCF) was established as the main operating entity of the Convention's financial mechanism. The measurable target was established to provide 30 billion USD by 2012, and 100 billion USD by 2020, to finance different climate change mitigation/adaptation activities, including REDD+. However, since no legally binding commitments to provide financial resources were established, the GCF fails to raise the announced funding.[1123]

As far as the present context is concerned, the Copenhagen Accord signalized a qualitative shift in the international climate change mitigation policies. It promoted the role of forest ecosystems as sinks of greenhouse gases. For the first time, climate change mitigation activities in developing States were addressed. Despite this, the documents adopted in Copenhagen did not deliver sufficiently precise policies, procedures, and tools for the implementation of REDD+.

1119 Ibid, paragraph 1(c).

1120 Ibid, paragraph 1(d.iii).

1121 See also *La Viña/de Leon/Barrer* [n. 1110], p. 17; *Abidin* [n. 1009], p. 26.

1122 UNFCCC COP Decision 2/CP.15 "Copenhagen Accord", in: United Nations FCCC/CP/2009/11/Add.1, Report of the Conference of the Parties on its fifteenth Session, held in Copenhagen from 7 to 19 December 2009, United Nations, 2010. – Internet resource – https://unfccc.int/resource/docs/2009/cop15/eng/11a01.pdf [Last visit: 30.04.2023]

1123 Section D.IV.7 of this Chapter.

3. From the Cancun Agreement to the Warsaw Framework for REDD+

UNFCCC COP 16, held in Cancun in 2010, further contributed to the development of REDD+. Particularly, it adopted comparatively detailed regulations in relation to the development and implementation of the new mechanism. These regulations include a list of necessary infrastructural elements to be established in developing States.[1124] Accordingly, developing States have to address a number of issues related to REDD+. They should address the drivers of deforestation and the lack of secure land tenure rights, and promote the environmental and social safeguards of paragraph 2, Appendix 1 of the Cancun Agreement.[1125] The safeguards deal with the risks for natural forests, forest biodiversity and local communities resulting from the implementation of REDD+.[1126] The establishment of the safeguards for REDD+ is another significant achievement of the Cancun Agreement.[1127] Thus, the Cancun Agreement delivered a comprehensive legal framework for REDD+. However, certain issues still had to be clarified. For example, the guideline for the development of the FREL/FRL was absent.

Several decisions related to REDD+ were also adopted by COP 17 in Durban in 2011. Decision 2/CP.17 further developed the international legal framework for REDD+, particularly regarding its implementation.[1128] It is noteworthy that the document in question directly links the REDD+-mechanism to the promotion of other development goals, including the alleviation of poverty and the strengthening of ecosystem resilience. Decision 3/CP.17, by which Green Climate Fund was launched, stipulates that the Fund will fully finance all incremental costs related to the implementation of REDD+.

1124 See Subsection D.IV.1 of this Chapter.

1125 UNFCCC COP Decision 1/CP.16 "Cancun Agreement", in: United Nations FCCC/CP/2010/7/Add.1, Report of the Conference of the Parties on its sixteenth Session, held in Cancun from 29 November to 10 December 2010, United Nations, 2011, Paragraph 72. – Internet resource – https://unfccc.int/resource/docs/2010/cop16/eng/07a01.pdf [Last visit: 30.04.2023]

1126 See Section V of this Chapter.

1127 UNFCCC COP Decision 1/CP.16 "Cancun Agreement", Appendix 1, Paragraph 2.

1128 UNFCCC COP Decision 2/CP.17 "Outcome of the work of the Ad Hoc Working Group on Long-term Cooperative Action under the Convention", in: United Nations FCCC/CP/2011/9/Add.1, Report of the Conference of the Parties on its seventeenth Session, held in Durban from 28 November to 11 December 2011, United Nations, 2012, Paragraphs 66 ff. – Internet resource – https://unfccc.int/resource/docs/2011/cop17/eng/09a01.pdf [Last visit: 30.04.2023]

The progress achieved in relation to REDD+ at UNFCCC COP 18 was modest. The REDD+-relevant issues were included in Decision 1/CP.18. Particularly, it was decided to "undertake a work programme on results-based finance" for REDD+ in 2013.[1129] The programme was deemed to promote fundraising and to enhance the effectiveness of finance for REDD+. For this purpose, the programme had to address different options, including the ways and means for the transfer of results-based payments, the ways to incentivize non-carbon benefits, and the ways to improve the coordination of finance.[1130] Among these issues, the attempt to incentivize non-carbon benefits is particularly important for the promotion of holistic approach to the forest conservation.[1131]

Finally, the 19th session of the UNFCCC COP has significantly contributed to the development of REDD+. The detailed Warsaw Framework for REDD+ is one of the most important outcomes of this session, as it finally contains detailed regulations for the implementation of REDD+. However, it does not contribute substantially to the promotion of compliance and does not deliver incentives for the implementation of REDD+.[1132]

4. *Paris Agreement and REDD+*

The Paris Agreement is deemed to guide international climate change mitigation/adaptation policies after the commitment period of the KP expired in 2012, and after the parties failed to agree on a second commitment period. As the main outcome of the Paris negotiations in 2015, the Paris Agreement is a binding international treaty in terms of the VCLT.[1133] Despite this fact, the treaty includes both mandatory and non-mandatory provisions, meaning that

---

1129 UNFCCC COP Decision 1/CP.18, "Agreed outcome pursuant to the Bali Action Plan", in: United Nations FCCC/CP/2012/8/Add.1, Report of the Conference of the Parties on its eighteenth Session, held in Doha from 26 November to 8 December 2012, United Nations, 2013. – Internet resource – https://unfccc.int/resource/docs/2012/cop18/eng/08a01.pdf [Last visit: 30.04.2023]

1130 Decision 1/CP.18, Paragraph 29.

1131 Decision 1/CP.18, Paragraph 34.

1132 See also *Lang, Chris*, The Warsaw Framework for REDD Plus: The Decision on Summary of Information on Safeguards, in: REDD Monitor, 2013. – Internet resource – http://www.redd-monitor.org/2013/12/17/thewarsaw-framework-for-redd-plus-the-decision-on-summary-of-information-on-safeguards/ [Last visit: 30.04.2023]

1133 *Bodansky, Daniel*, The Legal Character of the Paris Agreement, in: RECIEL, Volume 25, Issue 2, 2016, p. 142.

not every provision of the Paris Agreement establishes a legally binding obligation.[1134] The contribution of the Paris Agreement to climate change mitigation and, as a side effect, to the conservation of forests will depend not only on the legally binding force of its provisions but also on their clearness, preciseness, and non-ambiguity.[1135] Unlike the KP that contains precise emission reduction goals, the Paris Agreement obliges Parties to develop, submit, and maintain their individual nationally determined contributions (NDCs).[1136] By doing this, the Paris Agreement promotes a bottom-up approach to the climate change regulation.[1137] By promoting a bottom-up approach to climate change mitigation, the treaty provides a significant degree of flexibility in relation to its implementation. It should also be noted that not enough incentives are created for States to foster their forest emission reduction efforts. From this background, it can be argued that the Paris Agreement can hardly be seen as a sufficient solution to the global challenge of climate change.[1138] At the same time, considering the standstill of international negotiations prior to Paris, its adoption ought to be regarded as an important milestone in multilateral climate diplomacy.[1139]

The negotiations were inspired by the recognition of the fact that the emission reduction commitments under the KP had proven to be insufficient in achieving the main objective of the UNFCCC.[1140] Moreover, the lack of emission reduction commitments for developing States under the KP was seen as a major weakness.[1141] This is why the necessity emerged to establish emission reduction commitments for all States, while taking into account their differing

---

1134 Ibid.

1135 Ibid.

1136 Articles 3 and 4.2.

1137 See *Rayner, Steve*, How to Eat an Elephant: A Bottom-up Approach to Climate Policy, in: Climate Policy, Volume 10, Issue 6, 2010, pp. 615 ff.

1138 See *Rajamani, Lavanja*, Ambition and Differentiation in the 2015 Paris Agreement: Interpretative Possibilities and Underlying Politics, in: International and Comparative Law Quarterly, Volume 65, 2016, pp. 493 f.

1139 Ibid.

1140 *Klein, Daniel/Carazo, María Pía/Doelle, Meinhard/Bulmer, Jane/Higham, Andrew*, The Paris Agreement on Climate Change: Analysis and Commentary, 2017, pp. 51 ff.; *Savaresi, Annalisa*, The Paris Agreement: A New Beginning?, in: Journal of Energy & Natural Resources Law, Volume 34, Issue 1, 2016, pp. 16 f.; *Maljean-Dubois, Sandrine*, The Paris Agreement: A New Step in the Gradual Evolution of Differential Treatment in the Climate Regime?, in: RECIEL, Volume 25, Issue 2, 2016, p. 151.

1141 *Klein/Carazo/Doelle/Bulmer/Higham* [n. 1140], pp. 52 f.; *Savaresi* [n. 1140], pp. 17 f.; *Maljean-Dubois* [n. 1140], p. 151.

national circumstances and capabilities.[1142] Consensus was achieved that the legally binding emission reduction targets set for Annex I States must be replaced by voluntary contributions of each State.[1143] Against the backdrop of ongoing deforestation and forest degradation that is a major source of CO_2 emissions, especially in the tropics, the Paris Agreement had to include policies to promote the conservation of forests, re/afforestation, and SFM. After the re/afforestation projects under the CDM had proven to be a failure, the REDD+-mechanism was the main major instrument for reducing deforestation emissions and for the enhancement of forest carbon stocks. Thus, the circumstances were favorable for the inclusion of REDD+ in the Paris Agreement.

During the negotiations of the Paris Agreement, some observers argued that, in order to promote combating climate change mitigation, a new document had to go beyond the framework of the UNFCCC.[1144] One of the statements was that the UNFCCC was not adequately designed to combat the drivers of climate change.[1145] Despite all its benefits, the principle of Common but Differentiated Responsibilities as included in the UNFCCC did not allow the international climate change regime to be fully operational.[1146]

The fact that the Paris Agreement was negotiated and adopted under the umbrella of the UNFCCC limited its ability to incorporate new approaches to climate change mitigation/adaptation. As the UNFCCC failed to valorize the corresponding role of forest ecosystems, the Paris Agreement could hardly assign a higher priority to reducing emissions from deforestation. However, a profound reform of the UNFCCC was an unrealistic scenario, and such an option never emerged during the negotiations.

The Paris Agreement is characterized by several innovations. It is characterized by its so-called bottom-up approach. At it was mentioned above, this approach means that States are obliged to determine their voluntary contributions to combating climate changes and submit it as "nationally determined

---

1142 *Falkner, Robert*, The Paris Agreement and the New Logic of International Climate Politics, in: International Affairs, Volume 92, Issue 5, 2016, p. 1111.

1143 Ibid.

1144 *Scott, Schirley V.*, Does the UNFCCC Fulfil the Functions Required of a Framework Convention? Why Abandoning the United Nations Framework Convention on Climate Change Might Constitute a Long Overdue Step Forward, in: Journal of Environmental Law, Volume 27, Issue 1, 2015, pp. 71 f.

1145 Ibid, p. 80.

1146 Ibid, p. 80.

contributions" (NDCs).[1147] It is noteworthy that developing States also have to develop and implement their NDCs.

REDD+ is explicitly mentioned in Article 5.2 of the Paris Agreement. However, its wording has maintained the vague and ambiguous character of the previous Decisions of the COP. The Parties to the UNFCCC are only encouraged to implement REDD+ in developing States, including through results-based payments.[1148] The Paris Agreement stipulates that relevant guidelines and decisions adopted by the COP must be applied in this process, but it does not further develop the existing legal framework for REDD+.[1149] In particular, it does not strengthen the weak environmental and socio-economical safeguards for the implementation of this mechanism.[1150] While the explicit reference to the existing legal framework for REDD+ in the Paris Agreement delivers a political signal that there is a commitment to the implementation of the new mechanism, the provision concerned does not create any legally binding commitments concerning whether and how REDD+ must be implemented. Importantly, the Paris Agreement foresees the development and implementation of the Joint Mitigation and Adaptation (JMA)-mechanism as an alternative to REDD+. The JMA mechanism represents another innovative tool for reducing emissions from deforestation that is designed to promote environmental and social issues more effectively when compared to REDD+.[1151]

According to the Paris Agreement, all Parties have to "take actions" to conserve and enhance the sinks and reservoirs of greenhouse gases, including forests.[1152] This provision serves as a legal basis for the inclusion of LULUCF activities in NDCs. However, no relevant obligations are established.

REDD+ is currently included in a bulk of NDCs submitted by developing States.[1153] 56 of the 162 submitted NDCs foresee the application of this mechanism.[1154] Despite many concerns about the lack of environmental safeguards and of adequate and predictable finance, the adoption of the Paris Agreement

1147 *Klein/Carazo/Doelle/Bulmer/Higham* [n. 1140], pp. 92 ff.

1148 *Klein/Carazo/Doelle/Bulmer/Higham* [n. 1140], pp. 170 ff.

1149 Article 5.2 of the Paris Agreement.

1150 See Section V of this Chapter.

1151 See Subsection D.IV.6 of this Chapter.

1152 Article 5.1 of the Paris Agreement.

1153 *Hein, Johannes/Guarin, Alejandro,* Deforestation and the Paris Climate Agreement: An Assessment of REDD+ in the National Climate Action Plans, in: Forest Policy and Economics, Volume 90, 2018, p. 8.

1154 Ibid, p. 9.

has already contributed to the promotion of the new mechanism.[1155] The efforts to implement it are expected to be fostered in the future. For this reason, the provisions regulating the implementation of REDD+ must be further analyzed.

IV. Legal Provisions for Implementing REDD+

1. *Infrastructural Elements for REDD+ and Three Phases of Implementation*

UNFCCC COP Decisions provide a legal framework for the development and implementation of REDD+. The first step for any State seeking to participate in REDD+ projects is to identify the drivers of deforestation leading to CO_2-emissions and the means to combat them, as well as to determine forest-related activities under its jurisdiction that can contribute to climate change mitigation.[1156] After that, it must establish infrastructural elements for the implementation of REDD+. According to the Cancun Agreements, these infrastructural elements include a national strategy or action plan, a national FREL/FRL, and a robust and transparent national forest monitoring system for the monitoring and reporting of REDD+ activities.[1157] Furthermore, they include a system for providing information on how the environmental and social safeguards contained in Appendix I to the Cancun Agreements are being "respected and addressed".[1158] Altogether, these elements can provide an appropriate infrastructural base for the implementation of REDD+.

It is noteworthy that the establishment of all these crucial infrastructural elements in developing States must be supported by adequate and predictable financial and technological support.[1159] However, developed States have so far opposed any legally binding commitments to provide the support needed. Furthermore, the requirement to establish the crucial elements of REDD+ infrastructure is further relativized by the reference to the "national circumstances and respective capabilities" of developing States.[1160]

The implementation of REDD+ includes three main phases. It begins with capacity building and the development of national action plans, policies and

1155 Ibid, p. 10.
1156 Decision 4/CP.15, Paragraph 1.
1157 Decision 1/CP.16, Paragraph 71.
1158 Ibid.
1159 Ibid, Paragraphs 71, 74.
1160 Ibid, Paragraph 71.

measures. During the second phase, national action plans, policies and measures, which can already involve results-based demonstration activities, must be implemented. The third phase consists of results-based actions, which should be "fully measured, reported and verified".[1161] A developing State is able to bypass one of the three phases, if its national circumstances and available technical resources permit it to do so.[1162] The implementation of the three phases depends on the level of support received. For this purpose, developed States should provide financial resources, technical, technological and scientific support and advice.[1163]

In order to receive results-based payments for the achieved reduction in emissions from forests, developing States have to ensure that the relevant action under REDD+ are fully measured, reported, and verified, and that all infrastructural elements were established.[1164]. However, these provisions do not remedy the lack of specificity and clarity regarding the reference to national circumstances in the afore-mentioned provisions.

2. *FREL/FRL and the Definition of Forests*

A national FREL/FRL should be established in order to assess the progress achieved in every State while implementing REDD+.[1165] Although the COP did not clarify the difference between FREL and FRL, it is widely acknowledged that FREL encompasses only emissions from deforestation and forest degradation, whereas FRL also includes removals by sinks achieved through the enhancement of forest carbon stocks and sustainable forest management.[1166] This means that FRL can be a better choice for REDD+, as it is a more comprehensive indicator that allows to consider multiple forest functions.

1161 Ibid, Paragraph 73.

1162 Ibid.

1163 Ibid, Paragraphs 71, 73.

1164 Decision 2/CP.17, Paragraph 64.

1165 UNFCCC COP Decision 12/CP.17, Guidance on systems for providing information on how safeguards are addressed and respected and modalities relating to forest reference emission levels and forest reference levels as referred to in decision 1/CP.16, in: United Nations FCCC/CP/2011/9/Add.2, Report of the Conference of the Parties on its seventeenth Session, held in Durban from 28 November to 11 December 2011, United Nations, 2012, Paragraph 7. – Internet resource – https://unfccc.int/resource/docs/2011/cop17/eng/09a02.pdf [Last visit: 30.04.2023]

1166 *FAO*, Emerging approaches to Forest Reference Emission Levels and/or Forest Reference Levels for REDD+, FAO, United Nations, 2015, p. 4.

Relatively few provisions dealing with the establishment of FREL/FRL are contained in the decisions of the COP. It is stipulated, among other things, that national FREL/FRL should be expressed in tons of carbon dioxide equivalent per year,[1167] and must be established transparently, in accordance with the historical data contained in national greenhouse gas inventories.[1168] However, a State is allowed to take due regard to its national circumstances.[1169] It can be evaluated positively that States are invited to submit the information on how they establish FREL/FRL including details on national circumstances, to the COP.[1170] They should also submit details on how national circumstances are taken into account, but only "if adjusted".[1171]

Moreover, relevant guidelines developed by the IPCC should be used while submitting the information in question. States must also include the definition of forests in their reports.[1172] If the definition of forests applied for the purpose of a REDD+-project defers from the definition that is commonly used by the national forest authority and in the communications with other international institutions, the State has to explain why this definition was chosen.[1173] It must be observed critically that the COP did not agree on any definition of forests but rather authorized States to develop their own definition. An internationally recognized definition of forests could contribute to transparency and legal certainty. It can be argued that a holistic definition that would underline the intrinsic value of forests and their multiple ecological functions is needed. However, considering sovereignty concerns and the desire of States to extend the scope of application of REDD+ on all forestland categories, if any REDD+-relevant definition of forests were developed, it would not go beyond pure quantitative indicators such as the minimal tree height and minimal tree crown cover. This means that it would not contribute significantly to the protection of forest biodiversity. This absence of an adequate international forest definition that would go beyond quantitative indicators is one of serious shortcomings not only of the international legal framework for REDD+, but also of the international legal regime for forests as such.

1167 Ibid.

1168 Decision 12/CP.17, Paragraph 8; Decision 4/CP.15, Paragraph 7.

1169 Decision 4/CP.15, Paragraph 7.

1170 Decision 12/CP.17, Paragraph 9.

1171 Ibid, Paragraph 9.

1172 Ibid, Annex, Paragraph 1(d).

1173 Ibid.

Decision 12/CP.17 did not specify further upon which criteria the UNFCCC bodies should assess the information on FREL/FRL submitted by Parties. It is also unclear, which consequences it will have for a State, if the assessment will reveal that national FREL/FRL are not adequate to be applied for REDD+.

3. National Forest Monitoring System

REDD+ furthermore requires the establishment of National Forest Monitoring Systems (NFSM). Decision 11/CP.19 contains the relevant modalities. According to it, the establishment of NFSM is subject to the availability of financial and technical resources.[1174] However, in the absence of a legal obligation to provide these resources, the provision in question is a declaratory one.[1175] NFSM can be developed based on the existing forest monitoring systems, which enables the involvement of existing forest-related governmental institutions and monitoring agencies in the implementation of REDD+.[1176] Decision 11/CP.19 stipulates further that NFSM should be sophisticated enough in order to facilitate the assessment of different types of forests, including natural forests.[1177] It must be seen critically that the Warsaw Framework, a regulatory framework for the implementation of REDD+, does not explain further upon which criteria the conditions of forests should be assessed and monitored. Considering that the provisions in question refer to national circumstances, it can be concluded that States enjoy a significant level of flexibility while determining these criteria. States are expected to follow the IPCC guidelines, but only "as appropriate". Moreover, as mentioned above, the guidelines concerned only deal with technical issues, such as the estimation and monitoring of changes of forest carbon stocks and forest-related emissions.

The Warsaw Framework also underlines the evaluation and monitoring of carbon stocks must be undertaken. This means that NFSM are likely to incorporate the same criteria for the assessment of the conditions of all types of forestlands including natural forests and tree plantations. Moreover, the generally used

---

1174 UNFCCC COP Decision 11/CP.19 "Modalities for national forest monitoring systems", in: United Nations FCCC/CP/2013/10/Add.1, Report of the Conference of the Parties on its nineteenth session, held in Warsaw from 11 to 23 November 2013, United Nations, 2014, Paragraph 1. – Internet resource – https://undocs.org/FCCC/CP/2013/10/Add.1 [Last visit: 30.04.2023]

1175 See Subsection D.IV.7 of this Chapter.

1176 Decision 11/CP.19, Paragraph 4(a).

1177 Ibid.

quantitative indicators, including tree height, tree crown cover and the estimation of carbon stocks, are likely to be applied. Consequently, a national NFSM is not adequate for the conservation of forest biodiversity and for the promotion of multiple forest functions.

On the other hand, the existing reference to the above-mentioned safeguards in Decision 11/CP.19 can potentially shift the emphasis to the promotion of environmental issues.[1178] The COP acknowledged that NFSM may sustain national systems for submitting information in relation to REDD+-relevant safeguards. Nevertheless, States only "may" apply NFSM for sustaining national systems for submitting this information; the COP avoided using the stronger modal verb "shall". The provision in question is further relativized by the weak wording "as appropriate". In other words, there is no direct link between NFSM and the environmental safeguards. This approach can result in the lack of attention to environmental issues in the process of implementing REDD+. Consequently, more precise regulations for NFSM are needed in order to promote its effective application for monitoring the conditions of forests.

4. *Technical Assessment of a FREL/FRL*

Decision 13/CP.19 contains "guidelines and procedures for the technical assessment" of submissions of Parties on FREL/FRL. The technical assessment is designed to serve as a tool for the evaluation of transparency, adequacy, accuracy and consistency of establishing FREL/FRL, in particular its consistency with the information included in national greenhouse gas inventories.[1179] Furthermore, it serves to check whether and how the historical data was considered in FREL/FRL.[1180] The technical assessment also serves to verify whether all carbon stocks, forest-related activities, and greenhouse gases were included in FREL/FRL. In the case where some of them are not included, an explanation for their omission should be provided.[1181]

1178 Ibid, Paragraph 5.

1179 UNFCCC COP Decision 13/CP.19 "Guidelines and procedures for the technical assessment of submissions from Parties on proposed forest reference emission levels and/or forest reference levels", in: United Nations FCCC/CP/2013/10/Add.1, Report of the Conference of the Parties on its nineteenth session, held in Warsaw from 11 to 23 November 2013, United Nations, 2014 Annex, Paragraph 1. – Internet resource – https://undocs.org/FCCC/CP/2013/10/Add.1 [Last visit: 30.04.2023]

1180 Ibid, Annex, Paragraph 2(b).

1181 Ibid, Annex, Paragraph 2(f).

However, the positive impact of the technical assessment of FREL/FRL on the development and implementation of REDD+ is likely to be insignificant. The assessment team is only authorized to provide recommendations for the improvement of FREL/FRL, including identifying areas where further capacity building is necessary.[1182] However, the assessment team is not authorized to make any judgement on policies used for the elaboration of FREL/FRL.[1183] Thus, the assessment team is more likely to serve as a kind of technical advisory body than as a tool for the promotion of compliance within the international legal framework for REDD+. The respective provisions suffer from a lack of holistic and comprehensive approach that would promote the environmental integrity of forests. It is not clear whether results-based payments will still be performed in case a State neither applies a holistic definition of forests in its FREL/FRL nor distinguishes between natural forests and monoculture plantations. It is problematic that the failure to consider multiple functions of forests may result in serious damage or threat to natural forests and their biodiversity.

5. *Modalities for Measuring, Reporting, and Verifying*

A State that seeks to receive results-based payments has to ensure that REDD+-activities are "measured, reported and verified".[1184] The Parties to the Convention have to identify drivers of deforestation and the means to combat them and to establish NFSM, but no details on criteria and indicators for measuring, reporting, and verifying is provided.[1185]

Decision 14/CP.19 contains further general statements concerning transparency and consistency with an established national FREL/FRL.[1186] Furthermore, it specifies that forest-related emissions by sources and removals by sinks must be expressed in the "tons of carbon dioxide equivalent per year".[1187] It further

1182 Ibid, Annex, Paragraphs 3, 4.

1183 Ibid, Annex, Paragraph 4.

1184 UNFCCC COP Decision 14/CP.19, "Modalities for measuring, reporting and verifying", in: United Nations FCCC/CP/2013/10/Add.1, Report of the Conference of the Parties on its nineteenth Session, held in Warsaw from 11 to 23 November 2013, United Nations, 2014, Paragraph 1. – Internet resource https://undocs.org/FCCC/CP/2013/10/Add.1 [Last visit: 30.04.2023]

1185 Decision 4/CP.15, Paragraph 1(a).

1186 Decision 14/CP.19, Paragraph 2.

1187 Ibid, Paragraph 4.

indicates that developing States must submit biennial update reports that contain the information relevant to REDD+, including changes in forest area.[1188]

It can be evaluated positively that changes in forest areas also must be reported, but it can be problematic that States do not have to differentiate between natural forests and monoculture plantations in their reports. Noteworthy, States have to demonstrate that they used the same methodology to estimate the results of the implementation of REDD+ and to establish FREL/FRL.[1189] By applying the same methodology, States can enhance the transparency and accurateness of their estimations. The information provided by Parties should be analyzed by a technical team of experts.[1190] Upon request of the State hosting a REDD+ project, two experts on forest-related issues from the UNFCCC roster of experts have to be included in the technical team of experts.[1191] In the end, despite the lack of clarity on environmental issues, procedures for measuring, reporting, and verifying are adequate for the development and implementation of REDD+. However, with the vagueness of provisions on FREL/FRL in mind, it is submitted that the lack of transparency in relation to reporting on achieved emission reductions can hinder the implementation of this mechanism.

Noteworthy, the legal framework for REDD+ underlines the need for addressing the drivers of deforestation and forest degradation.[1192] As already pointed out, States that participate in REDD+ have to identify drivers of deforestation and the means to combat them. The COP did not go beyond the statement of the importance of addressing the drivers of deforestation in the process of the implementation and development of national strategies and action plans.[1193] States were only encouraged to take actions to reduce the negative impact of deforestation and to exchange relevant information.[1194] Consequently, the COP neither established a legal duty to address the drivers of deforestation, nor promoted the relevant voluntary commitments of Parties, nor delivered precise guidelines in

---

1188 Ibid, Paragraph 6.

1189 Ibid, Annex, Paragraph 3.

1190 Decision 14/CP.19, Paragraph 11 in conjunction with 2/CP.17, Annex IV, Paragraph 4.

1191 Decision 14/CP.19, Paragraph 10.

1192 UNFCCC COP Decision 15/CP.19, "Addressing the drivers of deforestation and forest degradation", in: United Nations FCCC/CP/2013/10/Add.1, Report of the Conference of the Parties on its nineteenth Session, held in Warsaw from 11 to 23 November 2013, United Nations, 2014, Paragraphs 1 ff. – Internet resource – https://undocs.org/FCCC/CP/2013/10/Add.1 [Last visit: 30.04.2023]

1193 Decision 15/CP.19, Paragraph 1; Desicion 1/CP.16, Paragraphs 2 and 6.

1194 Decision 15/CP.19, Paragraph 3.

this regard. It can be proposed to apply results-based payments for REDD+ in order to combat the drivers of deforestation. However, the COP has so far not adopted any such provision. The need to apply results-based payments for the conservation of forests can be derived from the concept of SFM, which foresees, among other things, the utilization of financial resources generated by forest management for the promotion of forest conservation. However, it is problematic that SFM is not an established principle of international law, and no legal duty to protect forests is established as of now. In the absence of legally binding provisions on the issue, States are not obliged to apply results-based payments to combat the drivers of deforestation.

6. *Joint Mitigation and Adaptation Approaches*

In the process of development of REDD+, the concept of Joint Mitigation and Adaptation Approaches for the Integral and Sustainable Management of Forests (JMA) emerged. It can be applied instead of the conventional REDD+ scheme that incorporates results-based payments. The idea of these mechanisms originates from proposal of Bolivia that was presented at UNFCCC COP 17 in Durban. The aim of the proposal consisted of the promotion of the "integrated and sustainable management of forests and forest lands".[1195] Bolivia expects that this mechanism will promote mitigation and adaptation activities on different levels, by enhancing the role of forests and forest-dependent people.[1196] It is supposed to deliver financial resources for promoting national efforts dedicated to

---

1195 Submission by the Plurinational State of Bolivia "Joint mitigation and adaptation mechanism: "Sustainable Forest Life", December 2011. – Internet resource – https://unfccc.int/files/meetings/ad_hoc_working_groups/lca/application/pdf/submission_bolivia_redd.pdf [Last visit: 30.04.2023]; See also: UNFCCC, The Joint Mitigation and Adaptation Mechanism for the Integral and Sustainable Management of Forests (JMA), December 2014. – Internet resource – https://unfccc.int/files/cooperation_and_support/financial_mechanism/standing_committee/application/pdf/annex_1._the_joint_mitigation.pdf [Last visit: 30.04.2023]; *Müller, Robert/Pacheco Pablo/Montero, Juan Carlos*, The context of deforestation and forest degradation in Bolivia: Drivers, agents and institutions, Center for International Forestry Research (CIFOR), Occasional Paper 108, 2014, pp. 49 f.; *Klein/Carazo/Doelle/Bulmer/Higham* [n. 1140], p. 174.

1196 UNFCCC, Submission by the Plurinational State of Bolivia "Joint mitigation and adaptation mechanism: Sustainable Forest Life". – Internet resource – https://unfccc.int/files/meetings/ad_hoc_working_groups/lca/application/pdf/submission_bolivia_redd.pdf [Last visit: 30.04.2023]

the protection and sustainable management of forests. The idea behind the JMA mechanism was to redirect financial and technological resources coming from public funds of developed States in the promotion of the mitigation and adaptation role of integrally and sustainably managed forests. The bedrock of the JMA mechanism consists of the avoidance of the conversion of forest ecosystems or of their integral parts into tradable commodities.[1197]

It is particularly underlined that these activities can contribute to the long-term sustainability of REDD+.[1198] In order to apply the JMA, a State has to identify the drivers of deforestation and the means by which to combat them, and to establish a NSFM. Moreover, the provisions in relation to the REDD+-relevant safeguards are to be applied in this case. The development and implementation of alternative policy approaches requires international support. In order to receive such support, the State in question "may", but not "must", undertake certain activities.[1199] Inter alia, the State concerned is expected to develop national strategies or action plans and identify the spheres and activities, where technological and financial support is needed.[1200] Furthermore, it is expected to develop proposals describing the positive effects of alternative policy approaches on activities within the scope of REDD+. It may also consider the results of the implementation of alternative policy approaches and the areas for improvement.[1201] The legal framework for REDD+ does not deliver precise provisions on financing alternative policy approaches.[1202] The Green Climate Fund (GCF), other organizations and foundations serving as entities financing result-based payments for REDD+ are only "encouraged to continue" to financially support the JMA.[1203] Consequently, these activities are not automatically eligible to funding. It is also not clear, which part of funds can be allocated for this scope. Noteworthy, the JMA is non-market-based, meaning that market mechanisms will not be used for supporting its development and implementation. Therefore, the JMA can provide a more sustainable funding, which is not influenced by the fluctuations

1197 Ibid.

1198 Ibid.

1199 UNFCCC COP Decision 16/CP.21, in: United Nations FCCC/CP/2015/10/Add.3, Report of the Conference of the Parties on its twenty-first session, held in Paris from 30 November to 13 December 2015, United Nations, 2016, Paragraph 5. – Internet resource – https://undocs.org/FCCC/CP/2015/10/Add.3 [Last visit: 30.04.2023]

1200 Ibid, Paragraph 5(a).

1201 Ibid, Paragraph 5(b).

1202 Ibid, Paragraph 5(c).

1203 Ibid, Paragraph 6.

in market prices. On the other hand, in situations where there is a lack of available financial resources, market-based approaches should also be applied, at least as an interim measure. Independently from that, it would be desirable to ensure that the financial resources that are provided are adequate in order to evaluate in monetary terms multiple benefits of forests, and to create incentives for the conservation of forest ecosystems.

The JMA can potentially have a more positive outcome for the protection of forests and for the promotion of local livelihoods in comparison to the conventional scheme for the implementation of REDD+, where deforestation emission reductions and results-based payments play a crucial role.

In the case of conventional REDD+ projects, the emphasis is put on reducing emissions and enhancing carbon stocks, whereas the promotion of ecological and social benefits of forests play a secondary role and can be overlooked. In contrast, the JMA is dedicated to the achievement of SFM, while enhancing the role of forest ecosystems in climate change mitigation and adaptation.

Consequently, the JMA is an important initiative that possesses a higher potential to promote a holistic approach to the conservation forests. However, its implementation can be hindered by the lack of incentives and by the vagueness of the relevant provisions on finance. Whereas in the case of REDD+ the option of market-based financing can potentially enhance the availability of funding, the lack of legally binding commitments to support the JMA can compromise the implementation of this mechanism.

7. REDD+ Finance

During CBD COP 15, developed States committed themselves to raise at least 100 billion USD of funds by the year 2020 to meet the needs of developing States.[1204] The Green Climate Fund (GCF) was established in order to manage these financial resources.[1205] It was further underlined that these funds can come from a wide range of sources, including "public and private, bilateral and multilateral, including alternative sources of finance".[1206] Consequently, multiple sources of finance are foreseen. However, developed States opposed to any legally binding commitments to provide financial resources.

1204 Decision 2/CP.15, Paragraph 8.
1205 Ibid.
1206 Ibid.

In the Cancun agreement, developing States reaffirmed their political commitment to mobilize 100 Billion USD by 2020 for the Green Climate Fund.[1207] However, once again, no incentives were created to provide financial resources, and no legally binding documents were adopted.

Subsequently, the COP explicitly included REDD+-activities on the list of "enhanced actions on adaptation, and mitigation" eligible to receive financial resources from the Fund.[1208] Considering that funding raised through the GCF will be used to finance all the imaginable climate change mitigation and adaptation activities, it can be seen critically that the COP does not indicate which part of the funds raised should be applied for the implementation of REDD+.[1209]

It is important that a State seeking to receive financial support from the GCF must fulfil "fiduciary principles and standards, and environmental and social safeguards", developed by the Board of the Fund.[1210] The environmental and social safeguards mentioned in these provisions are different from the relevant safeguards for REDD+ contained in Appendix I to Decision 1/CP.16. It should be noted that the introduction of specific safeguards for access to the GCF-resources can be seen as positive. Indeed, it can contribute to the environmentally sound implementation of REDD+ projects, but only if the safeguards developed by the Board of the GCF will provide a higher standard of environmental protection than the safeguards of Decision 1/CP.16. On the other hand, considering that the GCF will not fund every single REDD+-project, a discrepancy in the level of environmental protection may arise.

The Warsaw Framework for REDD+ calls States and other relevant actors to foster the raising of funds from all possible sources, including the alternative ones, and to channel them to developing States in "a fair and balanced manner".[1211] It is foreseen that the GCF will play a relevant role in this process.[1212] However, the availability of financial resources is not ensured.

Thus, the legal framework for REDD+ finance does not establish a financial instrument that could serve for the promotion of a holistic approach to the protection of forest ecosystems. It is important to mention that the COP decided to establish an information hub in order to increase transparency. All relevant

1207 Decision 1/CP.16, Paragraphs 95–98.
1208 Decision 3/CP.17, Annex, Paragraph 35.
1209 *Abidin* [n. 1009], p. 26.
1210 Decision 3/CP.17, Annex, Paragraph 40.
1211 Decision 9/CP.19, Paragraph 5.
1212 Ibid.

information on REDD+ should be published in the information hub, including the information on infrastructure FREL/FRL and a NFSM, and on the emission reductions achieved by REDD+. Obviously, the information hub can facilitate the public control over REDD+-activities and relevant funding. However, States and private stakeholders are not obliged to remedy the eventual lack of compliance.

V. Legal Provisions for the Sustainable Implementation of REDD+ in Forest Ecosystems

REDD+ can have unpredictable impacts on forest ecosystems. Although this mechanism is supposed to contribute to sustainable development and wildlife conservation, these impacts can be both positive and negative. This is mainly because REDD+ is primarily dedicated to the reduction of greenhouse gas emissions from forests and to the enhancement of forest carbon stocks, whereas other forest functions play only a secondary role. Moreover, there is a risk that States and private actors will seek to maximize their revenues from REDD+ projects at the conservation of forests. Correspondingly, non-carbon environmental functions and benefits of forest ecosystems can be undermined. This can result in the destruction of natural forests and the establishment of monoculture plantations. Moreover, this situation may also threaten forest-dependent ILCs, which can be evicted from their customary lands without compensation. In order to tackle these hazards that threaten to create new risks for the integrity and vitality of forest ecosystems, the legal framework for REDD+ establishes the environmental and social safeguards. Additionally, it seeks to promote cooperation with other environmental treaties in order to elaborate a holistic approach to forests.

1. Environmental Safeguards for REDD+

The safeguards for REDD+ are needed because these mechanisms are primarily dedicated to reducing CO_2-emissions and enhancing carbon stocks. Therefore, a desire to maximize forest emission reductions and associated revenues has negative impacts on the environment in the process of implementing REDD+.[1213]

1213 See *Rey, Daniela/Roberts, J./Korwin, S./Rivera, L./Ribet, U.*, A Guide to Understanding and Implementing the UNFCCC REDD+ Safeguards, 2013, p. 17; *Harvey, Celia A./Dickson, Barney/Kormos, Cyril*, Opportunities for Achieving Biodiversity Conservation through REDD+, in: Conservation Letters, Volume 3, Issue 1, 2010, pp. 54 f.; *Pistorius, Till/Schmitt, Christine/Entenmann, Stephen K.* B. Greening REDD+: Challenges and Opportunities for Forest Biodiversity Conservation, 2011,

For example, the introduction of invasive species can help to curb CO_2 levels in the atmosphere, but, at the same time, it can have adverse impacts on biodiversity.[1214] It is noteworthy that the lack of attention to social and environmental issues can also have adverse impact on the sustainable implementation of projects. If biological diversity, vitality, and regeneration capacity of forest ecosystems are weakened to a certain level, they can become a net source of greenhouse gas emissions. Moreover, displaced forest-dependent people are more likely to adopt unsustainable economical practices leading to higher CO_2-emissions, in order to sustain their livelihoods.

This is why several actors advocated the adoption of environmental and social safeguards. Developed States and NGOs particularly supported the development of the safeguards. However, at first, Parties tried to promote environmental and social issues by referring to other international environmental treaties.[1215] An indicative guidance delivered by Decision 2/CP.13 stipulates that REDD+ demonstration activities should be compatible with SFM, by considering particularly the provisions of the NLBI, the UNCCD, and the CBD.[1216] However, not all threats arising from the implementation of REDD+ can be addressed by other international agreements, whose scope is limited by their respective objectives. For this reason, it is necessary to include detailed provisions promoting the environmental and social sustainability of REDD+ into the legal framework for its implementation. From this point of view, the inclusion of the safeguards in the Cancun Agreements can be seen as positive.

The safeguards have the potential to promote forest conservation and to improve local livelihoods. They are dedicated to the protection of human rights

---

pp. 3 ff.; *Secretariat of the Convention on Biological Diversity*, Connecting Biodiversity and Climate Change Mitigation and Adaptation: Report of the Second Ad Hoc Technical Expert Group on Biodiversity and Climate Change. CBD Technical Series No. 41, 2009, pp. 5 ff.

1214 *Nelson, Eric/Polasky, Stephen/Lewis, David J./Plantinga, Andrew J./Lonsdorf, Eric/White, Denis/Bael, David/ Lawler, Joshua J.*, Efficiency of Incentives to Jointly Increase Carbon Sequestration and Species Conservation on a Landscape, in: Proceedings of the National Academy of Sciences, Volume 105, Issue 28, 2008, pp. 9471 ff.; *Savaresi, Anna*, The Role of REDD in the Harmonization of Overlapping International Obligations, in: *Hollo Erkki/Kulovesi Kati/Mehling Michael* (eds.), Climate Change and the Law. A Global Perspective. Ius Gentium: Comparative Perspectives on Law and Justice, Volume 21, 2013, pp. 393 ff.

1215 *Savaresi* [n. 1214], p. 404.

1216 Decision 2/CP.13, Annex, Paragraph 8.

and environmental goods, and not to the management of risks, as in the case with safeguards developed by economic institutions.[1217] In other words, the safeguards are not intended to avoid harm to forest ecosystems and local communities, but rather to take positive actions in order to improve their conditions.[1218]

The Cancun Agreement indicates only that States "should" "promote and support" the safeguards.[1219] At the same time, the Durban Accords stipulate that REDD+-activities "should be consistent" with the safeguards.[1220] In this case, the modal verb "should", which can be interpreted as "States are encouraged", is specified by the reference to "consistency". However, the wording is crucial for the interpretation of the provision in question. Moreover, further provisions clarifying the criteria for evaluating the consistency of REDD+ activities with the safeguards are needed. The legal value of the safeguards could be enhanced through the inclusion of a reference to the relevant documents adopted by the COP into Article 5.2 of the Paris Agreement. However, this provision does not establish any legal obligations.[1221]

The legal content of the wording "to address and respect safeguards" is unclear, but it can potentially be strengthened by recourse to the general international environmental legal principles, including the principle of prevention and the precautionary principle, from which the obligation to act with due diligence can be derived.[1222] These principles are also included in the UNFCCC.[1223] However, the legal content of these principles is also characterized by a lack of clarity.[1224]

Moreover, the safeguards for REDD+ can be enforced by way of collaboration with another environmental conventions, especially with the CBD. In fact, both the UNFCCC and the CBD enjoy an almost universal membership, meaning that States, when implementing REDD+, are also bounded by the provisions of the CBD and other international environmental treaties. At the same time, the conservationist provisions of these treaties are formulated in a vague manner,

1217 *Rey/Roberts/Korwin/Rivera/Ribet* [n. 1213], p. 18.

1218 Ibid.

1219 Decision 1/CP.16, Paragraph 69.

1220 Decision 2/CP.17, Paragraph 63.

1221 See Subsection D.III.4 of this Chapter.

1222 Legality of the Threat or Use of Nuclear Weapons, Advisory Opinion, ICJ Reports 1996, pp. 241–242, Paragraph 29; Pulp Mills on the River Uruguay (Argentina v. Uruguay), Judgment of 20.04.2010, ICJ Reports 2010, pp. 55 f., Paragraph 101.

1223 Article 3.3 of the UNFCCC.

1224 See Subsection C.II of Chapter 2.

and compliance mechanisms are weak or absent. This means that the obligations contained in the CBD are unlikely to remedy the lack of strong safeguards and compliance mechanisms in the legal framework for REDD+.[1225]

The wording of safeguards themselves is also vague and unspecified. However, they mention several important issues. Firstly, they reemphasize that REDD+-activities should be consistent with the objectives of other international legal documents and complementary to the objectives of National Forest Programmes.[1226] They emphasize the need to promote "transparent and effective national forest governance structures". Moreover, it is to be assessed positively that the safeguards stipulate explicitly that the implementation of REDD+ should be consistent with the preservation of "natural forests and biological diversity".[1227] Additionally, REDD+ should create incentives for the conservation of natural forests and for the promotion of their ecosystem services, and not be used for the promotion of logging activities, forestland conversion, and for the establishment of monoculture plantations.[1228] However, there is no clarity on how these incentives should be created, as States receive results-based payments only for the reduction of CO_2-emissions. Other multiple benefits of forests are not covered by the scope of REDD+. As higher amounts of CO_2 can be bound by the fast-growing monoculture plantations in a short-term perspective, there is a risk that REDD+ will have negative impacts on natural forests and forest biodiversity. This can happen because the establishment of monoculture plantations is not explicitly prohibited by the legal framework for REDD+. Moreover, it can be argued that the unclear reference to national circumstances can give States a possibility to justify certain negative impacts of REDD+ on forest ecosystems. Thus, further provisions and regulations are needed in order to ensure that REDD+ promotes the conservation of natural forests.

Concerning the permanency and leakage issues, the safeguards include provisions stipulating that States should take action in order to "address risk of

1225 However, there is still a possibility to promote a more holistic approach to forests by applying REDD+ in collaboration with other environmental agreements. ABS of the CBD deserves particular attention in this regard. The relevant issues concerning the possibility to strengthen environmental safeguards for REDD+ through collaboration with the CBD and by considering general obligations of States derived from prevention and precautionary principles, will be analyzed in the next Chapter.

1226 Decision 1/CP.16, Appendix 1, Paragraph 2(a).

1227 Decision 1/CP.16, Appendix 1, Paragraph 2(e).

1228 Decision 1/CP.16, Appendix 1, Paragraph 2(e).

reversals" and to "reduce displacement of emissions".[1229] However, these provisions are also weak and non-binding.

2. Reporting on the Implementation of the Safeguards

States have to report to the Conventions' bodies on how safeguards are "addressed and respected".[1230] The safeguards should be implemented on every stage of the implementation of REDD+, but only "when appropriate".[1231] Systems for providing information on how the safeguards are addressed and respected should be consistent with the guidance contained in Paragraph I, Appendix I to the Cancun Agreements.[1232] According to this guidance, the REDD+ activities should contribute to the achievement of the main objective of the UNFCCC.[1233] It is important that the REDD+-activities should be "consistent with the objectives of environmental integrity", while considering multiple functions of forests and promoting SFM.[1234] This guidance can contribute to the consistency and accurateness of reporting on how safeguards are addressed and respected. It can also promote the environmental sustainability of REDD+-projects. However, it is formulated in a vague and imprecise way.

National systems for providing information on how safeguards are addressed and respected should be established and maintained by the State concerned and should be based on the existing systems, which can permit to use the FAO reporting systems while submitting REDD+-relevant information.[1235] It was decided that a State "should" start to provide the information from the very beginning of the implementation of REDD+ and should continue to do so periodically.[1236] The intervals for submitting reports on how safeguards are addressed and respected should correspond to the frequency of submitting National Communications (NC) from non-Annex 1 UNFCCC Parties.[1237] The NC must be submitted by non-Annex I States every four years.[1238] Moreover, these States

1229 Decision 1/CP.16, Appendix 1, Paragraph 2(f-g).
1230 Decision 12/CP.17, Paragraph 1.
1231 Ibid.
1232 Ibid, Paragraph 2(a).
1233 Decision 1/CP.16, Appendix I, Paragraph 1(a).
1234 Ibid, Appendix I, Paragraph 1(d).
1235 Decision 12/CP.17, Paragraph 2.
1236 Decision 12/CP.19, Paragraphs 1, 2.
1237 Ibid, Paragraph 5.
1238 Decision 1/CP.16, Paragraph 60(b).

have to submit biennial update reports (BUR). However, submission of these documents is subject to national circumstances and the availability of international financial assistance. Moreover, more vague regulations concerning the frequency of submitting the NC and BUR exist for the Group of Least Developed States. Consequently, the provisions on the frequency of reporting on how safeguards are being respected and addressed are not sufficient in order to promote the environmental integrity of REDD+ activities. Even if a State provides the information on a biennial basis, the frequency of reporting will not be sufficient in order to avoid environmental harm in the process of implementing REDD+. Indeed, environmental damage can occur within a much shorter period of time. This is why the conditions of forest ecosystems and of forest biodiversity should be monitored constantly, and the relevant information should be regularly submitted to the Governing Bodies of the UNFCCC. If negative impacts on forests and forest-dependent people were detected, the remedial actions should be undertaken immediately. Thus, the safeguards and the system for providing the relevant information is insufficient in order to ensure the effective and constant monitoring of conditions of forest ecosystems in the process of implementing REDD+.

COP 21 in Paris decided that the summary of information on the safeguards should include the information on national circumstances that affect the safeguards.[1239] The information on how the safeguards have been addressed and respected should be included in the summary.[1240] Any further information concerning safeguards can be provided on a voluntary basis.[1241] Thus, the COP at least delivered a general guidance containing minimal requirements to the information on safeguards to be provided. It should be noted, however, that this "check-list" of information to be provided is insufficient in order to ensure the effective monitoring of impacts of REDD+ on forests. In order to be verifiable, the information concerning the conservation of biodiversity and natural forests should be gathered and analyzed based on established and empirically

---

1239 UNFCCC COP Decision 17/CP.21 "Further guidance on ensuring transparency, consistency, comprehensiveness and effectiveness when informing on how all the safeguards referred to in Decision 1/CP.16, Appendix I, are being addressed and respected", in: United Nations FCCC/CP/2015/10/Add.3, Report of the Conference of the Parties on its twenty-first session, held in Paris from 30 November to 13 December 2015, United Nations, 2016, Paragraph 5(a). – Internet resource – https://undocs.org/FCCC/CP/2015/10/Add.3 [Last visit: 30.04.2023].

1240 Ibid, Paragraph 5(c).

1241 Ibid, Paragraph 6.

demonstrable scientific criteria. Moreover, the pertinent provisions can be criticized for the lack of links to the information on FREL/FRL. If the information on the safeguards and on FREL/FRL were submitted simultaneously, the holistic approach to forests would be promoted more effectively. For this purpose, the achieved emission reductions should be analyzed together with the changes in status and trends of forest biodiversity. In order to promote compliance with the safeguards, it should be ensured that States would bear legal consequences in case if they failed to promote the conservation of forests in the course of REDD+ implementation.

3. Promotion of Compliance with the Safeguards

As stated above, States must provide information on how the safeguards are addressed and respected in order to receive results-based payments.[1242] However, the Warsaw Framework for REDD+ did not establish any modalities and procedures for the analysis and verification of the information. The legal consequences of insufficient implementation of safeguards are also unclear. Presumably, NGOs and other civil society institutions are supposed to play a significant role in the promotion of the environmental sustainability of REDD+-activities. This assumption is supported by the fact that the information on safeguards will be published on the information hub and be publicly available. This information hub was established in order to increase transparency.[1243]

Obviously, the information hub enables public control over REDD+-activities and over the relevant finance; however, Decision 9/CP.19 does not indicate how private or public entities could remedy a lack of compliance with the safeguards. It is unclear whether a State will receive results-based payments in the case when the implementation of relevant activities will result in the conversion of natural forests and in the loss of forest biodiversity.[1244]

1242 UNFCCC COP Decision 9/CP.19 "Work programme on results-based finance to progress the full implementation of the activities referred to in decision 1/CP.16, paragraph 70", in: United Nations FCCC/CP/2013/10/Add.1, Report of the Conference of the Parties on its nineteenth Session, held in Warsaw from 11 to 23 November 2013, United Nations, 2014, Paragraph 4. – Internet resource – https://unfccc.int/sites/default/files/resource/docs/2013/cop19/eng/10a01.pdf [Last visit: 30.04.2023].

1243 Ibid, Paragraphs 9–10.

1244 UNFCCC COP Decision 9/CP.19, Paragraphs 9–10; see also *Lyster, Rosemary,* International Legal Frameworks for REDD+: Ensuring Legitimacy, in: *Lyster, Rosemary/MacKenzie, Catherine/McDermott, Constance* (eds.), Law, Tropical Forests, and Carbon: The Case for REDD+, 2013, pp. 4 ff.; See also *Lang, Chris,* The Warsaw

4. *Current Impact of REDD+ on Forest Ecosystems and Forest Biodiversity*

While REDD+ has so far not been implemented on a full scale, numerous voluntary REDD+-projects are being currently implemented around the globe. The experience gained from such projects makes it possible to submit preliminary conclusions about the impacts of REDD+ on forest ecosystems and forest biodiversity. According to one piece of research, the analysis of 80 REDD+-projects implemented around the world has shown that biodiversity and other environmental issues can potentially be overlooked by these projects.[1245] Although several project's designs incorporate at least some environmental goals, a common feature of the projects is the lack of quantifiable, verifiable, and time-bound targets to be achieved within a specified period.[1246] The project documentation often fails to describe the conditions of forest ecosystems and of forest biodiversity at the initial phase of the project implementation, and to deliver the information on the projects' contribution to the achievement of environmental goals.[1247] If such information was included, relevant reports contained only a list of areas where some improvements had taken place without precisely describing their positive effects.[1248] If the environmental goals were precisely articulated, it would promote their achievement and would facilitate the demonstration of the benefits of REDD+-projects for forest ecosystems.[1249]

Other projects' designs do not incorporate environmental issues. Only 48% of afforestation/reforestation projects foresee that only native tree species are planted. The problem with such projects is also that only little or no information is available on the historical drivers of deforestation in the region.[1250] The

---

Framework for REDD Plus: The Decision on Summary of Information on Safeguards. REDD Monitor, 2013. – Internet resource – http://www.redd-monitor.org/2013/12/17/thewarsaw-framework-for-redd-plus-the-decision-on-summary-of-information-on-safeguards/ [Last visit: 30.04.2023].

1245 *Steven N. Panfil/Celia A. Harvey*, REDD+ and Biodiversity Conservation: A Review of the Biodiversity Goals, Monitoring Methods, and Impacts of 80 REDD+ Projects, in: Conservation Letters, Volume 9, Issue 2, 2016, pp. 145 ff.

1246 Ibid, p. 146.

1247 Ibid, p. 147.

1248 Ibid.

1249 *Caplow, Susan/Jagger, Pamela/Lawlor, Kathleen/Sills, Erin*, Evaluating Land Use and Livelihood Impacts of Early Forest Carbon Projects: Lessons for Learning About REDD+, in: Environmental Science & Policy, Volume 14, Issue 2, 2011, p. 164.

1250 Ibid.

project documentation does not provide the evidence that sufficient precautionary measures were taken in order to rule out the possibility of establishing monoculture tree plantations. A number of researches demonstrate that expansion of these plantations has already taken places in some areas where REDD+-projects were and are implemented.[1251] In other places, the implementation of REDD+ did not deliver any significant benefits for the reduction of the rates of deforestation. In other cases, positive impacts were scarce, such as in the Aceh Province of Indonesia.[1252] In this particular case, the conservation of forests and forest biodiversity was in the focus of the project's design.[1253] However, a change of government in Aceh resulted in the downgrading of the role of environmental issues.[1254] Consequently, the project included such activities as the expansion of monoculture plantations, granting mining concessions, and the infrastructure development in biodiversity hotspots. After the election of Irwandi Yusuf, who is committed to the protection of environment, to the post of the provincial governor, environmental issues were again incorporated into the project's design. This resulted in a notable reduction of deforestation rates in the region.[1255] This

---

1251 *Vanderhaegen, Koen/Verbist, Bruno/Hundera, Kitessa/Muys, Bart*, REALU vs. REDD+: Carbon and Biodiversity in the Afromontane Landscapes of SW Ethiopia, in: Forest Ecology and Management, Volume 343, 2015, pp. 22 f., 27 f.; *Huettner, Michael*, Risks and Opportunities of REDD+ Implementation for Environmental Integrity and Socio-Economic Compatibility, in: Environmental Science Policy, Volume 15, Issue 1, 2012, pp. 8 ff.

1252 *Anandi, Cut Augusta Mindry/Resosudarmo, Ida Aju Pradnja/Ekaputri, Andini Desita/Komalasari, Mella/Astri Pangestuti/Aryani, Riza*, Ulu Masen REDD+ Initiative, Aceh, Indonesia, in: *Sills, O Erin/Atmadja, Stibniati/de Sassi Claudio/Duchelle, Amy E/Kweka, Demetrius L/ Resosudarmo, Ida Aju Pradnja/Sunderlin, William D.* (eds.), REDD+ on the Ground: A Case Book of Subnational Initiatives Across the Globe, CIFOR, 2014, p. 380; *Lang, Chris*, After Seven Years, Norway's US$1 Billion REDD Deal in Indonesia Is Still Not Stopping Deforestation, REDD-Monitor, 2017. – Internet resource – http://www.redd-monitor.org/2017/12/28/after-seven-years-norways-us1-billion-redd-deal-in-indonesia-is-still-not-stopping-deforestation/ [Last visit: 30.04.2023].

1253 *Anandi/Resosudarmo/Ekaputri/Komalasari/Astri/Aryani* [n. 1252], p. 380; *Lang* [n. 1252].

1254 *Anandi/Resosudarmo/Ekaputri/Komalasari/Astri/Aryani* [n. 1252], p. 380.

1255 *Jong, Hans Nicholas/Hanafiah, Junaidi*, Deforestation Wanes in Indonesia's Aceh and Leuser Ecosystem, but Threats Remain, NGO Says, in: Mongabay, 2018. – Internet resource – https://news.mongabay.com/2018/02/deforestation-wanes-in-indonesias-aceh-and-leuser-ecosystem-but-threats-remain-ngo-says/ [Last visit: 30.04.2023].

example demonstrates that the implementation of REDD+ is highly dependent on local circumstances, which play a crucial role in the promotion of environmental benefits of these projects. It is somewhat frustrating to note that the Aceh province would have received results-based payments even if the environmentally detrimental version of the project had been implemented.

The analysis of environmental impacts of REDD+ projects on forest ecosystems is a topic for an independent research, which goes beyond the scope of this dissertation. However, the available sources demonstrate that a more robust international monitoring and compliance system is needed in order to avoid a damage or threat to forests in the context of project implementation.

5. Protection of the Rights of ILCs in the Context of Implementing REDD+

As far as the rights of forest-dependent people are concerned, the safeguards stipulate that "the knowledge and rights of indigenous peoples and members of local communities" should be respected. This provision refers to other international legal documents, including the United Nations Declaration on the Rights of Indigenous Peoples (UNDRIP).[1256] Moreover, it determines that the full and effective participation of ILCs should be promoted while implementing REDD+.[1257] This provision should also be read in the context of Paragraph 72 of the Cancun Agreements that requests States to ensure the participation of ILCs when developing and implementing national strategies or action plans. This request is a notable step forward in comparison to previous decisions that only recognized the need for consideration of the interests and rights of ILCs.[1258] In this regard, COP 15 has recognized the necessity for full and effective participation of ILCs in the monitoring and reporting of REDD+-related activities and encouraged the development of relevant guidelines.[1259]

In order to ensure a sufficient level of protection of the rights of ILCs, it would be desirable to promote the participation of their representatives at every stage of the development and implementation of REDD+-projects, and not only at the stage of monitoring and reporting.[1260] Moreover, it should be pointed out

---

1256 Decision 1/CP.16, Appendix 1, Paragraph 2(c).

1257 Decision 1/CP.16, Appendix 1, Paragraph 2(d).

1258 Decision 2/CP.13, Preamble.

1259 Decision 4/CP.15, Preamble, Paragraph 3.

1260 *Wallbott, Linda*, Indigenous Peoples in UN REDD+ Negotiations: "Importing Power" and Lobbying for Rights through Discursive Interplay Management, in: Ecology and

that the legal provisions for implementing REDD+ provide a lower level of protection of the rights of ILCs in comparison to the pertinent provisions of the Nagoya Protocol, which requires States to establish a procedure for acquiring a prior and informed consent or at least an approval of ILCs for access to genetic resources, in relation to which ILCs have established legal rights.[1261] It can be recommended to ensure the direct participation of the representatives of ILCs in those international activities that are related to REDD+. It is also necessary to stipulate explicitly that results-based payments can be provided only after the PIC of ILCs in relation to REDD+ projects was acquired.[1262]

With regard to the submission of information on how safeguards dealing with the rights of ILCs are addressed and respected, the same weak legal provisions as in the case of environmental safeguards for REDD+ are applied. Thus, it is submitted that no legal duty to protect the rights of ILCs has been established so far. Therefore, there is a need for a robust international system for monitoring the situation of ILCs while implementing REDD+. Once established, this system would serve to analyze the information related to the safeguards and decide on further actions in the case of any possible violation of the rights of ILCs. The lack of such system is another shortcoming of the Warsaw Framework for REDD+.

(a) *Promotion of the Rights of ILCs in the General Framework of the UNFCCC*

The general framework of the UNFCCC also includes a number of provisions, policies and procedures that deal with the rights of ILCs and can be applied while implementing REDD+.[1263] However, these measures are not sufficient in order to ensure a robust and effective protection of their rights. According to the provisions in question, States may include ILCs-related information in their national communications (NCs) and in biennial update reports (BURs). However, they are neither requested nor encouraged to do so.[1264] NCs have rarely

---

Society, Volume, 19, Number 1, 2014, pp. 22 ff.; See also: *Lyster, Rosemary*, REDD+, Transparency, Participation and Resource Rights: The Role of Law, in: Environmental Science & Policy, Volume 14, Issue 2, 2011, pp. 118 ff.

1261 See Subsection D.II.5 of Chapter 3.

1262 See also: *Anderson, Patrick*, Free, Prior, and Informed Consent: Principles and Approaches for Policy and Project Development, 2011, pp 10 ff.

1263 *Abidin* [n. 1009], p. 80.

1264 Ibid.

mentioned ILCs in their texts. Cases where the possible impact of climate change mitigation and adaptation on ILCs is included in NCs, are scarce.[1265]

It is worth noting that Ecuador's NC addresses issues that are relevant to ILCs.[1266] It describes the adverse impacts of deforestation on the indigenous population of the Amazon Basin. Subsequently, the document in question identifies the drivers of deforestation in the region and stresses the need for enhanced action for combating deforestation.[1267]

NCs contain only in rare cases the information on the role of ILCs in climate change mitigation and adaptation, as well as on the relationship between the central government and ILCs.[1268] For example, this information was provided in the NCs submitted by Australia and New Zealand.[1269] Some NCs describe also national legislation that is relevant to ILCs, as in the case of Mexico.[1270] Furthermore, Bolivia has underlined the need to ensure the full and effective participation of ILCs when implementing REDD+.[1271] It can be argued, however, that the inclusion of these issues in the NCs listed above can be partially explained by a comparatively high share of ILCs in the population of the States concerned. Consequently, these cases cannot be representative for other forest-rich States, where ILCs represent a minority, and their rights can be violated by REDD+ projects. As in the case of Indonesia, some forest-rich States have so far not included any information on ILCs in their NCs.[1272]

The NC serves as the main communication tool between States and the Governing Bodies of the Convention and as an important element of its compliance system. After the evaluation and assessment of NCs is accomplished, the expert review team can present a list of non-legally binding recommendations for the State concerned. In cases where non-Annex I States are involved, an ad-hoc expert group can sometimes be created in order to elaborate non-legally binding recommendations. It can be also expected that States providing financial support

---

1265 Ibid.

1266 *Ecuador*, "National Communication, Republic of Ecuador", (November 2000), p. 8. – Internet resource –https://unfccc.int/resource/docs/natc/ecuncl.pdf [Last visit: 30.04.2023].

1267 Ibid, pp. 11 f.

1268 *Abidin* [n. 1009], p. 81.

1269 Ibid.

1270 Ibid, 82.

1271 Bolivia, "Plurinational State of Bolivia, Second National Communication, Executive Summary", 2009, pp. 47–48, 60 ff.; See also *Abidin* [n. 1009], p. 82.

1272 *Abidin* [n. 1009], p. 83.

for mitigation and adaptation in developing States will consider the needs of ILCs. However, the general framework of the UNFCCC, including those COP Decisions that are relevant to REDD+, does not contain sufficient instruments and tools that could pursue States and other stakeholders to improve the conditions of ILCs while implementing REDD+. Neither a sanction mechanism nor incentives for the promotion of compliance have been created so far.

(b) REDD+ and International Treaties for the Protection of Human Rights

When implementing REDD+, States are also bound to observe the provisions of international legal treaties specifically dedicated to the protection of human rights. Furthermore, according to the safeguards for REDD+, a number of international documents dealing with the rights of indigenous people are to be considered while implementing REDD+.[1273] In this regard, the International Covenant on Civil and Political Rights (ICCPR),[1274] the International Convention on the Elimination of all Forms of Racial Discrimination (ICERD)[1275] and the International Covenant on Economic, Social, and Cultural Rights (ICESCR)[1276] should be taken in consideration. The systems of communication established by these treaties may serve as a channel for submitting information in case of violation of the rights of ILCs. In this regard, Article 1 of the First Optional Protocol to the ICCPR,[1277] Article 14 of the ICERD, and Article 2 of the Optional Protocol to the ICESCR[1278] foresee the possibility for individuals or groups of individuals to submit communications claiming that any of their rights protected by the Treaties in question were violated. However, indigenous people can submit

---

1273 Decision 1/CP.16, Appendix 1, Paragraph 2(c).

1274 International Covenant on Civil and Political Rights, adopted on 16 December 1966 and entered into force on 23 March 1976, in: 999 UNTS 171.

1275 International Convention on the Elimination of all Forms of Racial Discrimination, adopted on 21 December 1965 and entered into force on 4 January 1969, in: 660 UNTS 195.

1276 International Covenant on Economic, Social, and Cultural Rights, adopted on 16 December 1966 and entered into force on 3 January 1976, in: 993 UNTS 3.

1277 First Optional Protocol to the International Covenant on Civil and Political Rights, adopted on 16 December 1966 and entered into force on 23 March 1976, in: 999 UNTS 171.

1278 Optional Protocol to the International Covenant on Economic, Social, and Cultural Rights, adopted on 16 December 1966 and entered into force on 3 January 1976, in: UNTS No. 14531.

their complaints to the Governing Bodies of these conventions only in the cases where they are recognized as a "representative group of individuals".[1279] Such a procedure presupposes a wide interpretation of the term "indigenous people". In the past, the Human Rights Committee has already contributed to the protection of the rights of ILCs through its communications and concluding observations. For example, the Angela Poma Poma v. Peru case is remarkable in this regard.[1280] In this case, the Human Rights Committee not only proclaimed that Peru was in breach of Article 27 of the ICCPR and obliged this State to improve the situation of ILCs and pay them a compensation, but also underlined the need to prevent any future violations of the rights of ILCs.[1281] Theoretically, such a procedure could also be applied in the case of the violation of the rights of ILCs by REDD+-projects.

However, this procedure is only possible in the case where the State concerned has ratified the First Optional Protocol to the ICCPR.[1282] In the case when this State is not a Party to this Protocol, but only to the ICCPR, the Human Rights Committee can still play an important role as a monitoring body by analyzing periodic reports submitted by States.[1283] However, communications submitted by individuals and representative groups of individuals are not considered in this case.[1284] The procedure of submitting individual communications to the Human Rights Committee in accordance with the Optional Protocol to the ICESCR is also available only in the cases where a State concerned has ratified the Protocol.[1285]

1279 See Article 1 of the First Optional Protocol to the ICCPR, Article 14 of the ICESCR, and Article 2 of the Optional Protocol to the ICESCR.

1280 United Nations CCPR/C/95/D/1457/2006, Ángela Poma Poma v. Peru, Communication No. 1457/2006, Ninety-fifth Session of the Human Rights Committee, 16 March to 9 April 2009, 2009. – Internet resource – https://undocs.org/en/CCPR/C/95/D/1457/2006 [Last visit: 30.04.2023]; See also *Göcke, Katja*, The Case of Ángela Poma Poma v. Peru before the Human Rights Committee, The Concept of Free Prior and Informed Consent and the Application of the International Covenant on Civil and Political Rights to the Protection and Promotion of Indigenous Peoples' Rights, in: *Bogdandy, Armin von/Wolfrum, Rüdiger* (eds.), Max Planck Yearbook of United Nations Law, Volume 14, 2010, pp. 340 ff.; *Abidin* [n. 1009], p. 84.

1281 United Nations CCPR/C/95/D/1457/2006 [n. 1280], Paragraph 9.

1282 Optional Protocol to the ICCPR, Paragraphs 1–3.

1283 *Abidin* [n. 1009], p. 84.

1284 Optional Protocol to the ICCPR, Article 1.

1285 Optional Protocol to the ICESCR, Articles 1.1–2, and 2.

Furthermore, the submission of complaints to the Human Rights Committee is a time- and resource-consuming procedure. This circumstance reduces the ability of ILCs to protect their rights under the existing international legal framework. Thus, despite that they deliver a number of positive incentives, the above-mentioned international documents dealing with the protection of human rights cannot ensure the protection of the rights of ILCs in the process of implementing REDD+. For this reason, the legal framework for REDD+ should include effective and robust tools for the promotion of their rights.

(c) Current Impact of REDD+ on ILCs

As in the case with environmental safeguards, the experience with the implementation of REDD+ projects make it possible to submit preliminary conclusions concerning the impacts of this mechanism on ILCs.[1286] Accordingly, ILCs could receive more benefits when a multilevel governance system for implementing REDD+ would be established. The multilevel governance system permits to take decisions exactly at a level where they must be implemented,[1287] while taking into consideration the interests of the relevant stakeholders. However, in most States a centralized approach to the implementation of REDD+ continues to prevail. Actually, the evidence suggests that REDD+ is more likely to promote the centralization of forest governance than its decentralization.[1288]

In several cases, States have attempted to establish procedures for instructing ILCs on planned REDD+ projects and for acquiring their PIC. However, national legal frameworks for REDD+ rarely provide for the direct participation of ILCs in the decision-making process. In the case of Papua New Guinea, only chiefs of tribes were instructed about REDD+, and most forest-dependent people were not properly informed about the relevant projects.[1289] In contrast, in the case of the pilot REDD+-project in Lam Dong province in Vietnam, the government

---

1286 *Mucahid Mustafa, Bayrak/Marafa, Lawal Mohammed*, Ten Years of REDD+: A Critical Review of the Impact of REDD+ on Forest-Dependent Communities, in: Sustainability, Volume 8, Issue 7, 2016, pp. 620 ff.

1287 *Vijge, Marjanneke J./Brockhaus, Maria/Di Gregorio, Monica/Muharrom, Efrian*, Framing National REDD+ Benefits, Monitoring, Governance and Finance: A Comparative Analysis of Seven Countries, in: Global Environmental Change, Volume 39, 2016, p. 66.

1288 Ibid, pp. 65 ff.

1289 See *Leggett, Matthew/Lovell, Heather*, Community Perceptions of REDD+: A Case Study From Papua New Guinea, in: Climate Policy, Volume 12, Issue 1, 2012, p. 127.

acquired the PIC of ILCs.[1290] Nevertheless, even in this case, ILCs were not properly informed about all the aspects of REDD+, and only a limited time for consultancies and discussions was provided.[1291] Furthermore, forest-dependent people in Vietnam were not granted the right of recourse to independent courts against possible violations of their rights.[1292] All these facts demonstrate that the lack of an internationally established procedure for acquiring the PIC of ILCs is a serious obstacle to the promotion of the rights of ILCs while implementing REDD+.[1293]

The risk of "land grab" is another serious issue related to the implementation of REDD+. Currently, voluntary carbon markets, which, in contrast to compliance markets, are not thoroughly monitored, are dominating the process of its implementation. This results in the violations of the rights of ILCs.[1294] In this regard, the term "carbon piracy" is already applied by analogy with biopiracy. This term refers to cases when foreign investors take advantage of a less-favored position of developing States and ILCs in commercial negotiations and convince their representatives to sign unequal treaties. The treaties in question grant exclusive rights in relation to a forest area to foreign companies, whereas the forest-dependent people do not receive any adequate compensation.[1295] As a result, the REDD+-mechanism provides a new tool permitting private companies to

---

1290 *Nguyen, Tan Quang/Luong, Thi, Truong/Nguyen, Thi Hai Van/Enters, Thomas/Yasim, Yurdi/Vickers, Ben*, Evaluation and Verification of the Free, Prior and Informed Consent Process under the UN-REDD Programme in Lam Dong Province, RECOFTC – The Center for People and Forests, 2010, pp. 8 ff.

1291 Ibid.

1292 Ibid.

1293 *Mucahid/Marafa* [n. 1286], p. 625; *Griffiths, Tom/Martone, Francesco*, Seeing "REDD"? Forests, Climate Change Mitigation and the Rights of Indigenous Peoples and Local Communities; Update for Poznan (UNFCCC COP 14), Forest Peoples Programme, 2009, pp. 11 ff.

1294 *Mucahid/Marafa* [n. 1286], p. 625; *Asia Indigenous Peoples Pact*, REDD+ Implementation in Asia and the Concerns of Indigenous Peoples; Asia Indigenous Peoples Pact: Chiang Mai, Thailand, 2010. – Internet resource – http://www.iwgia.org/iwgia_files_publications_files/0654_REDD_Plus_Implementation_in_Asia_and_the_Concerns_of_Indigenous_Peoples.pdf [Last visit: 30.04.2021]

1295 *Llanos, Roberto Espinoza/Feather, Conrad*, The reality of REDD+ in Peru: Between Theory and Practice; Indigenous Amazonian Peoples' Analyses and Alternatives, Forest Peoples Forest Programme, 2011, p. 1. – Internet resource – http://www.forestpeoples.org/sites/fpp/files/publication/2011/11/reality-redd-peru-betweentheory-and-practice-website-english-low-res.pdf [Last visit: 30.04.2023]

appropriate community-owned forestlands and TRFK. One of the consequences is the reduction of the costs of access to forests and the absence of benefit-sharing with ILCs.

The impacts of REDD+ on ILCs will also depend on the general scope of specific projects. If the activities in question are narrowly focused on the reduction of forest-related emissions, social issues will be overlooked, and the main scope of safeguards will be to avoid negative impacts on ILCs. At the same time, if the project is socially and environmentally oriented, the improvement of local livelihoods and instructing indigenous people on environmentally sound agricultural and forestry practices will represent an indispensable part of the project implementation.[1296] Against this backdrop, it is submitted that REDD+ safeguards, no matter how vague they are, tend to favor positive actions for the improvement of the conditions of ILCs.

To sum up, a socially and environmentally sustainable project design should be preferred for implementing REDD+. Such project design can also contribute to the general sustainability of REDD+. Indeed, forest-dependent people must be provided with the resources and knowledge needed in order to apply sustainable agricultural and forestry practices. Otherwise, they are likely to be constrained to opt for such forestry and agricultural practices that result in enhanced greenhouse gas emissions in order to sustain their livelihoods. For this reason, it can be seen negatively that the existing REDD+-projects have rarely included a large-scale capacity building for ILCs.[1297] These projects have so far seldomly contributed to the improvement of the livelihoods of ILCs in a sustainable and comprehensive manner. At best, there were no negative impacts on their livelihoods.[1298] The equal and fair sharing of benefits generated by projects was not ensured in many cases. In this regard, it is submitted that small-scale projects are preferable in comparison to large-scale ones, because they can better promote the interests of ILCs.[1299] However, considering that larger project areas generate higher revenues, large-scale REDD+-projects are more attractive for investors.

REDD+-projects can contribute to the diversification of incomes.[1300] It is argued that REDD+ can create incentives for "shifting cultivators and extensive

1296 *Mucahid/Marafa* [n. 1286], p. 627.

1297 Ibid.

1298 Ibid.

1299 *Vijge/Brockhaus/Di Gregorio/Muharrom* [n. 1287], p. 60.

1300 *Campbell, Bruce. M.*, Beyond Copenhagen: REDD+, Agriculture, Adaptation Strategies and Poverty, in: Global Environmental Change, Volume 19, Issue 4, 2009, pp. 397 f.

cattle ranchers" to abandon their unsustainable agricultural practices.[1301] However, these incentives will contribute to the promotion of forest conservation only in the case when the monetary benefits generated by REDD+ will outweigh the revenues arising from unsustainable agricultural and forestry activities.[1302] Otherwise, REDD+ activities cannot effectively substitute commercial agricultural crops and tree plantations, unless additional financial and legislative instruments are provided. Even a small-scale intensive agricultural and forestry activity can be more profitable than REDD+.[1303] These issues can undermine not only the positive impact of REDD+ on ILCs, but also the general ability of these mechanisms to contribute to the conservation of forests. Considering that the financial benefits of REDD+ for ILCs are not guaranteed, it should be pointed out that these mechanisms can still contribute to the general improvement of socio-economic conditions of the forest-dependent people in question. This can be achieved by improving forest tenure issues, promoting ecosystem services, and strengthening forest governance.[1304] However, as it was already pointed out above, these co-benefits are likely to occur only if the national legislative framework favors them. Otherwise, REDD+-projects may have negative impacts on indigenous cultures and TFRK.[1305] In order to avoid these negative impacts on ILCs, community-based forest management (CBFM) should be applied.[1306]

1301 Ibid, p. 398.

1302 Ibid.

1303 Ibid, p. 399.

1304 *Chhatre, Ashwini/Lakhanpal, Shikha/Larson, Anne M., Nelson, Fred/Ojha, Hemant,* Social Safeguards and Co–Benefits in REDD+: A Review of the Adjacent Possible, in: Current Opinion in Environmental Sustainability, Volume 4, Issue 6, December 2012, p. 656.

1305 *Anderson, Nicholas,* REDDy or Not? The Effects on Indigenous Peoples in Brazil of a Global Mechanism for Reducing Emissions from Deforestation and Degradation, in: Journal of Sustainable Development, Volume 2, Issue 3, 2009, pp. 22 ff.; *Corbera, Esteve,* Problematizing REDD+ as an Experiment in Payments for Ecosystem Services, in: Current Opinion in Environment Sustainability, Volume 4, Issue 6, 2012, pp. 612 ff.

1306 *Ianni, Elena/Mattenet, Mauricio/Geneletti, Davide/Malizia, Lucio R.,* Community-Based Forest Management in the Yunga's Biosphere Reserve, Northern Argentina, in: Environment, Development and Sustainability, Volume 12, 2010, p. 632; *Johanne, Pelletier/Gélinas, Nancy/Skutsch, Margaret,* The Place of Community Forest Management in the REDD+ Landscape, in: Forests, Volume 7, Issue 8, 2016, p. 189; *Gilmour, Don,* Forty Years of Community-Based Forestry, a Review of Its Extent and Effectiveness, FAO Forestry Paper 176, 2016, pp. 70 ff.

However, CBFM has not yet been introduced in many developing States. In this regard, it can be seen critically that the scope of most of REDD+-projects does not include the promotion of CBFM.

Thus, the empirical evidence suggests that the vagueness of environmental and social safeguards for REDD+, combined with the lack of international monitoring and compliance mechanisms, results in negative impacts on ILCs. Despite this, there is still a potential for REDD+ to contribute to the improvement of livelihoods and deliver a number of benefits for ILCs. For this purpose, the system of safeguards should be strengthened, and robust and effective policies and procedures should be developed.

(d) Interrelations Between REDD+, Sustainable Forest Management, and Non-Carbon Benefits of Forests

UNFCCC COP 15 recognized the importance of SFM and of non-carbon benefits of forests, including biodiversity.[1307] Accordingly, REDD+ can contribute to the achievement of the objectives of national forest programmes and of other international environmental treaties.[1308] Theoretically, this provision can provide for the development of joint activities with other international environmental agreements.

States participating in REDD+ are not going to receive results-based payments for promoting non-carbon benefits. However, the promotion of benefits in question still can deliver a number of advantages for States that can motivate them to adopt the appropriate policies in order to promote a holistic approach to forests while implementing REDD+.[1309] It should be underlined that the promotion of non-carbon benefits can contribute to the sustainability of projects. Moreover, developing States can take advantage of the ecosystem benefits of forests. However, the existing legal framework for REDD+ does not contain robust provisions and incentives for the promotion of the non-carbon benefits of forests.

1307 Decision 4/CP.15, Preamble.

1308 Ibid.

1309 *Meyer, Chris*, Clarifying the Role of Non-Carbon Benefits in REDD+, Environmental Defense Fund Talks Global Climate, 2013. – Internet resource – http://blogs.edf.org/climatetalks/2013/08/19/clarifying-the-role-of-non-carbon-benefits-in-redd/ [Last visit: 30.04.2023]

E. Conclusions

Forests play a crucial role both in climate change mitigation and in adaptation. Although underestimated at first, climate-related forest functions have apparently found their way onto the agenda of the international legal regime for combating climate change. Several developed and developing States advocated the immediate inclusion of legally binding commitments on the reduction of emissions from deforestation and forest degradation and the enhancement of forest carbon stocks in the text of the UNFCCC. However, the majority of Parties, particularly the key forest-rich developing States, argued that the reduction of industrial and transport-related emissions in developed States should be in the focus of climate change mitigation activities. This is why only vague provisions encouraging the reduction of emissions from forests were included in the UNFCCC and the KP. Consequently, both documents could not serve as an effective tool for the promotion of forest protection. Forest-related projects implemented under the CDM and the JI played only a marginal role in the international efforts dedicated to the conservation of forests. The positive impact of these projects was particularly limited by the exclusion of avoided deforestation in tropics from the scope of the CDM.

However, the development of the REDD+ mechanism gave an opportunity to promote climate change mitigation activities in forests. Under certain circumstances, this mechanism may contribute both to climate change mitigation and to adaptation. REDD+ may potentially have positive impacts on forest conservation because this mechanism can create incentives for a more sustainable forest use in developing States. If the benefits of REDD+ are distributed equally and fairly, they will contribute to the sustainable development in forest-rich States by promoting the transfer of financial resources and technologies from donor States.

However, despite all these positive aspects, the current REDD+ mechanism is not sufficiently robust and effective in order to ensure the conservation of natural forests and their biodiversity, and to improve local livelihoods. Non-carbon benefits of forests are likely to be overlooked by REDD+ because developing States can receive results-based payments only for the emission reductions achieved while implementing REDD+. Consequently, these States are likely to promote the non-carbon benefits of forests while implementing REDD+ only if their domestic policies foresee it. Otherwise, they will do so if the non-carbon benefits permit to achieve higher emission reductions.

Consequently, the implementation of REDD+ can actually result in serious damage or threats to forest ecosystems and forest biodiversity. *Inter alia*, such risks include the introduction of invasive species, forestland conversions, and the establishment of artificial tree plantations. Moreover, REDD+-projects can potentially have negative impacts on ILCs. There is a serious risk of "land grab", when ILCs are evicted from their customary lands. Traditional forms of forest use can be denied to them, community resources and traditional knowledge can be expropriated. ILCs can be overlooked by foreign investors and local authorities in the process of sharing benefits arising from REDD+.

The existing environmental and social safeguards are not sufficiently robust to tackle these risks. The interplay with other international legal treaties, including those dealing with the protection of the rights of ILCs, has proven to be insufficient to remedy the vagueness of the social safeguards for REDD+.

Moreover, the REDD+-related provisions do not effectively tackle the risk of leakage, meaning that REDD+ can potentially encourage the deforestation outside of project areas. REDD+ cannot resolve the problem of deforestation on a global scale, while boreal and temperate forests are not covered by its scope.

It can be assessed positively that REDD+ was finally included in the legally binding Paris Agreement, even though Article 5.2 is framed in a vague and imprecise manner and does not establish any obligation to implement this mechanism in an environmentally sound way. A number of developing States are currently proceeding with the implementation of REDD+ and have already achieved a second or a third stage of its implementation. This means that results-based payments are likely to occur in the near future. Importantly, a number of positive incentives have emerged from the legal framework for REDD+, including the promotion of the non-carbon benefits of forests and, particularly, the concept of the JMA mechanism.

Chapter Five Promoting Synergies Between the Climate Change Mitigation/Adaptation Activities in Forests and the CBD

A. Introduction

As was shown in the previous chapters, the international legal regime for combating climate change and the one for the conservation and sustainable use of biodiversity are two areas of international environmental law where significant development of forest-related policies and mechanisms have recently taken place. Notably, the pertinent mechanisms and policies can potentially be implemented simultaneously in forest ecosystems. For example, a forest involved in a REDD+ project can also deliver genetic resources for bioprospecting in the context of ABS. A/reforestation projects under the CDM can constitute a part of degraded ecosystem rehabilitation strategies in terms of Article 8(f) of the CBD. Joint mitigation and Adaptation (JMA) mechanisms, whose concept was introduced in the course of the negotiations of REDD+, can support the promotion of biodiversity conservation targets under the CBD. Indeed, all strategies and measures adopted in the context of the climate change regime that foresee the protection of forest ecosystems and forest biodiversity can be designed and implemented in a way that contributes to the achievement of the CBD's objectives.

The possibility of simultaneous implementation of climate- and biodiversity-related projects in forests delivers opportunities for further development of the international legal regime for forests. As discussed in the previous chapters, the CDM, REDD+, ABS, and other mechanisms developed under the scope of the aforementioned legal regimes are likely to promote only a limited number of forest functions. Although the REDD+-related provisions foresee the promotion of multiple benefits of forests, States are not under any obligation to follow the relevant recommendations while implementing these mechanisms. For this reason, those characteristics and functions of forest ecosystems that are not covered by the main objectives of the UNFCCC are likely to be overlooked in the process of implementing REDD+. Consequently, it is desirable that the synergies between REDD+ and ABS be promoted. Joint projects can allow for the consideration of a wider range of forest functions in the course of their implementation.

The conservation and enhancement of forest carbon stocks can be promoted in tandem with the preservation and sustainable use of forest biodiversity. The REDD+ mechanism refers explicitly to sustainable forest management and to the conservation of forests, and in so doing it promotes the synergies between two regimes. These two aspects also reflect some of the core activities under the CBD, which is on its turn dedicated to the conservation and sustainable use of biodiversity. However, these issues play only a marginal role in the legal framework for REDD+ that contains precise regulations only in relation to technical issues, including measuring, reporting, and verifying forest-related emissions by sources and removals by sinks. For this reason, the promotion of the non-carbon benefits of forests can be strengthened by incorporating the mechanisms and policies adopted under the CBD into REDD+.

Strengthening the interplay between REDD+ and the CBD can also support the raising of funds for the promotion of forest conservation and sustainable forest management. In addition to the receipt of results-based payments for REDD+, engaged actors can also receive benefits from the participation in ABS. If the implementation of REDD+ does not bring sufficient economic benefits in order to promote the protection of forests, additional funding may be delivered by the mechanisms developed within the frameworks of the CBD and the Nagoya Protocol. Considering that funding for the implementation of the CBD is raised through voluntary contributions of developed States and other actors, and only scarce contributions were pledged until now, the availability of financial and technical assistance is not predictable. Notwithstanding this, the implementation of ABS can provide further stimulus for the protection of forest ecosystems. Even if the financial benefits of the synergetic implementation of REDD+ and ABS remain negligible, the CBD COP will still be able to promote the conservation of forests and of forest biodiversity. Moreover, it can contribute to the monitoring of environmental conditions in forest ecosystems while implementing joint projects.

The promotion of cooperation between the aforementioned forest-related regimes is a challenging process. For instance, institutional arrangements between the CBD and the UNFCCC should be adopted in order to establish and promote this cooperation. In order to maximize the benefits of synergies, the interplay between the two treaties should be organized and monitored based on specific regulations rather than occur spontaneously. In the absence of an arrangement regulating the relationship between the treaties in question, the collaboration between the UNFCCC and the CBD will be more challenging. Currently, the forest-related mechanisms and policies of both Conventions are developed independently in the course of two parallel international negotiation processes. Therefore, it is desirable for the competent international bodies and other stakeholders operating under one of the treaties to take due regard to the progress achieved under the other. As the legal relationship between the UNFCCC, the CBD, and their respective Governing Bodies is not directly clarified by the treaty texts, additional institutional arrangements should be adopted in order to provide the framework for enhanced collaboration.

The actors involved should be properly informed about the possibilities and benefits of simultaneously applying forest-related policies of two legal regimes. The relevant information and expertise should be shared by the Governing Bodies of both Conventions. Unnecessary overlaps between different mechanisms and policies should be avoided. Financial resources and implementation efforts should be adequately allocated in order to promote the contribution of relevant projects to the protection of forest ecosystems. Moreover, it is not completely clear how the synergies between the forest-related mechanisms and policies of both Conventions will affect the integrity of forest ecosystems and the rights of ILCs.

Against this background, this Chapter will analyze whether these synergies already take place and how they influence forest ecosystems. Another question is whether institutional arrangements for the promotion of cooperation between two Conventions exist. Moreover, the factual compatibility of the forest-related mechanisms and policies of both legal regimes must be examined. Approaches and strategies that may help to avoid overlaps between the two regimes need further consideration. The impact on the environmental conditions in forest ecosystems and on the conditions of ILCs is another important issue that must be clarified.

B. Relationship Between the UNFCCC and the CBD

I. Provisions for the Promotion of Synergies in the Treaty Texts

In order to proceed further with the analysis of synergies, the availability of conflict clauses in the treaty texts and their applicability to the cases when climate change mitigation/adaptation activities cause a damage or threat to forest ecosystems and forest biodiversity should be examined. It is also necessary to verify if the provisions of the CBD and the UNFCCC promote the synergetic implementation of the two Conventions.

In fact, both the UNFCCC and the CBD enjoy an almost universal membership, meaning that States are bound by the CBD provisions while implementing REDD+. What concerns the CBD is that, among important States, only the USA did not become a party to it; however, it has constantly supported its implementation through voluntary political commitments.[1310] Despite this, the obligations of the CBD to conserve natural habitats and biodiversity are vague and relativized by the wording "in accordance with its particular conditions and capabilities", and "as far as possible and as appropriate".[1311] The relevant stakeholders can justify the implementation of those measures that can cause a damage or threat to biodiversity by the urgent need to combat climate change. Moreover, no sanctions are foreseen for the violation of provisions in question. However, the risk of damage to biodiversity in the course of climate change mitigation/adaptation can be tackled by obliging States and private stakeholders to prioritize the conservation of forest ecosystems and biodiversity.

The relationship between two Conventions is not conclusively regulated in the treaty texts. It is to be evaluated positively that a special clause on the relationship with other treaties is included in Article 22.1 of the CBD. Accordingly, the rights and obligations of the Parties to the CBD derived from other existing international treaties shall not be affected by the provisions of the CBD, except for the cases when relevant activities under other treaties would cause a serious threat or damage to biodiversity. One commentator has argued that Article 22.1 follows the "reverse conflict clause" approach giving priority to the obligations arising from other Conventions but only upon condition of the absence of a

---

1310 For the explanation of the United States' choice not to join the CBD see: *Blomquist, Robert F.*, Ratification Resisted: Understanding America's Response to the Convention on Biological Diversity, 1989–2002, in: Golden Gate University Law Review, Volume 32, Issue 4 "Rio's Decade: Reassessing the 1992 Earth Summit", 2002, pp. 493 ff.; See also *Dickie, Glory*, The US is the only country that hasn't signed on to a key international agreement to save the planet, in: Quartz, December 25, 2016. – Internet resource – https://qz.com/872036/the-us-is-the-only-country-that-hasnt-signed-on-to-a-key-international-agreement-to-save-the-planet/ [Last visit: 30.04.2023]. Despite that the USA as a State, which has not yet ratified the CBD, cannot join the Nagoya Protocol, this international treaty influences also the domestic processes of utilization of genetic resources. For the impact of the Nagoya Protocol on bioprospecting in the USA see: *McCluskey, Kevin/Katharine B. Barker/Barton, Hazel/ Boundy-Mills, Kyria*, The U.S. Culture Collection Network Responding to the Requirements of the Nagoya Protocol on Access and Benefit Sharing, in: mBio, Volume 8, Issue 4, 2017. – Internet resource – https://mbio.asm.org/content/mbio/8/4/e00982-17.full.pdf [Last visit: 30.04.2023]

1311 CBD, Articles 6–10.

"serious damage or threat" to biodiversity.[1312] This provision is criticized for its ambiguity, taking into account that, in the absence of a precise definition of "a serious damage or threat" and of internationally standardized criteria and indicators for respective risk evaluation, it can be interpreted in different ways.[1313] Against this backdrop, this provision is sometimes labelled as deprived of any legal meaning.[1314] Furthermore, it has been argued that Article 22.1 of the CBD cannot be applied to the UNFCCC, as this treaty had not yet become effective at the time when the provision was written.[1315] Both treaties were elaborated and adopted independently from each other. While the CBD entered into force on 29 December 1993, the UNFCCC entered into force only on 21 March 1994. Following this logic, Article 22.1 of the CBD cannot produce any effects in relation to the Kyoto Protocol and the Paris Agreement, which both are autonomous legally binding international treaties and entered into force later than the CBD did.[1316] However, this statement can be contested by the *lex specialis* principle, which potentially could be applied in the case of disputes involving a significant threat to biodiversity regardless of the legal meaning of Article 22.1 of the CBD,[1317] if only the CBD was recognized as *lex specialis*. However, the application of the *lex specialis* principle is subject to a number of uncertainties and controversies. In this regard, similar to the case of the maritime transportation of hazardous

1312 *Morgera, Elisa*, Faraway, So Close: A Legal Analysis of the Increasing Interactions between the Convention on Biological Diversity and Climate Change Law, 2011, p. 4.

1313 *Pavoni, Riccardo*, Mutual Supportiveness as a Principle of Interpretation and Law-Making: A Watershed for the WTO-and-Competing-Regimes Debate? in: European Journal of International Law, Volume 21, Issue 3, 2010, p. 655.

1314 *Doelle, Meinhard*, Linking the Kyoto Protocol and Other Multilateral Environmental Agreements: From Fragmentation to Integration? in: Journal of Environmental Law and Practice, Volume 14, 2004, p. 86.

1315 *Hodas, David R.*, Biodiversity and Climate Change Laws: A Failure to Communicate, 3rd Colloquium of the IUCN Academy of Environmental Law, Macquarie University, Sydney, Australia, 10–15 July 2005, p. 17.

1316 See *Jacquemont, Frédéric/Caparrós, Alejandro*, The Convention on Biological Diversity and the Climate Change Convention 10 Years After Rio: Towards a synergy of the Two Regimes? in: RECIEL, Volume 11, Issue 2, 2002, pp. 169 ff.

1317 *Van Asselt, Harro*, Managing the Fragmentation of International Environmental Law: Forests at the Intersection of the Climate and Biodiversity Regimes, in: New York University Journal of International Law and Politics, Volume 44, 2012, pp. 1251 f.

substances discussed in the Report of the International Law Commission, none of both Conventions' provisions would enjoy an intrinsic priority of application over the others.[1318] Depending on who the actors are, whether they are state authorities, different environmental organizations, or forest-dependent people, the conservation of forests or climate change mitigation would assume higher priority.[1319] For this reason, in each specific case of conflict between these provisions, the final decision would have to consider both environmental problems as well as to determine the specific relationship between them. None of the provisions can be set aside in the process.[1320]

It is not conclusively clarified which of the Conventions should be regarded as *lex specialis*, as they both deal with very specific environmental issues. Arguably, the choice of a *lex specialis* provision in every particular case is left to the discretion of the States concerned.[1321] The settlement of conflicts between the provisions of the two Conventions is further complicated by the fact that the UNFCCC and the Paris Agreement will potentially enjoy the primacy of application according to *lex posterior* rule because they entered into force later than the CBD.[1322] The opposite conclusion could be made were the dates of adoption chosen as a threshold for the application of the *lex posterior* rule, as the CBD was adopted one day later than the UNFCCC.[1323] Notably, the time of the ratification of both treaties can vary for different States that joined them on different dates.[1324] Even if clear rules for the determination of a "successive treaty" were available, the applicability of the *lex posterior* maxim to these two Conventions would be questionable,[1325] as States did not intend to change the provisions of the UNFCCC while adopting and ratifying the CBD and vice versa.[1326]

1318 United Nations A/CN.4/L.682, Fragmentation of International Law: Difficulties arising from the Diversification and Expansion of International Law, Report of the Study Group of the International Law Commission, International Law Commission, Fifty-eighth Session, Geneva, 1 May–9 June and 3 July–11 August 2006, 2006, pp. 34 ff., p. 63. – Internet resource – https://undocs.org/en/A/CN.4/L.682 [Last visit: 30.04.2023]

1319 Ibid, p. 63.

1320 Ibid, pp. 63 ff.

1321 See *Van Asselt* [n. 1317], p. 1251.

1322 *Vierdag, E.W.*, The Time of the "Conclusion" of a Multilateral Treaty: Article 30 of the Vienna Convention on the Law of Treaties and Related Provisions, in: British Yearbook of International Law, Volume 59, Issue 1, 1988, pp. 75 ff.

1323 *Vierdag* [n. 1322], pp. 75 f.; see *Van Asselt* [n. 1317], p. 1251.

1324 *Vierdag* [n. 1322], p. 110.

1325 *Van Asselt* [n. 1317], p. 1251.

1326 Ibid.

Consequently, both maxims of the international law of treaties are not sufficient in order to resolve conflicts between the provisions of both Conventions.

Even if it is assumed that the UNFCCC is an "existent treaty" for the purpose of Article 22.1 of the CBD, one can hardly argue that the provision in question ultimately prohibits activities under the scope of the UNFCCC that can cause serious damage or threat to biodiversity.[1327] Arguably, this provision only establishes a minimal "code of conduct" urging States to constantly monitor the existence of a serious damage or threat originating from the exercise of their rights and obligations under other treaties.[1328] In practice, it is imaginable that the threat to biodiversity will be evaluated against the urgency of action to mitigate climate change. Nevertheless, Article 22.1 provides States with a legal tool to justify the lack of implementation of the UNFCCC in cases where there is a serious risk of the adverse impacts on biodiversity. However, it is difficult to derive a relevant obligation from the conflict clause of Article 22 (1), considering that this provision, if interpreted narrowly, refers only to the treaties already existing at the time of the CBD's entering into force. At any time, the provision in question can hardly be applied to the Kyoto Protocol and the Paris Agreement, which both entered into force later than the CBD.

Arguably, the relationship between both international legal regimes has slightly changed with the entry into force of the Nagoya Protocol. This document incorporates the reverse settlement clause of the CBD. Its Article 4.1 literally cites the wording of Article 22.1 of the CBD while emphasizing that it does not intend to regulate the hierarchy between international environmental agreements. At the time when the Nagoya Protocol was signed and entered into force, the UNFCCC and the Kyoto Protocol had already existed, which is not true for the Paris Agreement, however. Consequently, the provision, as ambiguous as it is, must also be applied in relation to at least two major agreements dealing with climate change.[1329] The purpose of the Nagoya Protocol is to further develop the ABS-regime in relation to genetic resources. However, it also refers to the conservation and sustainable use of biodiversity.[1330] Moreover, Article 4.1 clearly refers to the possibility that the implementation of other treaties can result in serious damage or threat to biodiversity. Its scope of application is not limited to cases where the implementation of other treaties limits access to genetic resources or

1327 But see *Morgera* [n. 1312], p. 5.
1328 Ibid.
1329 Ibid.
1330 Nagoya Protocol, Article 1.

otherwise undermines the application of ABS. Consequently, Article 4.1 of the Nagoya Protocol has the same legal meaning as Article 22.1 of the CBD, thus obliging States to take due regard to the provisions of the Nagoya Protocol in the cases where the activities under the UNFCCC and the KP can result in the serious damage or threat to biodiversity.

That said, the low number of ratifications of the Nagoya Protocol is a challenge for the application of its Article 4.1. Moreover, the provision in question is also not worded clearly, just like the original provision of the CBD.[1331] The literal interpretation of this provision supports that it cannot be applied to the Paris Agreement.[1332] It should be pointed out that the duties to conserve biodiversity as established by the Nagoya Protocol are even weaker than the corresponding duties of the CBD. Therefore, the CBD and the Nagoya Protocol encourage States to choose those climate change mitigation/adaptation options that do not cause a serious damage or threat to biodiversity, but no relevant obligation are set by the treaties in question. Even if such an obligation existed, assuring compliance with it would be challenging due to the absence of a sanction mechanism in both treaties.

It is noteworthy that a number of activities potentially damaging biodiversity, including REDD+, are not regulated by the UNFCCC, the KP, or the Paris Agreement. Their implementation is guided by UNFCCC COP Decisions. These Decisions are directly related to the treaty on which they are based; however, the conflict clauses in the treaty texts as well as the *lex specialis* and *lex posterior* principles cannot be automatically applied to them.[1333] If the decisions of the COP or of any other international organization were recognized as a treaty or a source of customary law, such legal techniques could also be applied to them.[1334] However, decisions of the COP can rarely constitute an independent source of legal obligations. The COP is able to adopt legally-binding decisions either in accordance with the explicit authorization in the respective treaty text or based on the doctrine of implied powers.[1335] This doctrine presupposes that the COP can adopt

---

1331 *Morgera, Elisa/Tsioumani, Elsa/Buck, Matthias*, Unraveling the Nagoya Protocol, A Commentary on the Nagoya Protocol on Access and Benefit-sharing to the Convention on Biological Diversity, Leiden, 2014, p. 87.

1332 Ibid.

1333 *Van Asselt* [n. 1317], p. 1253.

1334 *Wiersema, Annecoos*, The New International Law-Makers? Conferences of the Parties to Multilateral Environmental Agreements, in: Michigan Journal of International Law, Volume 31, Issue 1, 2009, pp. 247 f.

1335 *Churchill, Robin R./Ulfstein, Geir*, Autonomous Institutional Arrangements in Multilateral Environmental Agreements: A Little-Noticed Phenomenon in International

legally-binding decisions on the matters that are necessary for the fulfillment of its functions even if it is not explicitly laid down in its constituent treaty.[1336] However, even if the Decisions of the COP were considered legally binding, their binding force would be based on and subordinated to an authorizing provision in the treaty text, meaning that they could hardly be recognized as independent sources of international law.[1337] This is supported by the fact that the parties to the UNFCCC have constantly failed to agree on any provision that could create a legal obligation to implement REDD+.[1338] Consequently, neither Article 22.1 of the CBD nor the *lex posterior* and *lex specialis* maxims can be applied to conflicts arising between UNFCCC COP decisions and the CBD. The opposite conclusion could only be made if the parties explicitly assigned a character of an independent legal obligation to the decisions' provisions,[1339] which was not the case.

As far as the UNFCCC is concerned, it does not contain any provision regulating the relationship to other international treaties. Its Preamble does not contain any reference to the conservation of biodiversity. Moreover, the Convention does not oblige States to adopt those climate change mitigation policies that permit to promote the conservation of biodiversity and natural habitats.[1340]

As mentioned above, States which ratified both the CBD and the UNFCCC are automatically bound by the provisions of the two Conventions. However, as the legal relationship between two Conventions is unclear, and the legal obligations to conserve biodiversity under the CBD are vague and formulated in a weak and imprecise manner, there is no legal certainty that REDD+ and other forest-related mechanisms under the UNFCCC will be applied in a way that is consistent with the CBD provisions and contribute to the conservation and sustainable use of forest ecosystems and their biodiversity. The lack of strong environmental safeguards for REDD+ is thus not remedied by the fact that

---

Law, in: The American Journal of International Law, Volume 94, Number 4, 2000, pp. 632 ff.

1336 Legality of the Threat of Use of Nuclear Weapons, Advisory Opinion of 8.07.1996, ICJ Reports 1996, p. 242, Paragraph 25.

1337 Ibid.

1338 See Subsection D.III of Chapter 4.

1339 *Van Asselt* [n. 1317], p. 1253.

1340 *Morgera* [n. 1312], p. 5; *Doelle, Meinhard*, Integration among Global Environmental Regimes: Lessons Learned from Climate Change Mitigation, in: *Chircop, Aldo/ McDorman, Theodore/Rolston, Susan* (eds.), The Future of Regime-Building in the Law of the Sea: Essays in Tribute to Douglas M. Johnston, 2009, pp. 63, 75.

those States that implement this Mechanism are bound by the provisions of the CBD.[1341]

II. Promotion of Synergies Between the UNFCCC and the CBD by Their Governing Bodies

The objectives of all three Rio Conventions – the CBD, the UNFCCC, and the UNCCD – are closely interrelated, as each of these Conventions tends to regulate environmental issues that constitute integral parts of the planetary ecosphere. This interrelation of different natural phenomena and processes has its clear manifestation at global, regional and local levels. Against this backdrop, the adoption of a single all-embracing international treaty and the development of unified strategies for its implementation would be desirable. However, the elaboration of such a treaty can be extremely complicated due to its excessively complex character and the conflicting interests of States.[1342] As outlined in Chapter 2, these factors have also impeded the adoption of a legally binding treaty in relation to the conservation and sustainable utilization of forests. The option of adopting an all-embracing environmental convention has never emerged from the international negotiations until 2018, when the United Nations General Assembly adopted the resolution "Towards a Global Pact for the Environment."[1343] The Agenda 21 and the Rio Declaration on Environment and Development, which are both non-legally binding, serve as overarching arrangements that aim to promote the consolidation of international environmental policies.[1344] However detailed the provisions of these documents are, especially those of the Agenda 21, they represent only a set of non-legally binding political commitments. For that reason, they could not provide a sufficient legal basis for the collaboration between the Rio Conventions, each of which has at its disposal its

1341 *Van Asselt, Harro*, Integrating Biodiversity in the Climate Regime's Forest Rules: Options and Trade-offs in Greening REDD Design, in: RECIEL, Volume 20, Issue 2, 2011, p. 144.

1342 See: *de Lassus St-Geniès, Géraud*, The Outcome of the Negotiations on the Global Pact for the Environment: A Commentary, in: Sustainability, Volume 12, Issue 3, 2020, pp. 877 ff.; *Ling, Chee Yoke*, The Rio Declaration on Environment and Development: Assessment, 2012, pp. 6 ff.; *Sands, Phillipe*, Greening International Law, 1994, pp. 22 ff.

1343 *de Lassus St-Geniès* [n. 1342], p. 877; United Nations A/RES/72/277, Resolution adopted by the General Assembly on 10 May 2018: Towards a Global Pact for the Environment, United Nations, New York, 2018. – Internet resource – https://undocs.org/A/RES/72/277 [Last visit: 30.04.2023]

1344 *de Lassus St-Geniès* [n. 1342], p. 877.

own Governing Bodies and functional mechanisms. In this situation, there is a need for further regulations that can be adopted by the Governing Bodies of the treaties in question. These regulations can facilitate the mutual coordination of implementation efforts. They are likely to promote the synergetic implementation of REDD+ and ABS.

With the necessity for enhanced collaboration between the three treaties in mind, the Joint Liaison Group (JLG) between the secretariats of the Rio Conventions was launched in August 2001.[1345] The JLG serves as an informal forum for the promotion of synergies and collaboration between the Conventions and for the exchanging of information.[1346] The CBD SBSTTA has further underlined that the JLG delivers an important platform for enhanced collaboration on the international level.[1347] Its main aim is the enhancement of coordination between the Rio Conventions, particularly by collecting and sharing information in relation to the activities undertaken under each of the treaties.[1348] Each of the three Secretariats organizes the meetings of the JLG on a rotation basis. However, the mandate for the JLG has been limited, while major States preferred to maintain the independent nature of climate change related policies.[1349] The Secretariats of the three Conventions jointly identified areas where enhanced cooperation is likely to occur.[1350] These areas include the promotion of collaboration between the national implementation programmes, strategies and action plans aimed to achieve the goals of three Rio Conventions, including national biodiversity

1345 CBD, Joint Liaison Group, CBD official site: https://www.cbd.int/cooperation/liaison.shtml [Last visit: 30.04.2023]

1346 Ibid.

1347 CBD SBSTTA Information Note 9/10, Paragraph 9(d).

1348 CBD COP Decision XI/6 "Cooperation with other conventions, international organizations, and initiatives", Decision adopted by the Conference of the Parties to the Convention on Biological Diversity at its eleventh Session, held at Hyderabad, India, from 8 to 19 October 2012, United Nations, 2012, Paragraph 17(b). – Internet resource – https://www.cbd.int/doc/decisions/cop-11/cop-11-dec-06-en.pdf [Last visit: 30.04.2023]

1349 *Van Asselt, Harro/Sindico, Francesco/Mehling, Michael A.*, Global Climate Change and the Fragmentation of International Law, in: Law & Policy, Volume 30, Number 4, 2008, p. 432.

1350 CBD SBSTTA Information Note 9/10, Options for enhanced cooperation among the three Rio Conventions, SBSTTA, ninth Meeting, held at Bangkok from 7 to 11 February 2005, United Nations, 2004, Annex, Paragraph 9. – Internet resource – https://www.cbd.int/doc/meetings/sbstta/sbstta-10/information/sbstta-10-inf-09-en.pdf [Last visit: 30.04.2023]

strategies and action plans (NBSAPs) under the CBD, national adaptation programmes of action (NAPAs) for least developed States under the UNFCCC, and national action programmes (NAPs) under the UNCCD.[1351]

While CBD COP Decision VII/1 on forest biodiversity does not contain any reference to the provisions, policies and mechanisms of other Rio Conventions,[1352] CBD COP Decision VII/26 has provided further impulses for establishing such synergies. Besides the promotion of cooperation on the international level, particularly through the JLG,[1353] this Decision also encouraged cooperation between different national programmes, and requested all Parties and other States to support cooperation initiatives on the national level.[1354] However, the document in question does not specify the type and scope of these initiatives.

The UNFCCC COP has also outlined that climate change mitigation/adaptation policies have to be complementary to the national policies developed under other Conventions, but this provision uniquely concerns climate change adaptation in Least Developed States, meaning that the biggest share of forest-related projects is not included.[1355] The provision concerned does not take into account that there is an urgent need to promote the complementary approach in relation to the mitigation activities in forests. For example, the policy that foresees the utilization of biofuels produced from plants cultivated on intentionally converted forestlands as an emission reduction option results in a serious damage or threat to forest biodiversity. In order to tackle this risk, the links between mitigation

---

1351 Ibid, Annex Paragraph 9(a).

1352 CBD COP Decision VII/1 "Forest biological diversity", Decision adopted by the Conference of the Parties to the Convention on Biological Diversity at its seventh Session, held at Kuala Lumpur from 9 to 20 and on 27 February 2004, UNEP, United Nations, 2004. – Internet resource – https://www.cbd.int/doc/decisions/cop-07/cop-07-dec-01-en.pdf [Last visit: 30.04.2023]

1353 CBD COP Decision VII/26 "Cooperation with other conventions and international organizations and initiatives", Decision adopted by the Conference of the Parties to the Convention on Biological Diversity at its seventh Session, held at Kuala Lumpur from 9 to 20 and on 27 February 2004, UNEP, United Nations, 2004, Paragraphs 1–5. – Internet resource – https://www.cbd.int/decision/cop/default.shtml?id=7763 [Last visit: 30.04.2023]

1354 Ibid. Paragraph 6.

1355 UNFCCC COP Decision 28/CP.7, "Guidelines for the preparation of national adaptation programmes of action", in: United Nations FCCC/CP/2001/13/Add.1, Report of the Conference of the Parties on its seventh Session, held at Marrakesh from 29 October to 10 November 2001, United Nations, 2002. – Internet resource – https://undocs.org/FCCC/CP/2001/13/Add.4 [Last visit: 30.04.2023]

policies and national biodiversity strategies and action plans (NBSAPs) ought to be established. Of course, the discretion enjoyed by States when implementing their NBSAPs permits them to apply non-biodiversity-friendly climate change mitigation activities. The same would apply if synergies between different national programmes were established. However, these synergies would promote the integration of the main objective of NBSAPs that is to promote the conservation and sustainable use of biodiversity in the climate change mitigation activities. In fact, the CBD requires that NBSAPs must be integrated in any activities that can potentially influence biodiversity and its conditions.[1356] However, it should occur only "as far as possible and as appropriate". This wording makes it possible to ignore NBSAPs in certain cases. Therefore, there is a risk that climate change mitigation strategies that have negative impacts on forest ecosystems and on forest biodiversity will be implemented. Consequently, the collaboration between national implementation strategies and action plans foreseen by the Rio Conventions cannot ensure that the possible negative impacts of climate change mitigation/adaptation mechanisms on forest biodiversity will be avoided.

The JLG has also promoted collaboration between national focal points of the Rio Conventions.[1357] It can be expected that this kind of cooperation will contribute to a more effective implementation of conventions and will permit to avoid overlaps between them.[1358] In this regard, creating and strengthening coordination mechanisms between the national focal points in question could facilitate avoiding duplicates and achieving goals when working on common issues.[1359]

Despite the restrictive nature of provisions relating to synergies on the international level, the Secretariats of the Rio Conventions have continued to encourage the elaboration of projects that integrate their respective goals on the national

1356 CBD, Article 6(a) and 6(b) in combination with the Article 10(a).

1357 CBD SBSTTA Information Note 9/10, Annex, Paragraph 9(b).

1358 UNFCCC SBSTA 19 Conclusions, Report of the Subsidiary Body for scientific and technological advice on its nineteenth Session held in Milan from 1 to 9 December 2003, United Nations, 2004, Paragraph 44(d). – Internet resource – https://unfccc.int/resource/docs/2003/sbsta/15.pdf [Last visit: 30.04.2023]

1359 CBD COP Decision VII/1 "Forest biological diversity", Decision adopted by the Conference of the Parties to the Convention on Biological Diversity at its seventh Session, held at Kuala Lumpur from 9 to 20 and on 27 February 2004, UNEP, United Nations, 2004. – Internet resource – https://www.cbd.int/doc/decisions/cop-07/cop-07-dec-01-en.pdf [Last visit: 30.04.2023]

level.[1360] However, it is not specified whether the projects' goals have to integrate the objectives of other Conventions, or only to take due regard to non-climate-related issues. Even the implementation of projects that integrate the goals of all three Rio Conventions does not ensure that no conflicts will arise between the provisions in question. Due to the lack of provisions regulating the legal relationship between the CBD and the UNFCCC, and because of the vague character of the conservationist provisions of the CBD, it is not clear how the balance between the goals of these treaties can be achieved.

III. Promotion of the Collaboration Between the Rio Conventions on the International Level

The international level of cooperation should be promoted by strengthening collaboration between the respective scientific and technical bodies, including the SBSTTA of the CBD, the Committee for Science and Technology (CST) of the UNCCD, as well as the SBSTA of the UNFCCC.[1361] The development of joint work programmes or plans constitutes another option for creating synergies between the Conventions. As for the moment, a joint work programme was established only between the CBD and the UNCCD. It deals with the conservation and sustainable use of biological diversity of dry and sub-humid lands.[1362] On the contrary, no joint work programme has yet been developed between the CBD and the UNFCCC. In 2010, an attempt was undertaken to elaborate a trilateral Joint Work Programme between the CBD, the UNFCCC and the UNCCD.[1363]

1360 UNFCCC SBSTA Information Note 2, Cooperative activities with United Nations entities and other intergovernmental organizations that contribute to the work under the Convention, the Kyoto Protocol and the Paris Agreement, forty-eighth Session of the SBSTA, held in Bonn from 30 April to 10 May 2018, United Nations, 2018. – Internet resource – https://unfccc.int/sites/default/files/resource/inf02_2.pdf [Last visit: 30.04.2023]

1361 CBD SBSTTA Information Note 9/10, Paragraph 9(c).

1362 CBD SBSTTA Information Note 8/10 "Joint Work Programme between the Secretariat of the Convention on biological Diversity and the Secretariat of the Convention on the biological diversity of dry and sub-humid lands", eighth Meeting, held in Montreal from 10 to 14 March 2003, United Nations, 2003. – Internet resource – https://www.cbd.int/doc/meetings/sbstta/sbstta-08/information/sbstta-08-inf-10-en.pdf [Last visit: 30.04.2023]

1363 *Böhringer, Ayşe-Martina*, Die Kooperationsvereinbarungen der Sekretariate multilateraler Umweltschutzübereinkommen, 2014, pp. 147 f.; *Van Asselt, Harro*, The Fragmentation of Global Climate Governance, Consequences and Management of Regime Interactions, 2014, p. 146.

The impulse for this enhanced level of collaboration was given by the CBD Secretariat. However, the initiative did not find sufficient support from the Executive Secretariat of the UNFCCC.

The possible impact of another collaboration tool applied in relations between the conventions – joint workshops[1364] – is likely to be insignificant. Although it can promote the gathering and exchange of information, it cannot contribute to the elaboration of joint implementation activities. At any time, reports of workshops can help to determine options for enhancing synergies and – in this way – influence decision-making processes under the scope of the Conventions in question. The workshop on "Forests and Forest Ecosystems: Promoting Synergy in the Implementation of the Three Rio Conventions" should be specifically mentioned here. The event was held on 5–7 April 2004 in Viterbo, Italy.[1365] The impulse for the organization of a joint workshop on the topic came from the Secretariats of the CBD and the UNCCD, but later also the Secretary of the UNFCCC joined the workshop activities.[1366] The objective of the workshop was to promote on-ground activities in forest ecosystems and to elaborate new synergetic processes in this sector.[1367] During the workshop, the Working Group on Sustainable Livelihoods and Forest Resources recommended a bottom-up approach in order to increase synergies among the Rio Conventions.[1368]

Despite that numerous positive impulses and suggestions for enhanced collaboration in the forest sector were generated by the workshop, its final outcome was modest. The participants could not agree on a number of issues. The planned final statement suggesting that a lack of financial resources represents one of the major obstacles for synergies between the Conventions raised concerns on behalf of other participants.[1369] They argued that if such a conclusion had been taken seriously, projects that do not promote collaboration between the Conventions in question would have been deprived of financial resources needed for their implementation.[1370] Against this backdrop, if the promotion of synergies

1364 CBD SBSTTA Information Note 9/10, Paragraph 9(f).

1365 *Gaynutdinova, Tamilla/Juncurt, Stefan*, Workshop on "Forests and Forest Ecosystems: Promoting Synergy in the Implementation of the Three Rio Conventions", in: Earth Negotiations Bulletin, International Institute for Sustainable Development, Volume 4, Number 174, 2004, p. 3. – Internet resource – http://enb.iisd.org/download/pdf/enb04174e.pdf [Last visit: 30.04.2023]

1366 Ibid.

1367 Ibid.

1368 Ibid, p. 6.

1369 Ibid, p. 8.

1370 Ibid.

was recognized as a new major criterion for funding, it would undermine the independent nature of the Rio Conventions.[1371] For this reason, participants could not achieve a consensus on many important issues, and the outcome document is formulated in a vague and imprecise manner.[1372] Despite this, it contains points on which future consensus could be achieved. Among other things, these points focus on forest landscape restoration and the promotion of SFM and ecosystem services.[1373] Subsequently, the recommendations of the workshop were further considered by the Governing Bodies of the CBD. For example, the report of the workshop was further cited in Decision IX/16, which deals with the linkages between biodiversity and climate change.[1374] Particularly, it requested the Executive Secretary to use the report of the Workshop while elaborating its recommendations that can be relevant for the development and implementation of REDD+.[1375]

Joint capacity-building activities and the facilitation of information gathering and exchange can also contribute to enhanced collaboration between the Rio Conventions.[1376] Among other things, the Secretariats of the treaties in question conduct case studies on the interlinkages between different issues that play a crucial role under the three Conventions.[1377] The cooperation in question should include collaboration in communication, education, and public awareness (CEPA) programmes.[1378] However, it should be pointed out that a bulk of these activities has been promoted only under the CBD, whereas no major developments under the UNCCD and the UNFCCC have taken place.

It is important to note that the cooperation between the Secretariats of the Rio Conventions is also promoted by respective treaty COPs.[1379] The exchange

---

1371 Ibid.

1372 Ibid, p. 9.

1373 Ibid.

1374 CBD COP Decision IX/16 "Biodiversity and climate change", Decision adopted by the ninth Session of the COP, held in Bonn from 19 to 30 May 2008, United Nations, 2008, Paragraph 16. – Internet resource –https://www.cbd.int/doc/decisions/cop-09/cop-09-dec-16-en.pdf [Last visit: 30.04.2023]

1375 Ibid.

1376 CBD SBSTTA Information Note 9/10, Paragraph 9(g).

1377 CBD SBSTTA Information Note 9/10, Paragraph 9(h).

1378 CBD SBSTTA Information Note 9/10, Paragraph 9(j).

1379 UNCCD, Enhanced scientific cooperation and knowledge exchange between the Committee on Science and Technology and the scientific subsidiary bodies of the United Nations Framework Convention on Climate Change and the Convention on Biological Diversity, the Scientific and Technical Advisory Panel of the Global

of relevant knowledge between the scientific bodies of the Conventions should support decision-making processes in relation to their respective main goals.[1380] It should be taken into account, though, that the scientific bodies of the Conventions merely play an advisory and consultancy role, which is why they cannot directly influence the implementation of the treaties in question. States can be obliged to apply guidelines and tools elaborated by the respective Scientific Bodies only by decisions of the COPs of the Conventions. Consequently, the collaboration between these Scientific Bodies can support the elaboration of joint activities, but it is unlikely that it will encourage their implementation.

To sum up, it is submitted that the collaboration between the Governing Bodies of the three Rio Conventions is not sufficiently promoted. The particularly low level of institutionalization and the lack of legal regulations overshadow the collaboration between the UNFCCC and the CBD. The lack of an institutional arrangement between these treaties can be seen negatively in the light of the far-reaching interdependences of biodiversity issues and climate change mitigation/adaptation activities. Against this backdrop, it is submitted that a Memorandum of Cooperation (MOC) or a Memorandum of Understanding (MOU) should be adopted in order to promote synergies between the Conventions in question.

IV. Institutional Arrangements Between Governing Bodies of International Treaties: The Example of the CBD and the UNCCD

Which factors determine the importance of MoUs and MoCs as tools for the promotion of collaboration between international treaties? From a formal viewpoint, such agreements are characterized by their fixed written form. MoUs and MoCs are not legally binding.[1381] However, they give rise to specific commitments for respective Governing Bodies of the Conventions concerned only upon the signature by their Executive Secretariats – a practice that resembles the entrance into force of legally binding treaties,[1382] as in the case of the MoU

Environment Facility and relevant specialized agencies of the United Nations, UNCCD COP 10th session, Changwon, 11–13 October 2011. – Internet resource – https://www.unccd.int/sites/default/files/sessions/documents/ICCD_COP10_5/cst INF5eng.pdf [Last visit: 30.04.2023]

1380 Ibid, Paragraph 6.

1381 *Ulfstein, Geir*, Treaty bodies, in: *Bodansky, Daniel/Brunnée, Jutta/Hey, Ellen* (eds.) The Oxford Handbook of International Environmental Law, 2007, pp. 887 f.

1382 *Ulfstein* [n, 1381], pp. 887 f.; See also *Böhringer, Ayşe-Martina* [n. 1363], pp. 8 f.

between the CBD and the UNCCD. Therefore, they share certain characteristics of binding international treaties.[1383] It is argued that, from a substantive point of view, MoUs and MoCs seem to be able to create "internal" rights and obligations for the Governing Bodies of the Conventions concerned.[1384] For example, the MoU between the CBD COP and the Council of the GEF defined the modalities for the financial mechanism of the Convention and, among other things, established the commitment for the Council to present a report to each session of the COP.[1385] It is submitted that MoUs and MoCs can provide a political stimulus and a legal mandate for fostering cooperation between different international treaties and their respective Governing Bodies.[1386] For this reason, they can be prioritized over other forms of cooperation, such as joint working groups, at least in the case of treaties with highly divergent mandates.[1387]

The MoC between the CBD and the UNCCD, concluded in Montreal on 31 July 1998, stands out for its strong wording, indicating that the respective Executive Secretariats "will" undertake activities and implement measures, as foreseen by the treaty text.[1388] First, each of the Secretariats provides for the participation of the representatives of the other Convention in the sessions of the respective COPs and in other relevant meetings.[1389] Secondly, they ensure that information on corresponding implementation activities is exchanged.[1390] It is

---

1383 *Böhringer, Ayşe-Martina* [n. 1363], pp. 8 f.

1384 See Ibid, p. 8.

1385 CBD COP Decision III/8 "Memorandum of understanding between the Conference of the Parties to the Convention on Biological Diversity and the Council of the Global Environment Facility", Decisions adopted by the Conference of the Parties to the Convention on Biological Diversity at its eighth Session, held in Buenos Aires from 4 to 15 November 1996, United Nations, 1996, Paragraph 16. – Internet resource – https://www.cbd.int/decision/cop/?id=7104 [Last visit: 30.04.2023]; *Morgera, Elisa/Razzaque, Jona*, Biodiversity and Nature Protection Law, in: *Cottier, Thomas/Schefer, Krista Nadakavukaren* (eds.), Elgar Encyclopaedia of International Law, Volume 3, 2017, p. 479.

1386 *Van Asselt, Harro/Gupta, Joyeeta/Biermann, Frank*, Advancing the Climate Agenda: Exploiting Material and Institutional Linkages to Develop a Menu of Policy Options, in: RECIEL, Volume 14, Issue 3, 2005, p. 264.

1387 Ibid, p. 262.

1388 *Secretariat of the UNCCD/Secretariat of the CBD*, Memorandum of Understanding between the Convention on Biological Diversity and the United Nations Convention to Combat Desertification, 31 July 1998. – Internet resource – https://www.cbd.int/doc/agreements/agmt-unccd-1998-07-31-moc-web-en.pdf [Last visit: 30.04.2023]

1389 Ibid, Article 1.1.

1390 Ibid, Article 2.

important to note that the Secretariats ought to analyze their implementation activities in order to identify further options for collaboration and for the elaboration of joint work programmes.[1391] The MoC has four Annexes, which contain detailed regulations in relation to the promotion of scientific collaboration and joint implementation.

The Memorandum of Understanding between the CBD and the UNCCD, concluded in Geneva on 3 September 2011, delivers a further example of an institutional arrangement between the Secretariats of the two Conventions.[1392] It is primarily dedicated to the promotion of implementation of the Joint Work Plan 2011–2012. Apart from information gathering and exchange, the MoU foresees the development of joint actions, the elaboration of common indicators for the ten-year strategic plans of both Conventions,[1393] as well as the implementation of at least one pilot project for the adjustment of the NAPs and the NBSAPs.[1394] The Joint Work Plan, attached to the MoU, includes several goals that can be relevant also to the conservation and sustainable use of forest ecosystems. Inter alia, it foresees further research with the aim to identify areas with high potential for the conservation and enhancement of carbon stocks. The Joint Work Plan is also aimed to identify modalities for ecosystem management contributing to the achievement of climate change mitigation goals, which is highly relevant for the implementation of REDD+.[1395] It further promotes the development and implementation of joint actions that can reduce the risk of natural disaster through the promotion of ecosystem resilience.[1396]

Although the MoU and Joint Work Plan do not refer explicitly to forests, it is obvious that the aforementioned issues are directly related to their conservation and sustainable management. Obviously, in order to achieve progress on the relevant issues, participation of the Secretariat of the UNFCCC is needed. For this purpose, the Secretariat of the UNFCCC was identified as an external partner for promoting linkages between the respective national programmes, and for the

---

1391 Ibid, Article 3.

1392 *CBD/UNCCD*, Memorandum of Understanding between the Convention on Biological Diversity and the United Nations Convention to Combat Desertification Regarding the Joint Work Plan 2011–2012, 03 September 2011. – Internet resource – https://www.cbd.int/doc/agreements/agmt-unccd-2011-09-03-mou-en.pdf [Last visit: 30.04.2023]

1393 Ibid, Paragraph 13.

1394 Ibid, Paragraph 18.

1395 Ibid, Annex I.

1396 Ibid.

development of proposals for removing obstacles for the improved consideration of biodiversity issues in the climate change related activities.[1397]. However, as the Executive Secretariat of the UNFCCC is not a party to the MoU, it does not have to integrate the proposals elaborated under the Joint Work Plan in its activities. If the Executive Secretary joined the MoU, more synergies could be achieved, and consistent policies on joint activities could be elaborated. Such developments could contribute to the further development of the regulatory framework for REDD+.

It is submitted that if a MoC or a MoU between the CBD and the UNFCCC were adopted, it would promote the cooperation between the Conventions in question.[1398] The institutional arrangement could regulate the exchange of information and the elaboration of joint implementation activities, and promote the harmonization of activities under both Conventions. Adoption of an institutional agreement between the CBD and the UNFCCC could be especially beneficial for the development of the REDD+ mechanism, which incorporates, besides climate change mitigation/adaptation issues, also a considerable biodiversity dimension. Arguably, such an institutional agreement could serve as a basis for the development of a Joint Work Programme on REDD+ or at least on its biodiversity-relevant safeguards. As one of the options, common biodiversity targets to be achieved in the process of implementing REDD+ and a mechanism for monitoring their achievement could be established. Such an approach would permit to incorporate more effectively the conservation of biodiversity into the process of implementing REDD+. In doing so, the rights of forest-dependent people could also be promoted, as the CBD and the Nagoya Protocol deal with the rights of ILCs.[1399]

It is true that an institutional agreement would not resolve all controversial issues in relation to the absence of stronger safeguards for REDD+. For example, the legal framework for its implementation consists merely of the provisions which refer to the national circumstances of States, giving them the possibility to disregard biodiversity issues in the process of REDD+-implementation. Thus, a more profound reform of the legal framework for REDD+ would be needed, and the conclusion of an arrangement between the CBD and the UNFCCC would not remove this necessity.

1397 Ibid.

1398 *Van Asselt/Gupta/Biermann* [n. 1386], p. 264.

1399 See Subsection D.II.5 of Chapter 3.

On the other hand, the conclusion of an institutional arrangement between the UNFCCC and the CBD could contribute to the establishment of biodiversity-friendly forest carbon mechanisms. It could also facilitate the resolution of possible controversies and disputes that may arise in the process of implementing REDD+. As was pointed out above, the application of traditional legal techniques for conflict resolution, such as the *lex posterior* and *lex specialis* maxims, can be particularly challenging in the case of international environmental agreements that deal with factually inseparable elements of the planetary ecosphere and separate them artificially in order to elaborate applicable legal provisions.[1400] In this situation, every treaty can theoretically be seen as constituting a *lex specialis.*[1401] In the case of REDD+, which foresees the implementation of forest carbon projects, the collision between the two legal regimes is inevitable. For this reason, the institutional arrangement between the Governing Bodies of both treaties could strengthen the role of biodiversity issues in the implementation of REDD+.[1402]

Some factors pose serious obstacles for promoting the cooperation between the Rio Conventions. First, the development of collaboration between these treaties is overshadowed by controversial interests of the parties to these treaties. Every provision which promotes cooperation between these agreements tends to underline an independent character of the three Conventions and of their Governing Bodies.[1403] For this reason, such independence sets a threshold upon which any attempt to promote collaboration between them is destined to fail. Notably, the biggest part of initiatives concerning collaboration between the Rio Conventions originates from the Governing Bodies of the CBD. For example, the idea of a trilateral joint work programme was proposed exactly by the CBD Secretariat. Therefore, the leading role in the process of the development of synergies between the Rio Conventions is currently with the CBD, a fact which corresponds to the holistic nature of biodiversity-related issues. At the same time, the UNCCD engaged itself actively in the collaboration with the CBD, whereas the Governing Bodies of the UNFCCC were reluctant to conclude any institutional arrangement. This may be explained by the significant role of market-based and

---

1400 *Van Asselt, Harro*, Managing the Fragmentation of International Environmental Law: Forests at the Intersection of the Climate and Biodiversity Regimes, in: New York University Journal of International Law and Politics, Volume 44, 2012, pp 1250 ff.

1401 Ibid.

1402 Ibid, pp. 1255 ff.

1403 See CBD SBSTTA Information Note 9/10, Annex, Paragraph 5.

non-market-based financial instruments in the process of the implementation of this Convention. The market-based instruments of Joint Implementation introduced by the KP require high transparency and legal certainty based on the simplicity of accounting emissions and on traceability of determining of prices for certified emission reductions. For this reason, it was desirable for the actors involved to keep the relevant procedures and policies independent from the provisions of other international legal regimes. Additionally, the parties to the UNFCCC, especially the developing States, were deeply concerned that funding will be determined by the contribution of climate change mitigation/adaptation activities to the achievement of the objectives of the CBD and the UNCCD.

Arguably, the leading role in the most of international and national environmental policies was assigned to the international climate change regime. This assumption is confirmed by the allocation of significant amounts of financial resources for combating climate change in the past decades. In fact, major attention is currently being paid to the possibility of financial resources raised by market and non-market-based mechanisms developed under the UNFCCC to contribute to the conservation of biodiversity, and not to the ability of programmes and mechanisms implemented under the CBD to contribute to climate change mitigation/adaptation. The history of the development of REDD+ is a clear confirmation of this statement. In this situation, the Governing Bodies of the UNFCCC preferred to keep this Convention unbounded by the institutional arrangements with other Rio Conventions. It is noteworthy that the UNFCCC and the CBD enjoy an almost universal membership, but the cooperation between them is not effectively promoted. It should be noted that this discourse may serve as another piece of evidence that the international organizations and treaty bodies are not just a sum of opinions of their parties, but are able to act as autonomous actors of international public law.[1404]

In this regard, it can be argued that the termination of the reduction commitments of the Kyoto Protocol and the adoption of the Paris Agreement may serve as a game-changer. The Paris Agreement stipulates that all States bear the responsibility to combat global warming. The terrestrial and marine ecosystems play a more important role in this treaty in comparison to the Kyoto Protocol. Actions to conserve and enhance the sinks and reservoirs of greenhouse gases are foreseen, and REDD+ is explicitly mentioned in the treaty text.[1405] In contrast

1404 *Cançado Trindade, Antônio Augusto,* International Law for Humankind: Towards a New Jus Gentium, 2010, pp. 181 ff.

1405 See Subsection D.III.4 of Chapter 4.

to the Kyoto Protocol, the Paris Agreement should, according to its Preamble, be interpreted, in such a way that the integrity of all ecosystems and the conservation of biodiversity should be promoted. This does not only mean that climate change mitigation and adaptation measures should be undertaken in order to avoid adverse impacts of climate change on ecosystems, but also that the relevant mechanisms and policies should be implemented in a manner that ensures the integrity of all ecosystems. Indeed, the above-mentioned references to environmental issues in the treaty text can potentially promote the development of such an approach to climate change mitigation that would deal with the climate issues and the conservation of ecosystems in a holistic manner.

In terms of stronger reference to environmental issues contained in the Paris Agreement, the contribution of REDD+ and other ecosystem-related mechanisms to the conservation of forest ecosystems can be enhanced through closer collaboration between the CBD and the UNFCCC. The fact that developing States have committed themselves to implement climate change mitigation measures in forests is confirmed by the frequent inclusion of REDD+ in the NDCs submitted by these States. However, the respective provisions of NDCs rarely mention environmental and social safeguards.[1406] Therefore, the risk of a damage to forests through the implementation of REDD+ is going to increase. Moreover, there is growing scientific evidence that the role of forests within climate change mitigation is not limited to the absorption and storage of CO_2. Forests also influence regional and global precipitation formation and distribution and act in multiple other ways.[1407] Against this backdrop, the adoption of an institutional arrangement between the CBD, on the one hand, and the UNFCCC and the Paris Agreement, on the other, gains further importance. However, it should be noted that the current practice of the COPs of these treaties does not give much hope for the conclusion of such an institutional arrangement in the near future. The issue of cooperation with other Rio Conventions was not even raised at the latest UNFCCC COP meetings.

1406 Ibid.

1407 See Subsection A.II.3 of Chapter 1.

C. Place of Forests in the Collaboration Between the CBD and the UNFCCC

I. Climate Change and Forestry Issues under the CBD

The CBD COP has always discussed forest topics in relation to climate change issues. It is important to note that the Working Programme on Forest Biodiversity outlines the need for collaboration with the UNFCCC and the UNCCD.[1408] Furthermore, the COP urged the parties to the UNFCCC and the KP to ensure that future climate-related activities in forests will be compatible with the conservation and sustainable use of biodiversity.[1409]

Moreover, the CBD COP 5 established an Ad Hoc Technical Expert Group on Forest Biological Diversity in order to support the work of the SBSTTA on forestry issues. In its first analysis, this Expert Group stated that climate change represents a serious threat to forest biodiversity, although the estimation of the gravity of this threat was complicated by scarce scientific knowledge and the impossibility to distinguish in certain cases between negative impacts of climate change and harm caused by other human activities.[1410] However, the expert group made a conclusion that in any case climate change will lead to the reduction of forest biodiversity.[1411] Consequently, the Governing Bodies of the UNFCCC were invited to enhance monitoring of interlinkages between forest biodiversity and climate change and to consider an opportunity for the establishment of an international network for assessing the impacts of climate change on forests and forest biodiversity.[1412]

1408 CBD COP Decision IV/7 "Work Program for forest biological diversity under the CBD", Annex to the decision IV/7, Decisions adopted by the Conference of the Parties to the Convention on Biological Diversity at its Fourth Meeting, held at Bratislava from 4 to 15 May 1998, United Nations, 1998. – Internet resource – https://www.cbd.int/doc/decisions/cop-04/full/cop-04-dec-en.pdf [Last visit: 30.04.2023]

1409 Article 16.

1410 CBD SBSTTA Information Note 7/3, Review of the status and trends of, and major threats to the forest biological diversity, prepared by the Ad Hoc Technical Expert Group on Forest Biological Diversity at its Seventh meeting, held in Montreal from 12 to 16 November 2001, United Nations, 2001, p. 106. – Internet resource – https://www.cbd.int/doc/meetings/sbstta/sbstta-07/information/sbstta-07-inf-03-en.pdf [Last visit: 30.04.2023]

1411 Ibid.

1412 CBD COP Decision VI/22, "Forest biological diversity", Decisions adopted by the Conference of the Parties to the Convention on Biological Diversity at its sixth

Besides the impacts of climate change on biodiversity, the Secretariat of the CBD outlined the role of terrestrial and marine ecosystems as carbon sinks, while pointing out that this role was not recognized by emission reductions trading schemes that existed at that time.[1413] It has further underlined the benefits that the conservation of biodiversity can deliver for combating climate change.[1414] Such statements can be interpreted as an implicit signal to the Governing Bodies of the UNFCCC that the forests and other ecosystems should be integrated into climate change mitigation/adaptation activities. However, before the concept of REDD+ was first proposed by Papua New Guinea in 2005,[1415] the COP CBD never explicitly called upon the Executive Secretariat of the UNFCCC to promote the role of forests in the fight against global warming.

CBD COP 9 underlined the need for further linkages between biodiversity and climate change while putting major emphasis on the promotion of collaboration between the Conventions.[1416]

Notably, the required options for collaboration between the Conventions do not go beyond consultations and exchange of information, promotion of capacity building, research and monitoring.[1417] These options for enhanced collaboration are undoubtedly important, but they do not guarantee that the gathered information and recommendations will be transformed into real actions on the ground. However, it should be evaluated positively that the activities proposed by Decision IX/16 also include reviewing national plans in order to identify areas where gaps in synergies exist.[1418]

Session, held at the Hague from 7 to 19 April 2002, Annex, Paragraph 40. – Internet resource – https://www.cbd.int/doc/decisions/cop-06/full/cop-06-dec-en.pdf [Last visit: 30.04.2023].

1413 *Secretariat of the CBD*, Interlinkages between biological diversity and climate change. Ad hoc Technical Group on Biological Diversity and Climate Change, CBD Technical Series No. 10, October 2003, pp. 1 ff.

1414 Ibid, p. 3.

1415 See Subsection D.IV.1 of Chapter 4.

1416 CBD COP Decision IX/16, Paragraphs 1 ff.

1417 Ibid, Annex II, Paragraphs 1–12, 20–31.

1418 Ibid, Annex II, Paragraphs 4–5.

It is noteworthy that the Global Biodiversity Outlook 3, which was prepared and published in 2010 shortly before the CBD COP 10 took place,[1419] underlined that current predictions of a dramatic decline in terrestrial biodiversity are largely based on the assessment of current trends in tropical forests.[1420] Consequently, if the conditions of tropical forests were improved, and the tropical forest loss was stopped, most biodiversity loss could be avoided. In this regard, the assessment argues that the two main threats to tropical forests are land-use change (conversion of forest areas) and climate change.[1421] The document in question determines that the interaction of these two hazards will largely determine the destiny of forest biodiversity.[1422] In order to improve the situation in tropics, the report calls for changes in land-use patterns through the application of sustainable forest management.[1423]

The Global Biodiversity Outlook did not develop further the legal concept of SFM. However, it further underlined that the reduction of human pressure on forests should occur under consideration of the CO_2 emissions that result from deforestation and forest degradation.[1424] In this regard, the REDD+ mechanism is the only specific international initiative explicitly mentioned in the part of the CBD report dealing with forests.[1425] Nevertheless, it is pointed out that the possible impact of REDD+-projects on forests will largely depend on their design.[1426] The fact that the focus was made on REDD+ signalizes that great attention is currently paid to climate change issues and the associating opportunities and threats for terrestrial biodiversity. At the same time, the report was not limited to them, but also argued that significant incentives for the conservation of biodiversity can be delivered by those schemes that foresee fair and equitable sharing of benefits arising out of the utilization of genetic resources.[1427]

1419 *Secretariat of the CBD*, Global Biodiversity Outlook 3, CBD, United Nations, 2010, p. 74.

1420 *Secretariat of the CBD 2010* [n. 1419], p. 75; see also *Long, Andrew*, The Convention on Biological Diversity, in: *Voigt, Christina* (eds.), Research Handbook on REDD+ and International Law, 2016, pp. 191 f.

1421 *Voigt* [n. 1420], pp. 191 f.

1422 Ibid.

1423 Ibid.

1424 *Secretariat of the CBD 2010* [n. 1419], p. 75.

1425 Ibid.

1426 Ibid.

1427 Ibid, p. 83.

However, the report remained blind on the promotion of synergies between REDD+ and ABS.

1. *Outcome of the COP CBD 10*

In terms of the conclusions of the Global Biodiversity Outlook 3, CBD COP 10 developed a guidance on the promotion of conservation and of sustainable use of biodiversity that would contribute simultaneously to climate change mitigation and adaptation.[1428] According thereto, Parties are invited to identify, assess, monitor and combat the adverse impacts of climate change on biodiversity and ecosystems, while great attention is paid to the promotion of the climate change resilience of species and ecosystems.[1429] For this purpose, the most sophisticated vulnerability assessments should be undertaken.[1430] The reduction of negative impacts of climate changes on biodiversity should be achieved through the enhanced conservation and sustainable utilization of biodiversity and ecosystem resources, meanwhile the ability of species and ecosystems to adapt to climate change should be further enhanced.[1431] Obviously, these goals can be achieved by avoiding other negative impacts on biodiversity, including, *inter alia*, chemical pollution, the destruction and fragmentation of natural habitats as well as the introduction of invasive species.[1432] That way, the CBD COP considered the fact that the natural resilience and vitality of forest ecosystems is higher when the natural level of biological diversity is maintained. Thus, this provision is important for the conservation of natural forests, whose ability to withstand unfavorable environmental conditions is significantly higher.

States are also invited to tackle specific climate-related hazards, for example, through better planning and management of water resources.[1433] This provision can contribute to the mitigation of changes in precipitation patterns. However, the COP did not mention the direct influence of forest ecosystems and

---

1428 CBD COP Decision X/33 "Biodiversity and climate change", Decisions adopted by the Conference of the Parties to the Convention on Biological Diversity at its tenth Session, held in Nagoya, Japan, from 18 to 29 October 2010, United Nations, 2010. – Internet resource – https://www.cbd.int/doc/decisions/cop-10/cop-10-dec-33-en.pdf [Last visit: 30.04.2023]

1429 Ibid, Paragraph 8(a).

1430 Ibid.

1431 Ibid, Paragraph 8(c–d).

1432 Ibid, Paragraph 8(d, i).

1433 Ibid, Paragraph 8(d.ii).

biodiversity on these patterns. It failed to consider that the IPCC report refers to the unclear impact of deforestation on the formation and distribution of atmospheric precipitation.[1434] The adaptation of natural ecosystems can also be promoted through expanding and strengthening protected areas, which should be integrated into the specific networks through wider involvement of natural linkages between habitats, such as ecological corridors.[1435]

An ecosystem-based approach to climate change adaptation is also promoted. In accordance with this approach, natural and near-natural forests are characterized by the highest level of resilience against climate change; consequently, States and other relevant actors should promote the integrity and vitality of forests.[1436] Importantly, ecosystem-based adaptation should be integrated not only in the overall adaptation strategies, but also in other national programmes, including national action plans to combat desertification, poverty reduction strategies, natural hazard mitigation programmes and strategies for promoting the long-term sustainability of agricultural activities.[1437]

The COP did not only underline the need for an ecosystem approach for adaptation, but also promoted this approach in relation to climate change mitigation,[1438] which can be relevant for the REDD+ mechanism and other forest-related projects. In this regard, the COP argued that a range of activities for restoring and conserving ecosystems, including protection of natural forests and mangroves, sustainable management of other forests and scientifically based approaches to a/reforestation, based on the utilization of native forest species in their natural composition, should be applied.[1439] The COP presumed that an ecosystem-based approach to climate change mitigation/adaptation is likely to contribute to the simultaneous achievement of the goals of the UNFCCC, the CBD and the UNCCD as well as of the Ramsar Convention.[1440]

Importantly, the CBD COP's guidance underlined that only forest areas of low ecological value or already degraded forest areas should be chosen for

---

1434 IPCC, Climate Change: The 1990 and 1992 IPCC Assessments, Intergovernmental Panel on Climate Change, 1992, p. 82.

1435 CBD COP Decision X/33 [n. 1428], Paragraph 8(d.iii).

1436 See *Chong, Joanne*, Ecosystem-based Approaches to Climate Change Adaptation: Progress and Challenges, in: International Environmental Agreements, Volume 14, Number 4, February 2014, p. 392.

1437 CBD COP Decision X/33 [n. 1428], Paragraph 8(k).

1438 Ibid, Paragraph 8(m–t).

1439 Ibid, Paragraph 8(n).

1440 Ibid, Paragraph 8(o).

conversion for expanding agricultural areas or for other purposes.[1441] At the same time, it can be criticized that the option of restoration of degraded forest ecosystems instead of their conversion was not prioritized in this case. As it was already pointed out above, the native or acclimatized tree species should be preferably chosen for planting, meaning that introducing invasive species should be avoided.[1442] However, the guidance did not oblige States to avoid planting invasive tree species. It remains unclear if the decision regards monoculture tree plantations as restored forests. Despite this, the COP promoted those approaches to re/afforestation that prioritize the maintenance and promotion of forest biodiversity. Planted or restored forests should also be located properly in order to increase the connectivity between forest ecosystems and to promote multiple forest benefits.[1443]

It is also noteworthy that the COP underlined that in the light of the promotion of ecosystem-based approaches for mitigation, the REDD+ mechanism should be designed properly in order to enhance benefits for biodiversity, and to avoid negative impacts on forests, whereas the representatives of ILCs should be widely included in the process of the elaboration of relevant strategies and projects.[1444]

The guidance further indicated that in order to determine the possible risks for biodiversity and the ways to address them, strategic environmental assessments (SEAs), as well as environmental impact assessments (EIAs), should be conducted.[1445] SEAs and EIAs are two schemes for the assessment of impacts of different activities on biodiversity that can be designed based on a voluntary guideline contained in Decision COP VIII/28.[1446] Whereas SEAs serve to assess environmental impacts of entire programmes and action plans, and to reconcile conflicting interests of different stakeholders, EIAs are conducted in order

1441 Ibid, Paragraph 8(i, p).

1442 Ibid, Paragraph 8(p.ii–iii).

1443 Ibid, Paragraph 8(p, v).

1444 Ibid, Paragraph 8(q).

1445 Ibid, Paragraph 8(u).

1446 CBD COP Decision VIII/28, "Impact assessment: Voluntary guidelines on biodiversity-inclusive impact assessment", Decisions adopted by the Conference of the Parties to the Convention on Biological Diversity at its eighth Session, held at the Curitiba, from 20 to 31 March 2006, United Nations, 2006. – Internet resource – https://www.cbd.int/doc/decisions/cop-08/cop-08-dec-28-en.pdf [Last visit: 30.04.2023]

to verify the environmental sustainability of specific projects.[1447] It is important that, in cases when negative impacts on biodiversity are discovered in the course of an EIA, the possible solution can be to abandon the project concerned, to develop a new one, and/or to shift the project implementation area.[1448] Another option can be to strengthen the environmental safeguards included in the project design and to develop the strategies for compensation of the harm to biodiversity.[1449]

The voluntary guidelines for EIAs and SEAs represent useful tools that can potentially facilitate the assessment of environmental impacts of REDD+-projects. However, the pertinent provisions of the CBD do not stipulate the duty to conduct these assessments before the implementation of climate change mitigation and adaptation activities. Notwithstanding this, the duty to conduct the EIA can be derived from the customary law principles of due diligence and prevention that address transboundary environmental damage.[1450] In the case of environmental damage within a State's borders, the applicability of these principles can be affirmed if the relevant obligation is established by any international treaty.

1447 *Iyer, Vijayan Gurumurthy*, Strategic Environmental Assessment (SEA) Process for Sustainable Mining and Mineral Management Development, in: Open Access Library Journal, Volume 4, Issue 04, 2017. – Internet resource – https://m.scirp.org/papers/75449 [Last visit: 30.04.2023]

1448 CBD COP 8, Voluntary guidelines on biodiversity-inclusive environmental impact assessment, Annex to the Decision CBD COP Decision VIII/28, Paragraphs 5(b) and 20–27.

1449 Ibid.

1450 Pulp Mills on the River Uruguay (Argentina v. Uruguay), Judgment of 20.04.2010, ICJ Reports 2010, pp. 55 f., Paragraph 101; Legality of the Threat of Use of Nuclear Weapons, Advisory Opinion of 8.07.1996, ICJ Reports 1996, p. 242, Paragraph 29; See also: *Proelß, Alexander*, Prinzipien des Internationalen Umweltrechts, in: *Proelß, Alexander* (Hrsg.), Internationales Umweltrecht, 2017, pp. 75–84; *Proelß, Alexander*, Raum und Umwelt im Völkerrecht, in: *Proelß, Alexander/Graf Vitzthum, Wolfgang* (Hrsg.), Völkerrecht, 7. Auflage, 2016, pp. 424, 427; *Brownlie, Ian*, Principles of Public International Law, 2008, pp. 275–285; *Birnie, Patricia/Boyle, Alan/Redgewell, Catherine*, International Law and the Environment, 3rd ed., 2009, pp. 143–152; *Chen, Ling*, Realizing the precautionary principle in due diligence, in: Dalhousie Journal of Legal Studies, Volume 25, 2016, pp. 1 ff.; *Yotova, Rumiana*, The Principles of Due Diligence and Prevention, in: International Environmental Law, Volume 75, Issue 3, 2016, p. 445; *Soto, Valverde Max*, General Principles of International Environmental Law, in: ILSA Journal of International & Comparative Law, Volume 3, Issue 1, 1996, pp. 195 ff.

In this regard, the parties to the CBD are obliged to conduct the EIA of any proposed project or policy that is likely to have "significant adverse impacts on biodiversity", even though the provision concerned is relativized by the wording "as far as possible and as appropriate".[1451] Despite the vagueness of this provision, REDD+-projects can be regarded as activities that can potentially have "significant adverse impacts on biodiversity". Therefore, the parties to the CBD are arguably obliged to conduct the EIA while implementing REDD+.

It is important to note that the CBD also incorporates the precautionary principle in its Preamble, stipulating that in the case of a threat of significant reduction or loss of biodiversity, a lack of full scientific certainty should not be used as a reason for postponing measures to avoid or minimize this threat.[1452] Combined with the general obligations established by the CBD, this provision means that if there is a threat that the implementation of REDD+ and other climate change mitigation/adaptation activities is likely to result in the loss of biodiversity, a State should consider an alternative project design or move a project into another area. Arguably, this principle is also relevant in relation to the construction of wind parks in biodiversity-rich forest regions of Europe.[1453] Notwithstanding this, the need for further interpretation of the wording "significant adverse impacts" and "as far as possible and appropriate" makes the practical application of these provisions uncertain. This is proved by the fact that the current experience with REDD+-projects demonstrates only a low level of attention to biodiversity issues in States-Parties to the CBD. Thus, stricter international legal provisions are still needed in order to further develop the duty of States to conduct EIAs before implementing REDD+.

It is important to note that the COP's voluntary guidelines underline that the conversion and degradation of areas that are important for the conservation and maintenance of biodiversity should be avoided.[1454] This provision also refers explicitly to the installation of sources of renewable energy and is thus important for the construction of hydro, wind- and solar power stations in forest areas.[1455]

1451 Article 14.

1452 See also *McIntyre, Owen/Mosedale, Thomas*, The Precautionary Principle as A Norm of Customary International Law, in: Journal of Environmental Law, Volume 9, Number 2, 1997, p. 228.

1453 *Baier, Tina*, Wie Klimaschutz der Natur schadet, in: Süddeutsche Zeitung, 20. August 2019. – Internet resource – https://www.sueddeutsche.de/wissen/klimawandel-tierschutz-energie-1.4558398?reduced=true [Last visit: 30.04.2023]

1454 CBD COP Decision X/33, Paragraph 8(v).

1455 Ibid, Paragraph 8(v.i,v).

In summary, Decision X/33 contains an extensive voluntary guideline on the promotion of conservation of biodiversity and natural habitats, and on the maintenance and enhancement of ecosystem services in the process of development and implementation of climate change mitigation and adaptation activities. It promotes a comprehensive approach that establishes links between climate change-related activities and an ecosystem approach. It refers explicitly to climate change mitigation activities in forests, including the REDD+ mechanism. Notwithstanding this, it is only a voluntary guideline, and the CBD COP did not deliver further regulations in relation to the promotion of its implementation. The corresponding provisions are vague and do not indicate which measures should be undertaken in order to create incentives for implementing biodiversity-friendly climate projects in forests.

The CBD COP 10 also adopted a decision on forest biodiversity,[1456] which promotes linkages between different forest-related policies, while requesting the Executive Secretary to strengthen the implementation of the Expanded Programme of Work on Forest Biodiversity and of the NLBI.[1457] Despite the declaratory nature of the provisions in question, it ought to be assessed positively that the COP requested the SBSTTA to develop a specific document in relation to the application of biodiversity-relevant safeguards for REDD+.[1458]

2. *CBD COP 11 and 12*

CBD COP 11 encouraged States to recognize the role of natural ecosystems and of biodiversity in climate change-related policies, and to enhance the synergies between climate change and biodiversity, by promoting traditional knowledge as well as scientific innovation.[1459] It also called for

---

1456 CBD COP Decision X/36, "Forest biodiversity", Decision adopted by the Conference of the Parties to the Convention on Biological Diversity at its tenth Session, held in Nagoya, Japan, from 18 to 29 October 2010, United Nations, 2010. – Internet resource – https://www.cbd.int/doc/decisions/cop-10/cop-10-dec-36-en.pdf [Last visit: 30.04.2023]

1457 Ibid, Paragraphs 5, 8.

1458 Ibid, Paragraphs 12–15.

1459 *CBD COP Decision XI/21*, "Biodiversity and climate change: integrating biodiversity considerations into climate-change related activities", Decisions adopted by the Conference of the Parties to the Convention on Biological Diversity at its eleventh Session, held at Hyderabad, India, from 8 to 19 October 2012, United Nations, 2012. – Internet resource – https://www.cbd.int/doc/decisions/cop-11/cop-11-dec-21-en.pdf [Last visit: 30.04.2023]

enhanced mobilization of resources for the purpose of scientific research on interrelations between biodiversity and climate, but failed to specify further how these resources can be activated.[1460]

CBD COP 12 further reiterated that an ecosystem-based approach must be integrated into climate change mitigation/adaptation and disaster risk reduction,[1461] while requesting that the Executive Secretary elaborate a summary of information on the issue.[1462] Subsequently, the Executive Secretary underlined multiple benefits of ecosystem-based approach to climate change mitigation and adaptation. It stated that certain limitations to its applicability exist, particularly from the point of view of ecosystem vulnerability,[1463] while arguing that the full extent of the benefits of this approach cannot be evaluated because a number of them, including cultural, educational, spiritual and scientific factors, cannot be expressed in monetary terms.[1464] It also mentioned that the lack of fair and equitable sharing of benefits obstacles the enhanced implementation of an ecosystem-based approach. The reason for this is that States and forest-dependent people are not motivated to promote the conservation of biodiversity as a part of climate change mitigation/adaptation and disaster risk reduction policies.[1465]

The Executive Secretariat also mentioned that the forestland conversion and other changes in land use such as the replacement of broadleaf forests with conifer tree plantations result in the changes in evapotranspiration and surface roughness.[1466] Despite mentioning that the impacts of such

---

1460 Ibid, Paragraph 5.

1461 CBD COP Decision XII/20, "Biodiversity and climate change and disaster risk reduction", Decisions adopted by the Conference of the Parties to the Convention on Biological Diversity at its twelfth Session, held at PyeongChang, Korea, from 6 to 17 October 2014, United Nations, 2014. – Internet resource – https://www.cbd.int/doc/decisions/cop-12/cop-12-dec-20-en.pdf [Last visit: 30.04.2023]

1462 Ibid, Paragraph 7(c).

1463 *SBSTTA CBD*, Synthesis report on experiences with ecosystem-based approaches to climate change adaptation and disaster risk reductions, 2 May 2016, Paragraphs 63, 77, 354, 387.

1464 Ibid, Paragraph 6.

1465 Ibid, Paragraph 7.

1466 *SBSTTA CBD 20*, Managing Ecosystems in the Context of Climate Change Mitigation: A Review of Current Knowledge and Recommendations to Support Ecosystem-Based Mitigation Actions that Look Beyond Terrestrial Forests, 31 March 2016, p. 7; *Myhre, Gunnar/Shindell, Drew/Bréon, François-Marie/Collins, William/Fuglestvedt, Jan/Huang, Jianping/Koch, Dorothy/Lamarque, Jean-François/Lee, David/Mendoza, Blanca/Nakajima, Teruyuki/Robock, Alan/Stephens, Graeme, Takemura, Toshihiko/*

changes on global climate are unknown, the Executive Secretariat erroneously concluded that they are likely to be insignificant. Finally, the Executive Secretariat adopted guidelines on integrating the issues related to genetic resources in climate change adaptation.[1467] It indicated that both issues are interdependent and that the conservation and enhancing of genetic resources should be integrated in all climate change-related activities.[1468]

3. *CBD COPs 13 and 14*

CBD COP 13 appreciated the adoption of the Paris Agreement,[1469] especially its provisions dealing with biodiversity,[1470] and encouraged Parties to fully consider the significance of safeguarding the integrity of ecosystems and to incorporate an ecosystem-based approach into the process of elaboration of the NDCs.[1471] However, once again these provisions are formulated in a vague manner. The COP only encouraged States to consider biodiversity issues but did not request them to do it; it referred to the importance of ensuring the integrity of natural ecosystems and not to their role in climate change adaptation/mitigation. That

---

Zhang, Hua, Anthropogenic and Natural Radiative Forcing, in: *Stocker, Thomas F./ Qin, Dahe/Plattner, Gian-Kasper/Tignor, Melinda M. B./Allen, Simon K./Boschung, Judith/Nauels, Alexander/Xia, Yu/Bex, Vincent/Midgley, Pauline M.* (eds.) Climate Change 2013: The Physical Science Basis, Contribution of Working Group I to the Fifth Assessment Report of the Intergovernmental Panel on Climate Change, 2013, pp. 687 ff.

1467 *SBSTTA CBD 20*, Voluntary guidelines to support the integration of genetic diversity into national climate change adaptation planning, 2 May 2016. – Internet resource – https://www.cbd.int/doc/recommendations/sbstta-20/sbstta-20-rec-10-en.pdf [Last visit: 30.04.2021]

1468 Ibid, Paragraphs 8–11.

1469 CBD COP Decision XIII/4, "Biodiversity and climate change", Decisions adopted by the Conference of the Parties to the Convention on Biological Diversity at its thirteenth Session, held at Cancun, Mexico, from 4 to 17 December 2016, United Nations, 2016, Paragraph 1. – Internet resource – https://www.cbd.int/doc/decisions/cop-13/cop-13-dec-04-en.pdf [Last visit: 30.04.2021]
Ibid, Paragraph 1.

1470 Besides already mentioned Article 5 dealing with the conservation and enhancement of carbon sinks in the natural ecosystems, the relevant provisions of the CBD include Article 7, which promotes linkages between the climate change adaptation and the conservation of biodiversity; and Article 8 dedicated to minimizing the loss and damage associated with climate change.

1471 CBD COP Decision XIII/4 [n. 1469], Paragraph 2.

way, the COP avoided any wording that could assign a legally binding character to its provisions or give rise to political commitments. The COP also encouraged States to ensure the application of information tools and guidelines developed by the Governing Bodies of the CBD, but only where appropriate.[1472] Moreover, the provision according to which these guidelines and tools have "to be used"[1473] is ambiguous, as it can be interpreted as the promotion of compliance with these documents as well as a simple call to consult these instruments and guidelines while developing NDCs.

However, it can be evaluated positively that CBD COP 14 encouraged States to avoid activities that can contribute to ecosystem degradation and biodiversity loss when designing and implementing an ecosystem-based approach to climate change adaptation and risk reduction,[1474] by citing, inter alia, the unsustainable utilization of natural resources as a major threat for biodiversity.[1475] The COP also adopted voluntary guidelines for the design and implementation of an ecosystem-based approach to climate change adaptation and disaster risk reduction, but these guidelines are not to be applied to climate change mitigation.[1476] They provide a non-exhaustive list of examples of ecosystem-based activities that should be implemented as a response to different hazards related to climate change.[1477] For example, forest and pasture restoration is proposed as one of the options to combat hazards such as drought, erratic rainfall and soil erosion in mountain and forest ecosystems.[1478] The restoration of mangrove forests is proposed as one of the options for reducing the risks associated with storm surges and sea-level rise.[1479] It is important to note that the voluntary guidelines also contain a list of safeguards, which should guide the implementation of ecosystem-based approaches.[1480] They foresee conducting environmental risk

---

1472 Ibid.

1473 Ibid.

1474 CBC COP Decision XIV/5, "Biodiversity and Climate Change", Decisions adopted by the Conference of the Parties to the Convention on Biological Diversity at its fourteenth Session, held at Sharm El-Sheikh, Egypt, from 17 to 29 November 2018, United Nations, 2018, Paragraph 3(c). – Internet resource – https://www.cbd.int/doc/c/9860/44b3/042fbf32838cf31a771bb145/cop-14-l-23-en.pdf [Last visit: 30.04.2023]

1475 Ibid, Annex, Paragraph 8.

1476 Ibid.

1477 Ibid, Annex, Paragraph 5.

1478 Ibid, Annex, Paragraph 5, table.

1479 Ibid, Annex, Paragraph 5, table.

1480 Ibid, Annex, Paragraph 8.

assessments, avoiding adverse impacts or "shift" of risks from one area to another; and enhancing and promoting biodiversity and ecosystem functions and services.

Notably, all these safeguards can also be relevant for the implementation of REDD+, although the document does not refer to this mechanism. This holds true also for three overarching considerations that should guide the implementation of an ecosystem-based approach, namely (1) wide application of traditional knowledge, technologies, experiences and efforts of ILCs, (2) mainstreaming the ecosystem-based approach in climate change adaptation, and (3) disaster risk management, raising of awareness and capacity-building.[1481] Moreover, the implementation of an ecosystem-based approach should be based on a stepwise approach. The first two steps include the understanding of the principles of ecosystem functioning and of the interrelations between nature and society. The next steps are the identification of the best options for the implementation of an ecosystem-based approach and the selection of the most suitable options based on different criteria including environmental compatibility, cost-effectiveness and social justice. Finally, a final project design is required to be developed and implemented. The implementation process has to be constantly monitored and evaluated.[1482]

To sum up, the international legal regime for biodiversity actively deals with the problem of climate change. The potential negative impacts on wildlife, as well as the role of biodiversity and natural ecosystems in climate change mitigation and adaptation is a major component in the measures adopted by the Governing Bodies of the CBD. Emphasis is put on adaptation activities, including both the adaptation of natural ecosystems to climate change and the importance of their ecological and socio-economic functions in the process of adaptation of human societies to climate change. Notably, the CBD also links these two major issues with disaster risk reduction. In this way, it promotes a holistic approach to the management of biodiversity and natural ecosystems. The role of biodiversity in climate change mitigation has also found its place in Decisions of the COP, although the relevant provisions play only a secondary role in these documents. However, it should be noted that, when following the wider interpretation of the relevant provisions, REDD+ should be used also for the promotion of climate change adaptation. Consequently, one should also recognize the major

---

1481 Ibid, Paragraph 9.

1482 Ibid, Paragraphs 23 ff.

significance of the CBD COP decisions in relation to climate change adaptation for the implementation of these mechanisms.

4. CBD Provisions Dealing with the Development and Implementation of REDD+ and the JMA

In 2008, directly after REDD+ was included in the Bali Action Plan, CBD COP 9 invited all relevant stakeholders to ensure that the implementation of REDD+ does not run counter to the objectives of the CBD but supports the implementation of its programme of work on forest biodiversity.[1483]

Subsequently, CBD COP 11 provided advice on the promotion of compliance with biodiversity-related safeguards for REDD+.[1484] The COP indicated that the safeguards for REDD+ should be implemented in the context of experiences gained from existing initiatives and policies focused on the sustainable use of forest biodiversity within the framework of SFM.[1485] However, this provision does not take into regard the risks and uncertainties related to the concept of SFM.

The COP indicated that the safeguards should be developed in order to mitigate the risks associated with implementing REDD+ and to deliver multiple benefits for the conservation of biodiversity and the protection of human rights. It is important to note that the COP noted that negative impacts of REDD+ include the conversion of natural forests to monoculture plantations of low environmental value and the displacement of deforestation to forest ecosystems with lower ability to sink CO_2 and of higher biodiversity value. Moreover, REDD+ can lead to the increased pressure on non-forest ecosystems.[1486] Other possible

---

1483 CBD COP Decision IX/5, "Forest Biodiversity", Decisions adopted by the Conference of the Parties to the Convention on Biological Diversity at its ninth Session, held in Bonn from 19 to 30 May 2008, United Nations, 2008, Paragraph 2(a). – Internet resource – https://www.cbd.int/doc/decisions/cop-09/cop-09-dec-05-en.pdf [Last visit: 30.04.2023]

1484 CBD COP Decision XI/19, "Biodiversity and climate change related issues: advice on the application of relevant safeguards for biodiversity with regard to policy approaches and positive incentives on issues relating to REDD+", Decisions adopted by the Conference of the Parties to the Convention on Biological Diversity at its eleventh Session, held at Hyderabad, India, from 8 to 19 October 2012, United Nations, 2012. – Internet resource – https://www.cbd.int/doc/decisions/cop-11/cop-11-dec-19-en.pdf [Last visit: 30.04.2023]

1485 Ibid, Paragraph 12.

1486 Ibid, Annex, Paragraph 4(a–c).

adverse impacts include the eviction of ILCs from their traditional territories and the restriction of their customary rights in relation to forests that can be aggravated by a lack of fair and equitable benefit sharing.[1487] Thus, the CBD COP mentioned the main concerns of the relevant stakeholders about environmental and social impacts of REDD+. The COP pointed out that the safeguards were intentionally developed in order to address these risks.[1488] However, the relevant provisions failed to strengthen the safeguards in question.[1489]

5. *Aichi Biodiversity Targets and REDD+*

The Aichi Biodiversity Targets, adopted by CBD COP 10, are relevant for the development and implementation of REDD+. They include five strategic goals and 20 targets. The first goal, Strategic Goal A, calls for combating the drivers of biodiversity loss and degradation by mainstreaming biodiversity issues in the decision-making processes in all relevant sectors.[1490] This goal is highly relevant for the REDD+ mechanism, as the relevant projects are more likely to contribute to the conservation of forests if biodiversity issues are directly integrated in the process of their development. In this way, the developers of REDD+ can more effectively combat the underlying causes of deforestation.

Strategic Goal B foresees the reduction of human pressure on biodiversity resources and the promotion of their sustainable utilization.[1491] This aspect is also relevant for REDD+: In cases when the biodiversity safeguards are properly implemented, these mechanisms can help to reduce anthropogenic pressure on forest resources. In the same manner, only the strict observation of safeguards can strengthen the links between REDD+ and Strategic Goal C, which foresees promoting a network of protected areas and strengthening biodiversity conservation efforts.[1492]

1487 Ibid, Annex, Paragraph 4(e–f).

1488 Ibid, Annex, Paragraph 3.

1489 See Ibid, Annex, Paragraph 6.

1490 CBD COP Decision X/2, "The Strategic Plan for Biodiversity 2011–2020 and the Aichi Biodiversity Targets", Decisions adopted by the Conference of the Parties to the Convention on Biological Diversity at its tenth Session, held at Nagoya, Japan, from 18 to 29 October 2010, United Nations, 2010. – Internet resource – https://www.cbd.int/doc/decisions/cop-10/cop-10-dec-02-en.pdf [Last visit: 30.04.2023]

1491 Ibid, Strategic Goal B.

1492 Ibid, Strategic Goal C.

It can also be evaluated positively that CBD Decision XI/19 contains a proposition to use the indicative list of indicators for evaluation of progress towards the achievement of the Aichi Biodiversity Targets in order to assess the contribution of REDD+ to the achievement of the CBD objectives.[1493] However, the indicators in question are not sufficiently elaborated and must be further interpreted by national authorities.[1494]

Thus, the documents adopted by the Governing Bodies of the CBD include a number of instruments and tools for promoting compliance with the safeguards for REDD+. However, the relevant provisions, all of which are to be applied on voluntary basis, suffer from a lack of clarity. The COP has provided guidance and certain positive incentives for the promotion of biodiversity conservation in the context of implementing REDD+, but the relevant documents are unlikely to strengthen the weak environmental and social safeguards for REDD+. It remains problematic that the contribution of REDD+-projects to the conservation and sustainable use of forest ecosystems and forest biodiversity is not a criterion to be considered while providing the results-based payments for REDD+.

II. Interlinkages Between the CDM, REDD+, the JMA and Biodiversity-Related Mechanisms

The legal frameworks for the development and implementation of REDD+ on the one hand and of ABS on the other do not explicitly promote the establishment of joint forest-related projects. As was shown in the previous chapters, biodiversity issues only play a marginal role in the Marrakesh Agreement, in the "Warsaw Framework for REDD+", in the Paris Agreement and in the relevant UNFCCC COP Decisions, just like how the Nagoya Protocol mentions the role of biodiversity in climate change mitigation/adaptation only in its Preamble.[1495]

Nevertheless, at least some provisions of the corresponding international legal documents can be interpreted in such a way that they deliver indirect impulses and positive incentives for the development of synergies between the mechanisms that promote valorization of ecological functions of forests.[1496] For

1493 CBD COP Decision XI/19 [n. 175], Paragraph 5.

1494 SBSTTA CBD Recommendation XV/1, Indicator framework for the Strategic Plan for Biodiversity 2011–2020 and the Aichi Biodiversity Targets, 7 December 2011, Annex I, Paragraph 5 ff. – Internet resource – https://www.cbd.int/doc/recommendations/sbstta-15/sbstta-15-rec-01-en.pdf [Last visit: 30.04.2023]

1495 See also *Van Asselt* [n. 1317], pp. 1229–1231.

1496 Ibid.

example, it is submitted that by encouraging States to promote the co-benefits of REDD+ and to address and support the safeguards while implementing REDD+, the UNFCCC COP has opened the way for the integration of ABS into forest carbon projects. States will have incentives to promote multiple forest benefits in the context of implementing REDD+ because the promotion of forest biodiversity may deliver additional monetary benefits.

On the other hand, the CBD COP has encouraged States to ensure that REDD+ is implemented consistently with its objectives, though without specifying how this should be achieved. The synergies between forest-related mechanisms implemented under the two regimes can represent a means for achieving this goal. The options for collaboration between the Rio Conventions proposed by the JLG also foresee the strengthening of synergies between the activities on the national level.[1497] With this provision in mind, the synergies between forest-related mechanisms should also be promoted.

1. Compatibility of the CDM, REDD+ and the JMA with ABS

The forest-related climate change mitigation/adaptation mechanisms are not equally compatible with ABS and other biodiversity-related mechanisms. For example, the CDM is limited to re/afforestation activities.[1498] Afforested areas cannot serve as a source of genetic resources. The biodiversity value of such young forests is far lower when compared to natural and near-natural forests. Moreover, the number of the forest-related CDM projects around the world is merely insignificant.[1499]

As the activities eligible under the REDD+ mechanism include reducing emissions from deforestation and forest degradation, such projects can be easier combined with activities undertaken within the frameworks of the CBD and the Nagoya Protocol. However, these mechanisms will be compatible with the biodiversity-related activities only if compliance with the safeguards for REDD+ is ensured. The same applies to the overall impact of REDD+ on forest ecosystems and forest biodiversity. This conclusion can be contested by the statement made by one commentator suggesting that the concept of REDD+ corresponds to the framework of the CBD rather than to the one of the UNFCCC.[1500] However, the lack of sufficiently effective safeguards as well as the current practice concerning

---

1497 See Subsection B.III of this Chapter.
1498 See Subsection C.III.1 of Chapter 4.
1499 See Subsection B.III of Chapter 4.
1500 *Long* [n. 1420], p. 202.

the implementation of REDD+ and its inclusion in the NDCs without taking due regard to biodiversity issues contest this statement.[1501]

Unless compliance with the safeguards in question is ensured, the implementation of REDD+ can have negative impacts on forest ecosystems. This is why the relevant activities may run counter to the main objectives of the CBD. Moreover, it should be noted that the implementation of REDD+ is subordinated to the main objective of the UNFCCC. Consequently, the above-mentioned statement, according to which REDD+ automatically fits to the scope of the CBD, cannot be supported. Their compatibility with ABS and other biodiversity-specific mechanisms is subject to the availability of favorable legislative and institutional frameworks on the national level. Of course, the potential enhancement of economic benefits through synergies between different forest-related mechanisms can motivate States to modify their policies guiding the implementation of REDD+. Notwithstanding this, a lack of certainty on the compatibility between different mechanisms demonstrates that the REDD+ mechanism is not necessarily compatible with the framework of the CBD.

The JMA mechanism is likely to have much greater potential for synergies with biodiversity-related activities, as these alternative policy approaches are aimed at achieving a sustainable and holistic management of forests. By putting emphasis on both mitigation and adaptation, they also incorporate by definition biodiversity-related objectives. In fact, Bolivia outlined in its proposal concerning the establishment of the JMA mechanism that these alternative policy approaches would not only contribute to the achievement of the UNFCCC objectives but would also promote the achievement of the CBD's objectives.[1502]

On the other hand, the lack of fundraising can become a major problem. Indeed, the concept of JMA foresees the availability of financial resources for their implementation according to the projected needs before any verifiable results are achieved.[1503] On the contrary, in the case of the ABS regime, the majority of payments are provided only after bioproducts are developed and commercialized. This practice corresponds to the concept of results-based payments for REDD+.

1501 See Subsection D.V.4 of Chapter 4.

1502 The Plurinational State of Bolivia, Proposal for the development of Joint Mitigation and Adaptation Mechanisms for the integral and sustainable management of forests, the UNFCCC. – Internet resource – https://unfccc.int/files/bodies/awg-lca/application/pdf/3_background_information_mecanismo-medium_(1)_bolivia.pdf [Last visit: 30.04.2023]

1503 See Subsection D.V.6 of Chapter 4.

Additionally, low attention is currently payed to the JMA mechanism, which results in a low number of developing States which choose this mechanism as an alternative to REDD+. This fact limits the possibilities for establishing synergies between the JMA and ABS. Despite this, States still have the option to combine the JMA mechanism with ABS.

It is submitted that in case of combination of JMA projects with the mechanisms for monetary evaluation of biodiversity, the reciprocal influence of both mechanisms will defer from the one in case of REDD+. In the second case, the main contribution of activities under the CBD and the Nagoya Protocol would be a shift from a carbon-centered approach to a comprehensive ecosystem-based approach. On the contrary, the concept of the JMA incorporates on its own a holistic approach to the conservation and sustainable management of forests by putting emphasis on the well-being of forest ecosystems and of forest-dependent people. The role of ABS would consist of delivering further financial resources for the conservation of forest biodiversity. Thereby, relevant stakeholders may receive more incentives for the implementation of the JMA. Thus, biodiversity-related market mechanisms can deliver further funding for the implementation of the JMA mechanism, whose implementation can be further supported by the activities under the CBD. Differently, in the case of synergies between REDD+ and ABS carbon projects can serve as a tool for the financing of biodiversity conservation.

It can also be assumed that States, when choosing the JMA mechanism as an alternative to REDD+, may be reluctant to integrate ABS into forest projects. Indeed, this choice is often inspired by the lack of acceptance of the utilitarian approach to environmental protection based on commoditization and commercialization of nature.[1504] The ABS regime can be perceived as the manifestation of the anthropocentric approach to the conservation of biodiversity that tends to overlook the intrinsic value of nature. Thus, it can be argued that those States that choose the JMA mechanism are more likely to reject the idea of synergies with the ABS regime and other similar biodiversity-related mechanisms and activities. At the same time, one should not neglect the fact that the policies and strategies under the CBD do not only deal with the commoditization of biodiversity resources, but also with the rights of ILCs. Therefore, States that oppose the utilitarian market-based approaches to the conservation of biodiversity may

1504 See also: *Wilson, Mark*, The Green Economy: The Dangerous Path of Nature Commoditization, Consilience, in: The Journal of Sustainable Development, Volume 10, Issue 1, 2013, pp. 85 ff.

be still willing to implement the ABS regime, as it can strengthen their position in international negotiations. Additionally, participation in these biodiversity-related market mechanisms can deliver further benefits for ILCs that are also in the focus of the JMA mechanism. As the CBD tends to promote a holistic attitude to the conservation of nature and introduces an ecosystem-based approach, the mechanisms developed under its framework are likely to be compatible with the conservationist mechanisms and approaches that promote the intrinsic value of nature.

Moreover, those strategies for the conservation of forests that entirely reject market-based approaches can be difficult to implement. Financial resources for the implementation of these strategies must come from public funds. However, developed States must be further motivated to participate in the implementation of the JMA mechanism.[1505] Indeed, the partnership between private and public stakeholders can stimulate fundraising.[1506] The benefits that domestic companies can retrieve from ABS and similar mechanisms can motivate these States to provide financial resources. It should not be expected that States and other stakeholders acting under the scope of the UNFCCC are sufficiently aware of the intrinsic value of nature to push them to choose the JMA mechanism as an alternative to the conventional REDD+-schemes. This desirable shift to alternative holistic policy approaches can be inspired by strengthening the linkages with the CBD and the Nagoya Protocol. Against this backdrop, it is submitted that non-market-based approaches such as the JMA mechanism can be further strengthened by promoting the synergies in question. Consequently, the JMA mechanism can also deliver an adequate basis for synergies with the CBD. Moreover, the synergies between the JMA mechanism and the CBD can represent a suitable way for intensifying forest conservation efforts.

2. *Promotion of Synergies on the National Level*

As was already mentioned, the promotion of synergies between REDD+- and ABS is not specifically supported by the relevant international legal documents.

1505 See also: *Casey, Frank/Vickerman, Sara/Hummon, Cheryl/Taylor, Bruce*, Incentives for Biodiversity Conservation: An Ecological and Economic Assessment, 2006, pp. 1 ff.

1506 *Casey/Vickerman/Hummon/Taylor* [n. 196], pp. 5, 11; *Wilson, George R./Hayward, Matt W./Wilson, Charles*, Market-Based Incentives and Private Ownership of Wildlife to Remedy Shortfalls in Government Funding for Conservation, in: Conservation Letters, Volume 10, Issue 4, 2017, pp. 485–492.

This means that there is a clear need for enhancing synergies between the two mechanisms on a national level.

With due regard to the above-mentioned differing levels of mutual compatibility of different mechanisms, it is submitted that the forest-related mechanisms for combating climate change are deeply related to those initiatives that deal with the monetary evaluation of the genetic resources of forest ecosystems.

In this regard, the CBD COP has outlined the need for strengthening efforts for promoting the positive effect of REDD+ on the achievement of the CBD objectives, by simultaneously providing benefits for the wildlife and indigenous people. This is to be achieved by promoting synergies between the national biodiversity strategies and action plans and the national strategies or action plans for the development and implementation of REDD+.[1507] Developed States are encouraged to contribute to the establishment of these synergies in developing States by promoting capacity building and technology transfer.[1508] Consequently, the synergies between REDD+ and ABS are more likely to occur on the national level. However, under the current conditions, States will be interested in such joint projects only if biodiversity-related initiatives will deliver higher economic benefits than the establishment of monoculture carbon plantations.

As the documents adopted by the Governing Bodies of the CBD and the UNFCCC stipulate,[1509] and as supported by the example of Cameroon,[1510] synergies on the national level are likely to occur at first through the exchange of information and other communicative interaction between the national institutions and bodies. However, as the next step, it is necessary to promote closer cooperation between institutions and strengthen the development of joint implementation strategies, such as joint projects. It is argued that at least four elements should be established and promoted on the national level in order to achieve an effective integration of biodiversity issues in climate-related forest projects and vice versa, namely, financial mechanisms, national programmes, national and regional projects and national actors (institutional level of synergies).[1511] For example,

---

1507 CBD COP Decision XI/19, Paragraph 7(a).

1508 CBD COP Decision XI/19, Paragraph 7(b).

1509 See Section B of this Chapter.

1510 *Somorin, Olufunso O./Visseren-Hamakers, Ingrid J./Arts, Bas/Tiani, Anne-Marie/Sonwa, Denis, J.*, Integration through Interaction? Synergy between Adaptation and Mitigation (REDD+) in Cameroon, in: Environment and Planning C: Government and Policy, Volume 34, Issue 3, 2016, pp. 429 ff.

1511 *Morita, Kanako/Matsumoto, Ken´ichi*, Synergies among Climate Change and Biodiversity Conservation Measures and Policies in the Forest Sector: A Case Study of

research addressing the promotion of synergies in the Southeast Asian States has revealed that in this region Thailand has the best possibilities for strengthening linkages between different mechanisms on the institutional level. It is achieved due to a single national ministry dealing with the promotion of environmental services of forests.[1512]

The example of Thailand highlights the significance of national actors for the achievement of synergies between forest-related climate change mitigation/adaptation and biodiversity conservation. In fact, the Ministry of Natural Resources and Environment is a single national authority dealing both with climate change mitigation/adaptation and the conservation of biodiversity, as well as with the protection of forests and reforestation. Forests play a major role in the 20-Year Strategic Plan for the Ministry of Natural Resources and Environment for the years 2017–2036.[1513] Thailand's second national communication to the UNFCCC has particularly outlined the role of REDD+ in combating climate change.[1514] The same national ministry also deals with the conservation and sustainable use of biodiversity as demonstrated by the Master Plan for Integrated Biodiversity Management for the period 2015–2021.[1515] At the same time, despite all these opportunities for synergies, implementation of REDD+ in Thailand is characterized by a low level of the consideration of biodiversity issues. It has been argued that the national land use policy is particularly favorable for the promotion of large-scale monoculture plantations.[1516] This situation hinders

Southeast Asian Countries, in: Forest Policy and Economics, Volume 87, February 2018, pp. 59 ff.

1512 Ibid, p. 62.

1513 Ministry of Natural Resources and Environment, 20-Year Strategic Plan for the Ministry of Natural Resources and Environment (B.E. 2560 – 2579), MONRE. – Internet resource – http://www.mnre.go.th/en/about/content/1065 [Last visit 30.04.2023]

1514 Ministry of Natural Resources and Environment, Thailand's second national communication under the UNFCCC, 24 Mar 2011. – Internet resource – https://unfccc.int/resource/docs/natc/thainc2.pdf [Last visit 30.04.2023].

1515 Ministry of Natural Resources and Environment, Master Plan for Integrated Biodiversity Management B.E. 2558–2564 (2015–2021). – Internet resource – https://www.cbd.int/doc/world/th/th-nbsap-v4-en.pdf [Last visit: 30.04.2023].

1516 *Sukhotanang, Anawat/Iadpala, Phanumard/Kamuttachart, Panuwat*, Thailand Experience with REDD+, CBD Asia-Pacific Workshop on REDD+, Singapore, 15 – 18 March 2011. – Internet resource – https://www.cbd.int/doc/meetings/for/wscbredd-apac-01/other/wscbredd-apac-01-thailand-en.pdf [Last visit: 30.04.2023]; *Boulay, Axelle/Tacconi, Luca*, The Drivers of Contract Eucalypt Farming in Thailand, in: International Forestry Review, Volume 14, Issue 1, 2012, pp. 1 ff.

the promotion of biodiversity-relevant safeguards for REDD+. In fact, national policies and regulations promoting the diffusion of monoculture plantations can also be found in other Southeast Asian States as well as in other tropical regions of the world.[1517] It can be argued that such policies can also have negative impacts on the promotion of synergies between the forest-related mechanisms for climate change mitigation/adaptation and the conservation of biodiversity. For this reason, the availability of national policies favoring the inclusion of biodiversity issues in REDD+-related activities should be added to the aforementioned four elements that are necessary for the promotion of synergies.

Arguably, States can potentially be motivated to promote synergies between both forest-related mechanisms, while these synergies can deliver further economic benefits for the developing States and ILCs, and can strengthen the position of developing States on the international stage. Moreover, the synergetic implementation of REDD+ and ABS can be more cost-effective. Notably, the benefits could be further enhanced if REDD+ were combined with mechanisms that foresee the valorization of biodiversity resources beyond the issues related to genetic resources. Such mechanisms can include a wider range of goods and services delivered by forest biodiversity. In this regard, some incentives and positive impulses are already delivered by international organizations.

3. *Joint BioTrade and REDD+-Projects*

The BioTrade initiative can serve as an example of these mechanisms. It was launched by the United Nations Conference on Trade and Development (UNCTAD) in 1996.[1518] According to the definition of the UNCTAD, BioTrade comprises all those activities that are related to the "collection, production, transformation and commercialization of goods and services" originating from biodiversity and natural ecosystems based on environmental and socio-economical sustainability.[1519] By developing the BioTrade initiative, the UNCTAD intended

---

1517 *Ching Liu, Corsa Lok/Kuchma, Oleksandra/Krutovsky, Konstantin V.*, Mixed-species versus Monocultures in Plantation Forestry: Development, Benefits, Ecosystem Services and Perspectives for the Future, in: Global Ecology and Conservation, Volume 15, 2018, pp. 1 ff.

1518 *Jaramillo, Lorena/Kane, Malick/Rimmer, Lawrence/Garcia, Beatriz/Frondisi, Isaura/Siqueira, Fernanda*, Training Manual on Developing Joint BioTrade and REDD+ Projects, 2016, p. 2.

1519 UNCTAD, Biotrade Initiatives, BioTrade Principles and Criteria for terrestrial, marine and other aquatic biodiversity-based products and services, 2020, p. 2.

to promote the implementation of three main objectives of the CBD on the level of ecosystems and species.[1520] In correspondence to the PES concept, the BioTrade is also deemed to serve as a stimulus for the conservation and sustainable use of biodiversity in accordance with the CBD provisions. The provisions in question foresee that every party shall implement measures that are economically and socially sustainable and at the same time can incentivize the conservation and sustainable use of biodiversity.[1521] In addition to the CBD, the BioTrade initiative is also guided by the documents adopted by the COP, among which the Strategic Plan for Biodiversity with its Aichi Biodiversity Targets and the Nagoya Protocol ought to be specifically mentioned.[1522] The BioTrade initiative has also found its way into a number of the CBD COP decisions.

Inter alia, CBD COP 8 explicitly mentioned the BioTrade initiative and invited the UNCTAD to continue to support the relevant programme of work.[1523] Moreover, CBD COP 9 invited the BioTrade initiative to continue to promote the trade in sustainably harvested biodiversity-based products in accordance with the objectives of the CBD.[1524] The CBD COP 12 described the BioTrade initiative as an "engine for the sustainable use of biodiversity and its conservation".[1525]

Thus, although the BioTrade initiative has not been launched by the COP, it is factually deeply interrelated with the provisions of the CBD. It has already found its way into a number of COP decisions. The implementation of this initiative is guided to a great degree by the guidelines and other documents developed by the COP. Moreover, the BioTrade initiative is recognized as one of the means for the achievement of the CBD's objectives. This means that it can be regarded as one of

1520 Ibid, p. 3.
1521 CBD, Article 11.
1522 *Jaramillo/Kane/Rimmer/Garcia/Frondisi/Siqueira* [n. 1518], p. 3.
1523 CBD COP Decision VIII/26, "Incentive measures: preparation for the in-depth review of the programme of work on incentive measures", Decisions adopted by the Conference of the Parties to the Convention on Biological Diversity at its eighth Session, held at the Curitiba, from 20 to 31 March 2006, United Nations, 2006, Paragraph 9. – Internet resource – https://www.cbd.int/doc/decisions/cop-08/cop-08-dec-26-en.pdf [Last visit: 30.04.2023]
1524 CBD COP Decision IX/6, "Incentive measures", Decisions adopted by the Conference of the Parties to the Convention on Biological Diversity at its ninth Session, held in Bonn from 19 to 30 May 2008, United Nations, 2008, Paragraph 13. – Internet resource – https://www.cbd.int/doc/decisions/cop-09/cop-09-dec-06-en.pdf [Last visit: 30.04.2023]
1525 COP 12 Decision XII/6, Paragraph 18.

the market-based mechanisms for the conservation and sustainable use of forest biodiversity that is relevant under the framework of the CBD.

The BioTrade initiative has already delivered a number of benefits across the globe. For example, it has favored the strengthening of community ownership over forests in Namibia. It has also contributed to enhancing the population of wild species in community-owned lands and has promoted more environmental-friendly harvesting methods and practices.[1526] Overall, BioTrade has contributed to the shift from a less effective centralized governance of forest ecosystems that does not consider the interests and concerns of local residents to a community-based management of natural resources that tends to promote sustainable livelihoods strategies.[1527]

Notably, the UNCTAD recently proposed to develop further synergies between REDD+ and the BioTrade initiative.[1528] According to this idea, the previously mentioned interrelationship between forests, biodiversity and climate justifies the need for an integral approach to their management.[1529] The development and implementation of projects that integrate both REDD+ and BioTrade can better valorize numerous functions of forest ecosystems and deliver greater benefits for local communities.[1530] However, while promoting the interlinkages between REDD+ and the BioTrade initiative, the UNCTAD is guided by the wrong assumption that the objectives and principles of the two initiatives always complement each other.[1531] It is affirmed that REDD+ will always promote the protection of forests from deforestation and forest degradation, while the conservation of biodiversity is considered to be a positive side outcome of relevant activities.[1532] However, as was mentioned above, if compliance

---

1526 ***Suleman, Kassahun Kelifa*, Enhancing Biotrade in an era of biodiversity mainstreaming, South African Institute of International Affairs, 18 May 2016. – Internet resource – https://saiia.org.za/research/international-day-for-biological-diversity-enhancing-biotrade-in-an-era-of-biodiversity-mainstreaming/ [Last visit: 30.04.2023]**

1527 ***Chevallier, Romy*, The State Of Community-Based Natural Resource Management in Southern Africa, South African Institute of International Affairs, 13 Sep 2016. – Internet resource – https://saiia.org.za/research/the-state-of-community-based-natural-resource-management-in-southern-africa-assessing-progress-and-looking-ahead/ [Last visit: 30.04.2023]**

1528 *Jaramillo/Kane/Rimmer/Garcia/Frondisi/Siqueira* [n. 1518], pp. 7 ff.

1529 Ibid, p. 24.

1530 Ibid

1531 Ibid.

1532 Ibid.

with environmental safeguards is not assured, REDD+ may also have negative impacts on biodiversity.

The UNCTAD has developed an extensive guidance on development and implementation of Joint BioTrade and REDD+ projects.[1533] This guidance can serve as a basis for the synergetic implementation of other climate change mitigation/adaptation activities and biodiversity-related projects in forest ecosystems. According to the guidance, it is necessary in a first step to identify ecosystems and species that will be covered by the project in question.[1534] Interestingly, the wording of the guidelines leads to the conclusion that only forest areas under risk of deforestation and forest degradation, but not areas designated for afforestation or reforestation, are proposed as an eligible project areas for the scope of the implementation of joint projects.[1535] Such an approach can create further incentives for the conservation of natural forests. It can be also argued that the limitation of joint projects to forest areas under the risk of deforestation and forest degradation will deliver further incentives for States to promote the conservation of forest ecosystems and forest biodiversity in the process of implementing REDD+, as it delivers an opportunity to obtain further benefits from the economic evaluation of non-carbon benefits of such ecosystems. Furthermore, project developers ought to verify that the selected forest area is not subject to any land tenure disputes.[1536] Not only land ownership, but also land use rights must be assured before the project implementation can be launched.[1537] Beyond the land, this also refers to carbon and biodiversity. Considering that in some States the use and ownership rights on biodiversity and carbon stocks are not explicitly recognized,[1538] it is necessary to identify the holder of these rights.[1539] Indeed, secure tenure rights play a crucial role in the context of implementing REDD+.[1540] In fact, relevant stakeholders may be reluctant to invest

1533 Ibid, pp. 29 ff.

1534 Ibid, p. 29.

1535 Ibid, p. 26.

1536 Ibid, p. 30.

1537 Ibid.

1538 See, for example *Vhugen, Darryl/Aguilar, Soledad/Peskett, Leo/Miner, Jonathan*, REDD+ and Carbon Rights: Lessons from the Field, Property Rights and Resource Governance Project (PRRGP), 2012. – Internet resource – https://www.land-links.org/wp-content/uploads/2016/09/USAID_Land_Tenure_REDD_and_Carbon_Rights_Lessons_from_the_Field.pdf [Last visit: 30.04.2023]

1539 *Jaramillo/Kane/Rimmer/Garcia/Frondisi/Siqueira* [n. 1518], p. 30.

1540 *Sunderlin, William D./De Sassi, Claudio/Sills, Erin O./Duchelle Amy A./Larson, Anne M./Resosudarmo, Ida Aju Pradnja/Awono, Abdon/Kweka, Demetrius Leo/Huynh, Thu*

in REDD+-projects in the cases where the ownership and user rights are not insured. Moreover, small landowners and ILCs can be disadvantaged in the process of project implementation.[1541] Concerns over the lack of secure land tenure rights can be further fueled by the fact that there is the clear evidence that REDD+-projects are strengthening social injustice.[1542] In order to avoid negative outcomes for forest-dependent people, documentary assured land tenure rights ought to be secured for the entire period of project implementation and verification.[1543]

At the stage of validation, a validating entity has to verify if the joint REDD+ and BioTrade project is developed according to the rules and guidelines established by the CCB standard or other standard chosen by project developers. If in the process of validation serious deviations from the chosen standard are discovered, a list of non-conformity responses (NCR) must be submitted to the project developers in order to provide them with a guidance for the improvement of a project design.[1544] The project should not be validated until the project design will be modified.[1545]

However, it is not clear which entity should be responsible for the project validation. The guidelines in question indicate that it should be an "independent body"[1546] with experience with similar climate or biodiversity-related projects. However, a number of important issues remain unclarified. The possibility is not excluded that the project developers will influence the process of designating the validating entity. In fact, the guidance explicitly determines that the project developers ought to sign a contract with the validation body, meaning that their approval is likely to be decisive for the selection of this validating entity.

The same ambiguity exists in relation to the determination of the standard according to which the project should be validated. How transparent and accurate this standard should be is not clarified in the manual. However, such a lack

---

B, Creating an appropriate tenure foundation for REDD+: The record to date and prospects for the future, in: World Development, Volume 106, June 2018, pp. 376 ff.

1541 *Jaramillo/Kane/Rimmer/Garcia/Frondisi/Siqueira* [n. 1518], p. 30; *Chomba, Susan/ Kariuki, Juliet/Lund, Jens Friis/Sinclair, Fergus*, Roots of inequity: How the implementation of REDD+ reinforces past injustices, in: Land Use Policy, Volume 50, June 2018, pp. 202 ff.

1542 *Chomba /Kariuki/Lund/Sinclair* [n. 1541], pp. 202 ff.

1543 *Jaramillo/Kane/Rimmer/Garcia/Frondisi/Siqueira* [n. 1518], p. 30.

1544 Ibid.

1545 Ibid.

1546 Ibid.

of transparency and legal certainty is unlikely to be resolved within voluntary guidelines such as those provided by the UNCTAD. The question arises whether and to what extent the relevant issues can be regulated by the non-legally binding framework for joint projects. It can be argued that international strong legally binding regulations are needed in order to promote the protection of forests through joint projects.

Thus, voluntary approaches for the development of joint projects such as the one promoted by the UNCTAD give hope for a breakthrough in the establishing of synergies between REDD+ and biodiversity-related mechanisms. The guidelines elaborated by the UNCTAD are far more precise and far-reaching than the provisions of the CBD and UNFCCC COPs' decisions. They deliver an important example of a framework for the development of joint projects. With certain modifications, it can also be applied to joint REDD+ and ABS projects. However, voluntary and legally non-binding documents cannot deliver robust and transparent regulations.

4. *Impact of the Synergies Between Forest-Related Mechanisms on the Environment and on the Rights of ILCs*

The actual impact of synergies between REDD+ and biodiversity-related mechanisms on forests, biodiversity, and ILCs will depend on their contribution to the enforcement of the respective environmental and social safeguards. In relation to the promotion of interests of ILCs, the ABS regime with its requirements of MAT and PIC ensures a significantly higher level of protection of their rights than the REDD+-mechanism.

For this reason, the simultaneous implementation of REDD+ and ABS can potentially improve the situation of ILCs and provide them with further means of protecting their rights and freedoms. In addition, the BioTrade initiative can contribute to the promotion of the rights of ILCs which occupy a prominent place on its agenda.

Arguably, the requirement of PIC and MAT for the ABS regime of genetic resources is unlikely to be automatically extended on REDD+ in the case when joint projects will be developed and implemented. Notwithstanding this, the promotion of synergies between both mechanisms presupposes the elaboration of unified policies and procedures on the national level. At any time, forest-dependent people will be able to deny their PIC for joint project activities if their rights are not addressed by the REDD+-related part of the project concerned. However, this strategy presupposes awareness of their rights on the side of ILCs. Consequently, synergies between both forest-related legal mechanisms will substantially improve the conditions of forest-dependent people. Notwithstanding this, they will hardly provide a remedy for the lack of stronger provisions in the international legal framework for REDD+.

Synergies between both forest-related legal regimes can also promote the conservation of forest biodiversity and of natural forest ecosystems. Although the Nagoya Protocol does not establish actual duties to preserve forest ecosystems, the concept of ABS presupposes the conservation and sustainable use of biodiversity resources in order to ensure the availability of genetic resources in the future. Additionally, the implementation of ABS is subordinated to the main objectives of the CBD. In other words, the promotion of biodiversity conservation is to certain extent inherent to the provisions of the Nagoya Protocol. Incorporating environmental issues in ABS through the reference to the main objectives of the CBD can be more effective than declaratory and vaguely formulated environmental and social safeguards for REDD+.

At the same time, the major weak point of the ABS regime is its unclear contribution to the promotion of forest finance. In fact, there are no clear obligations to redirect the revenues from bioprospecting into the financing of forest conservation. So far, only a small number of ABS projects have been implemented. On the other hand, the idea to use REDD+ as a tool for financing the conservation of biodiversity was discussed from the very beginning of negotiations of this mechanism.[1547] However, it is unlikely that the benefits from REDD+ projects will flow automatically in the conservation of forest biodiversity, as this issue does not belong directly to the goals of climate change mitigation/adaptation mechanisms. This does not hold true for the JMA mechanism, whose scope is to promote the conservation of forest ecosystems, as the mechanism in question does not prioritize the enhancement of carbon stocks over biodiversity issues as the conventional REDD+-scheme does.

Moreover, the joint implementation of both mechanisms can favor integration of biodiversity concerns in all stages of implementing REDD+, including the development, implementation, evaluation, and verification. Such integration is crucial for the achievement of environmental sustainability of these mechanisms.[1548] The integration of both climate change and biodiversity concerns

1547 *Loft, Lasse*, Erhalt und Finanzierung biologischer Vielfalt – Synergien zwischen internationalem Biodiversitäts- und Klimaschutzrecht, 2009, pp. 211–212.

1548 *Gardner, Toby Alan/Burgess, Neil David/Aguilar-Amuchastegui, Naikoa/Barlow, Jos/Berenguer, Erika/Clements, Tom/Danielsen, Finn/Ferreira, Joice/Foden, Wendy/Kapos, Valerie/Khan, Saiful M./Lees, Alexander C./Parry, Luke/Roman-Cuesta, Rosa Maria/Schmitt, Christine B./Strange, Niels/Theilade, Ida/Vieira, Ima C. G.*, A framework for

within a joint implementation scheme can help to reduce the overall costs of monitoring the projects' impacts on biodiversity.[1549]

It should be also noted that the simultaneous implementation of climate change-related and biodiversity-related projects on the national scale could help to avoid the displacement of deforestation into areas with low forest carbon stocks and with high biodiversity value. However, certain forest ecosystems will still be disadvantaged in this process. For example, boreal forests have a lower biodiversity value when compared to tropical forests. Moreover, some researchers speculate that the contribution of boreal forests to climate change mitigation is relativized by their albedo warming effect.[1550] However, it is necessary to avoid a situation where REDD+ and other mechanisms implemented in developing States would result in higher deforestation rates in temperate and boreal forests, both of which also fulfil important environmental functions on the global and regional scale. This is why it is problematic that the forest mechanisms implemented in the tropics does not consider the risk of enhanced hazards for boreal forests. Arguably, the wording of the relevant provisions of the CBD and the UNFCCC leads one to conclude that they deal predominantly with the risk of displacement of deforestation only in tropical regions. For example, the CBD COP invited States and international organizations to reduce the leakage into forest ecosystems of low carbon value and of high biodiversity value.[1551] Nevertheless, if read in the context of the entire Decision XI/19, this provision can be interpreted as referring primarily to developing tropical States. Avoiding a shift of deforestation and forest degradation from tropical regions to boreal ones and

---

Integrating Biodiversity Concerns into National REDD+ Programmes, in: Biological Conservation, Volume 154, October 2012, p. 62.

1549 Ibid.

1550 For overview of different estimates of the effects of afforestation in boreal and temperate areas see: *South, David B./Lee, Xuhu/Messina, Michael G.* Will Afforestation in Temperate Zones Warm the Earth? in: Journal of Horticulture and Forestry, Volume 3, Number 7, 2011, pp. 195–199; For the importance of preservation of old-growing boreal forests for the climate change mitigation/adaptation see: *Olsson, Roger*, To Manage or Protect? – Boreal Forests from a Climate Perspective, in: Air pollution and climate series, Volume 26, 2011, pp. 28 ff.; Particularly for the role of intact Canadian boreal forests see: *Ireland, Perin*, Canada's Boreal Forest Is a Carbon Bomb Unless We Keep It Intact, Natural Resource Defense Council, October 12, 2018. – Internet resource – https://www.nrdc.org/stories/canadas-boreal-forest-carbon-bomb-unless-we-keep-it-intact [Last visit: 30.04.2023]

1551 Decision XI/19, Paragraph 14.

vice versa is another topic that should be covered by the legal framework of both treaties in order to ensure the protection of forest ecosystems on a global level.

However, the ABS regime cannot completely resolve the problem of leakage, because biodiversity samples can also be gathered in forest areas of small size. On the other hand, REDD+ usually requires larger forest areas for its implementation. Therefore, there is a risk that States will establish higher standards for the conservation of biodiversity on smaller areas whereas outside of these areas REDD+ will be implemented in an unsustainable manner. This risk will be reduced if States receive higher revenues when participating in the ABS regime, BioTrade initiative, or other biodiversity-related mechanisms. However, in order to avoid the conversion of natural ecosystems into monoculture plantations, the projects in question must be highly profitable.

Consequently, biodiversity-related mechanisms can strengthen the environmental safeguards, whereas REDD+ can serve as a tool for financing forest conservation activities.[1552] In the case of the JMA mechanism, the overall outcome of synergetic implementation will be also positive for forest ecosystems and forest-dependent people, but in this case, the role of ABS would be to deliver further incentives for the implementation of alternative policy approaches. For this reason, it can be argued that both the conservation of forests and the rights of ILCs can benefit from the synergetic implementation of the mechanisms in question.

5. *Avoiding Overlaps Between the Two International Legal Regimes When Applied to Forests*

The issue of unnecessary overlaps between REDD+ and ABS has not yet been discussed in legal literature. It is submitted that this issue should not represent a major challenge. In fact, as already demonstrated, the promotion of synergies between two types of forest-related mechanisms is likely to be beneficial for forest ecosystems and forest-dependent people. Consequently, avoiding overlaps between the two mechanisms should be understood as a more efficient distribution of financial and technical resources in order to ensure the best possible overall outcome for forest ecosystems, for developing States, and for ILCs. For example, the facilities used for raising public awareness on REDD+ can also be used for strengthening the capacity building on ABS. The benefits delivered by both mechanisms should also be allocated and distributed properly in order to

1552 See also *Loft* [n. 1547], pp. 211–212.

assure that the conservation efforts and co-benefits of projects are promoted in an equal manner in different forest ecosystems and local communities. Arguably, these issues can be resolved through better coordination and exchange of information between entities that implement different projects. At the same time, better coordination is likely to contribute substantially to the promotion of synergies between different mechanisms.

In conclusion, the developing States are primarily responsible for avoiding unnecessary overlaps between projects. Moreover, overlaps should not be perceived as a serious obstacle for the promotion of synergies between REDD+ and ABS.

D. Conclusions

The establishment and promotion of synergies between different forest-related policies and mechanisms can contribute to the development of a holistic approach to the conservation and sustainable utilization of forests and their resources. Those ecological functions of forests that are not currently covered by REDD+ can be valorized by enhancing synergies with the mechanisms and initiatives developed and implemented under the scope of the CBD. More financial resources can be raised, and a shift to a more biodiversity-friendly implementation of climate change mitigation/adaptation activities can be promoted. Moreover, there is the possibility that the development of synergies between both regimes may contribute to the shift from those projects that favor the establishment of monoculture tree plantations to those initiatives that promote the conservation of natural and near-natural forests. This assumption can be supported by the fact that natural forests deliver a perfect basis for both the conservation and sustainable use of biodiversity resources and for climate change mitigation/adaptation.

However, the promotion of links and synergies has so far not received sufficient attention in international documents. The cooperation between the three Rio Conventions is basically restricted to the exchange of information and to the promotion of capacity building, whereas the development and implementation of joint projects only plays a marginal role in the texts of the CBD and the UNFCCC and in the respective documents adopted by the COPs. The main contribution to strengthening synergies and links between different mechanisms has been made by the Governing Bodies of the CBD; however, the documents in question are deprived of any legally binding force. The idea to use indicators developed in order to evaluate the progress in application of the Aichi Biodiversity Safeguards for verification of the promotion of safeguards for REDD+ can be evaluated positively. At the same time, these indicators are too vague and unspecified. Moreover, there is no legal duty to apply them while evaluating the implementation of REDD+.

It can be argued that the promotion of synergies must occur on the national level of developing States. The development of synergies on the national level represents a plausible and realistic option. While most States are parties both to the CBD and the UNFCCC, in order to achieve the synergetic implementation of different mechanisms, States must establish synergies between the relevant institutions, policies, and procedures. This process requires serious efforts on the national level. Moreover, the promotion of synergies can be hindered by national policies that cause a damage or threat to biodiversity, such as the expansion of monoculture plantations.

The promotion of synergies between different policies could be further enhanced if States were aware of the benefits that potentially arise from the simultaneous implementation of climate change- and biodiversity-related mechanisms. However, the lack of synergies on the international level can have a negative impact on the strengthening of links between REDD+ and ABS on the national level. For this reason, it is submitted that the parties to the CBD and the UNFCCC should elaborate stronger, precise, and detailed regulations in relation to the development and implementation of joint projects. The efforts of the international community should be mainstreamed, *inter alia*, into the promotion of synergies between different forest-related mechanisms. However, it would be a mistake to limit the further development of the international legal regime for forests only to the promotion of synergies. A lack of strong safeguards for REDD+ cannot be completely resolved by promoting synergies. Thus, there is a need to consider other approaches to the development of the international legal regime for forests that can contribute to the promotion of forest protection.

Chapter Six Alternative Approaches to the Development of the International Legal Regime for Forests

A. Introduction

In the course of the analysis of the international legal regime for forests in previous chapters, it was demonstrated that there is a lack of effective mechanisms that could significantly reduce deforestation rates and promote forest conservation efforts. Despite progress, the problem of deforestation and forest degradation remains unresolved in many regions around the world.[1553] Some of the most valuable primary forests are still subjected to forest clearances, industrial logging, forestland conversion, and other threats caused by human activities. The international trade in unsustainably harvested timber and agricultural products that are cultivated on converted forestlands or in tree plantations through equally unsustainable agricultural techniques contributes to the rapid rate of deforestation. States, forest owners, and multinational corporations still maintain unsustainable forest management practices, while seeking to reap higher short-term revenues. At the same time, poverty, underdevelopment, and a lack of sustainable forestry and agricultural practices in the forest-rich States remain as major driving forces of deforestation; in fact, all these could be at least partially remedied through international cooperation.

Thus, the problem of deforestation is still directly linked to the global governance. The issue of the conservation of forests remains to a great extent within the realm of international environmental law. At the same time, the ratification and implementation of international forest-related documents are proceeding with considerable difficulties. As demonstrated in the previous chapters, REDD+, JMA, ABS, and other similar climate- and biodiversity-related mechanisms can promote the protection of forests; however, these mechanisms are plagued by a slew of uncertainties, including financial issues, promotion of environmental conservation, and the protection of ILCs' rights. In the case of REDD+, the pertinent international provisions are not robust enough to ensure that its

1553 See *Rayner, Jeremie/Buck, Alexander/Katila, Pia*, Embracing Complexity in International Forest Governance: A Way Forward, Policy Brief, IUFRO, 2010, pp. 8.

implementation will contribute to the protection of natural forests and forest biodiversity, rather than to forestland conversion and the establishment of monoculture plantations. Against this backdrop, it should be underscored that the international legal regime for forests still lacks a holistic approach that could comprehensively take into regard major forest-related issues.

Thus, this research would be incomplete without an analysis and reconsideration of the approaches that could enhance the effectiveness of the international legal regime for forests. These approaches should effectively tackle the challenges posed by the complex issue of deforestation. The need for action is also determined by the fact that the consequences of deforestation on global, regional, and local levels are still not completely clear, even though they are likely to be detrimental for the human civilization as a whole.[1554]

Adopting a legally binding forest convention has been on the table since the early 1990s. For legal clarity and transparency, which could be strengthened by this treaty, the relevant issue deserves to be revisited in this chapter. A further inquiry into the possibility of the adoption of a forest convention and its potential contribution to the protection of forests is necessary. This need is also dictated by the fact that the attempts to resolve related problems, such as the conservation of forest ecosystems, by compartmentalizing and handling them individually often prove to be misleading. For instance, various conflicting international legal documents have shown ineffective results in the protection of the indigenous people's cultural heritage.[1555] Moreover, the fourth largest decline ever in tropical forests around the world in 2018 – with 12 million hectares of forests destroyed – had demonstrated how inconsistent and insufficient today's international forest-related policies are.[1556]

It can also be argued that even if the international community had failed to reach a consensus on the adoption of a treaty, negotiations on the global scale should resume as soon as the international political situation modifies or the key stakeholders' stand on forest issues changes.

1554 See Subsection B.II of Chapter 1.

1555 *Gilbert, Jérémie*, Indigenous Peoples, Human Rights, and Cultural Heritage: Towards a Right to Cultural Integrity, in: *Xanthaki, Alexandra/Valkonen, Sanna/Heinämäki, Leena/Nuorgam, Piia Kristiina* (eds.), Indigenous Peoples and Cultural Heritage: Rights, Debates, Challenges, 2017, pp. 20–38.

1556 *Suliman, Adela*, World's Forests "in Emergency Room" After Years of Losses, in: Reuters, 25 April 2019. – Internet resource – https://www.reuters.com/article/us-global-environment-climatechange/worlds-forests-in-emergency-room-after-years-of-losses-idUSKCN1S10DZ [Last visit:30.04.2023]

Notwithstanding the possible benefits of a legally binding forest convention, the possibility of its adoption is considered low. This means that it is also necessary to focus on other approaches to develop the international legal regime for forests, going beyond climate change and biodiversity-related approaches discussed in previous Chapters. These approaches include the adoption of legally binding international documents or in the strengthening of existing soft law documents dedicated to forests. In previous Chapters, the regulation of forest issues through other conventions with forest conservation as a side effect, as in the case of UNFCCCC, was analyzed. The further development of specific law dedicated to forests is, therefore, in focus of this chapter. As alternatives to adopting the forest convention, the relevant approaches can include implementing forest-specific protocols to the existent conventions or strengthening the functions of international coordinative bodies dealing with the forest issues, including the UNFF and the Collaborative Partnership on Forests. Clearly, the international bodies exercising only coordinative and consultative functions can hardly contribute to the forest conservation. To be able to promote the conservation of forests in a consistent and holistic manner, the UNFF and other relevant bodies ought to be adequately empowered. Otherwise, their increased coordinative function will hardly contribute to a more holistic approach. However, the enhancement of the UNFF's role must be analyzed as a plausible interim measure until robust approaches will be available. In addition, the effectiveness of the combination of different approaches will be also analyzed in this Chapter.

The international community must nonetheless ensure that different policies, procedures and practices won't have adverse impacts on the environment and forest-dependent people. For this reason, possible impacts on forest ecosystems as well as local communities will also be considered in this chapter's analysis.

B. Need for the Further Development of International Law Specific to Forests

By analyzing the existing international legal regime for forests, it is possible to shed light on the question of whether there is a need for further development of international law specific to forests. It also evaluates whether the problem of deforestation and forest degradation can be at least partially resolved through the development of forest-related mechanisms of the existing international environmental treaties, and enforcing implementation of their relevant provisions.[1557]

1557 See *MacKenzie, Catherine P.*, Future Prospects for International Forest Law, in: The International Forestry Review, Volume 14, Issue 2, 2012, pp. 250 f.; *Gießen, Lucas*,

As it was demonstrated in previous Chapters, forest-related mechanisms and policies are now developed and implemented under the scope of other international environmental legal regimes, with the particular emphasis on the REDD+-mechanism. In addition, higher attention is now given to the regulation of relationships between and among different forest-related treaties. In this regard, it is a move towards the right direction: efforts are now mainly devoted to the promotion of coordination and cooperation between different forest-related regimes, rather than credit the resolution of conflicts resulting from the provisions of different conventions. Nevertheless, fundamental controversies remained unresolved. An example of this is the fact that States are allowed to implement those climate change mitigation and adaptation mechanisms, which can be harmful for forest ecosystems and for forest biodiversity.[1558]

However, the cooperation and coordination between and among different conventions will unavoidably be limited by the independent nature of these treaties. States are deeply concerned that their rights and obligations set by one of the conventions will be undermined by the provisions of another treaty.[1559] In fact, by signing any international convention, the States agree to a number of legal provisions that determine their corresponding rights and duties. Every provision represents a part of a larger international consensus that has made the adoption of a treaty possible. The promotion of cooperation with other treaties that can alter the independent character of the first convention is a threat for this consensus.[1560] Such considerable modification of rights and duties set by the treaty, which makes their fulfillment dependent on the provisions of other conventions, would require the renegotiation of the treaty itself.[1561] Hence, the modification of any treaty through its internal negotiation process or adoption of a new treaty can represent a more viable approach than establishing such wide-going cooperation that alters the independent character of international treaties.

Reviewing the Main Characteristics of the International Forest Regime Complex and Partial Explanations for its Fragmentation, in: International Forestry Review, Volume 15, Issue 1, 2013, p. 60.

1558 See Subsection D.V.1 of Chapter 4.

1559 See Subsection B.I of Chapter 5.

1560 *Gaynutdinova, Tamilla/Juncurt, Stefan*, Workshop on "Forests and Forest Ecosystems: Promoting Synergy in the Implementation of the Three Rio Conventions", in: Earth Negotiations Bulletin, International Institute for Sustainable Development, Volume 4, Number 174, 2004, p. 3. – Internet resource – http://enb.iisd.org/download/pdf/enb04174e.pdf [Last visit: 30.04.2023]

1561 Ibid.

The latter option is plausible only if the rights and duties established by two conventions exhibit a purely complementary character without contradicting each other at all. Therefore, the different highly specific mandates of the conventions pose clear limits on the promotion of cooperation between them.[1562]

Thus, the strengthening of coordination and cooperation between and among existent forest-related conventions cannot be seen as a viable way for the resolution of the present deforestation crisis.

Another argument that militates in favor of a further development of international environmental law specific to forests is the complex nature of deforestation and forest degradation. It makes impossible to consider all major forest-related issues under the framework of any legal regime established by any convention, which is in the first place not primarily aimed at ensuring the preservation of world forest ecosystems. The current practice with the REDD+-mechanism and the ABS regime of genetic resources proves this: both mechanisms fail to consider multiple functions of forests and numerous forest-related topics.[1563] So long as the convention deals with only a single aspect of forest management, it will be ineffective in establishing holistic and comprehensive policies and mechanisms to combat deforestation and forest degradation – never mind if the convention's title refers to forest-specific issues, such as the ITTA.[1564] In the specific case of the ITTA, credit should be given for that the text of this commodity agreement dealing primarily with the regulation of the international timber trade tends to also promote sustainable forest management in tropical forest ecosystems.[1565] However, forest conservation requires the development and implementation of those policies and strategies that go beyond the proper management of timber harvesting practices and international timber trade. The regulations cannot combat other driving forces of deforestation, such as the ongoing conversion of forest areas into the farm fields and monoculture tree plantations.[1566] Moreover,

---

1562 See *Van Asselt, Harro*, Managing the Fragmentation of International Environmental Law: Forests at the Intersection of the Climate and Biodiversity Regimes, in: New York University Journal of International Law and Politics, Volume 44, 2012, pp. 1265 ff.

1563 See Conclusions to Chapters 3 and 4.

1564 See *Wang, Sen*, Towards an International Convention on Forests: Building Blocks versus Stumbling Blocks, in: The International Forestry Review, Volume 3, Number 4, 2001, p. 258; *Flejzor, Lauren*, Reforming the International Tropical Timber Agreement, in: RECIEL, Volume 14, Issue 1, 2005, pp. 19 ff.

1565 See Subsection B.V of Chapter 2.

1566 See Subsection B.I of Chapter 1.

the mandate of the ITTA is limited to tropical forests and completely neglect boreal and temperate forests. Thus, the ITTA still falls short in promoting a holistic approach to the protection of forests.

Furthermore, the existing fragmented system of international treaties dealing with forest issues cannot substitute effective legally binding documents specific to forests. This is despite the legitimate concerns about the feasibility, effectiveness, and enforceability of the hypothetical forest convention. At this juncture, it is important to point out one of the significant disadvantages inherent to the present system of forest-related conventions and non-legally binding documents dealing with forests. Accordingly, it can be argued that the international community is heading the wrong way when they attempt to manage forest issues through the provisions, mechanisms, and instruments based on other international environmental agreements – which are not specifically designed to tackle the problem of deforestation and forest degradation in the first place. Even though the relevant stakeholders may be temporarily satisfied with the scarce achievements through the implementation of these regulations, efforts and attention are in fact diverted from the core issues of the deforestation crisis. As a result, the development of specific forest law is not prioritized on the agenda of international negotiations. After having faced the opposition of certain States or groups of States in the adoption of a legally binding forest treaty, activities are immediately shifted to areas covered by other conventions. Consequently, there will be few or no attempts made to understand the conflicting interests of different stakeholders; by logical extension, there can be no discussions to reach a consensus.

There is perhaps another explanation to the lack of will among States to invest more efforts in drafting and implementing more stringent forest-specific regulations. Indeed, the States often have other agenda and objectives when advocating the adoption of the forest convention or other legally binding forest-specific documents. Such agenda may include promoting local timber production; supporting domestic climate change mitigation policies by including REDD+-activities; combating illegal logging with the primary goal to strengthen domestic industries relying on the legal timber harvesting; obtaining international financial aid and technological assistance; and, finally, protecting the rights of forest-dependent people or regulating access to genetic resources of forests.[1567] On the one hand, the existence of these multiple interests can be justified, as they may

1567 See also *Humphreys, David,* The Elusive Quest for a Global Forests Convention, in: RECIEL, Volume 14, Issue 1, 2005, pp. 3 ff.

encourage agreement and consensus on forest issues. However, in practice, as soon as the States achieved their short-term goals through other conventions, they tend to abandon the negotiations of forest-specific legally binding documents. Thus, it can be argued that if the primary interest of States advocating the adoption of legally binding documents specific to forests – such as an independent instrument or a protocol to other convention – were to be in the promotion of the conservation of world forest ecosystem, they would be more likely to continue negotiations.

The inadequacy of the current approach to the development of the international legal regime for forests manifests also in the following fact. Indeed, it is impossible to achieve the secondary benefits of forest conservation without maintaining and promoting flourishing forest ecosystems. For instance, it is impossible to sustainably reduce the emissions from forests without conserving the forests and forest biodiversity. The rights of indigenous and local communities and their forest-related knowledge also cannot be effectively protected when they are detached from the forest ecosystems.

Of course, one of the benefits of REDD+ and other similar mechanisms is that their implementation allows progress to be achieved now – albeit a limited one. On the other hand, the consistently futile negotiations of legally binding documents do not contribute to the protection of forests at all. However, on a long-term perspective, it will be highly unlikely that mechanisms such as REDD+ can resolve the problem of deforestation and forest degradation. At the same time, there is a serious risk that the absence of a holistic approach to forests will be an obstacle to the primary objectives of REDD+ and other similar mechanisms, as well as the conventions on which they are based.

Nevertheless, when existing international legal regime for forests fails to resolve the problem of deforestation and the negotiations of legally binding documents specific to forests proceed with huge difficulties, the soft-law market-based approach could be an alternative to the development of international law for forests.[1568] Such soft-law market-based approaches can include both the Payments

1568 See conclusions to Chapter 4; See also: *Farley, Joshua/Costanza, Robert*, Payments for Ecosystem Services: From Local to Global, in: Ecological Economics, Volume 69, Issue 11, 2010, pp. 2065 f.; *Wang, Pu/Poe, Gregory, L./Wolf, Steven A.*, Payments for Ecosystem Services and Wealth Distribution, in: Ecological Economics, Volume 132, 2017, pp. 63–64.; *Börner, Jan/Baylis, Kathy/Corbera, Esteve/Ezzine-de-Blas, Driss/ Honey-Rosés, Jordi/Persson, U. MartinPersson/Wunder, Sven*, The Effectiveness of Payments for Environmental Services, in: World Development, Volume 96, 2017, pp. 359 ff.; *Beyerlin, Ulrich*, Internationaler Waldschutz am Scheideweg? Von den

for Ecosystem Services schemes and the voluntary certification schemes. As it was demonstrated in the previous chapters, implementing this approach has significant difficulties, due to the need to ensure the integrity of forest ecosystems and the well-being of forest dependent people. It is, therefore, necessary to investigate if this soft-law market-based approach can finally remedy the absence of legally binding forest-specific international documents.

I. Market-Based Soft Law Approach to Forest Governance Versus Legally-Binding Treaties

Since the early 2000s, the international community directed its attention to the possibility of developing the international legal regime for forests by establishing soft-law market-based instruments for promoting the conservation of forests.[1569] *Inter alia*, these activities were guided by the assumption that the forest convention or other forest-specific legally binding documents would hardly contribute to the reduction of deforestation rates on their own, unless significant financial and technical resources that are to be properly managed and allocated are provided to forest-rich States. From this background, forest protection was directly linked to development assistance issues and establishment of new, predominantly market-based sources of new, additional forest finance.[1570] From this point of view, the international community should focus on the development of those market- or non-market based mechanisms that deliver further incentives for tropical States to conserve and manage in an environmentally friendly manner the forest ecosystems within their sovereign territories.[1571] It was argued

---

"Rio-Waldprinzipien" zu "REDD plus", in: *Hecker, Bernd/Henlder, Reinhard/Proelß, Alexander/Reiff, Peter* (eds.), Jahrbuch des Umwelt- und Technikrecht, Schriftenreihe des Instituts für Umwelt- und Technikrecht der Universität Trier, 2012, pp. 190 ff.

1569 *Humphreys*, The Elusive Quest [n. 1567], pp. 1 ff.

1570 *Pagiola, Stefano/Bishop, Joshua/Landell-Mills, Natasha*, Selling Forest Environmental Services: Market-based Mechanisms for Conservation and Development, 2002, pp. 1–14, 16–37; *Streck, Charlotte/O'Sullivan, Robert/Toby Janson-Smith, Toby/Tarasofsky Richard G.*, Climate Change and Forests: Emerging Policy and Market Opportunities, 2008, pp. 11 ff.; *Adhikari, Bhim*, Market-Based Approaches to Environmental Management: A Review of Lessons from Payment for Environmental Services in Asia, Working Paper 134, Asian Development Bank Institute, Tokyo, 2009, pp. 7 ff.; *MacKenzie*, Future Prospects [n. 1557], pp. 252 ff.; *Humphreys, David*, Discourse as Ideology: Neoliberalism and the Limits of International Forest Policy, in: Forest Policy and Economics, Volume 11, Issues 5–6, October 2009, pp. 319 ff.

1571 *MacKenzie, Future Prospects* [n. 1557], pp. 252 ff.; *Streck, Charlotte/O'Sullivan, Robert/Toby Janson-Smith, Toby/Tarasofsky Richard G.* [n. 1570], p. 12.

that negotiating and adopting a legally binding forest treaty can be perceived as redundant and even counter-productive, considering that time and resources spent could have been more efficiently deployed elsewhere.[1572]

The regulations ought to be clearly laid down and implemented in relation to the provision of financial and technical resources, as well as to their proper utilization for the benefit of forest ecosystems and forest dependent people. Otherwise, a major risk exists that there will not be sustained forest protection even if further financial resources were to be provided.[1573] The ongoing attempts to negotiate an effective international mechanism to deliver additional finances from public funds and market-based mechanisms – which would have ensured the conservation of global forest ecosystems through the implementation of clear, transparent and robust legally binding regulations – have not yet been successful.[1574] Sometimes, the need for these regulations was also challenged, with some arguing that the focus ought to be placed on establishing market-based mechanisms.[1575]

That said, the attempts to strengthen forest conservation through effectively implemented soft law market mechanisms have delivered only few benefits. It is true that certain progress has been achieved in this regard and significant efforts have already been put in place for the development of mechanisms for the valorization of ecosystem functions of forests with REDD+ in focus. Nonetheless, there is growing frustration over the ability (or the lack thereof) of these mechanisms to halt and reverse deforestation and forest degradation worldwide.[1576] The initial idea was to pull together forest- and climate-related policies to tackle both issues effectively through what were perceived as highly promising mechanisms; this was thought to facilitate the introduction of tougher regulations into the international legal regime for forest. Contrary to that, however, REDD+ is evolving into a voluntary instrument without clear regulation and deprived of a robust international oversight.[1577] Consequently, if current trends ensue, there will be a

1572 *MacKenzie*, Future Prospects [n. 1557], p. 252;
1573 See *Humphreys*, Discourse as Ideology [n. 1570], pp. 319 ff.
1574 See also *Eikermann, Anja*, Forests in International Law: Is There Really a Need for an International Forest Convention? 2015, pp. 150 ff.; *Humphreys*, Discourse as Ideology [n. 1570], pp. 319 ff.
1575 *Humphreys*, Discourse as Ideology [n. 1570], pp. 319 ff.
1576 *Wiersema, Annecoos*, Climate Change, Forests, and International Law: REDD's Descent into Irrelevance, in: Vanderbilt Journal of Transnational Law, Volume 47, Issue 1, 2014, pp. 5 ff., 66.
1577 See conclusion to Chapter 4; See also *Wiersema* [n. 1576], pp. 5 ff., 66; *Kill, Jutta*, REDD+: A Lost Decade for International Forest Conservation, Heinrich Böll

high risk that this mechanism will result in low emission reductions, unsatisfactory forest conservation, and violation of ILCs' rights; it will even harm natural forests and their biodiversity.[1578] If REDD+ does not meet the initial high expectations of its development, the implementation of soft-law approaches to forest conservation focusing on the promotion of financial incentives would face further challenges. To date, no other similar globally relevant international mechanism has been or is being developed. Moreover, the ABS regime of the CBD and Nagoya Protocol is unable to replace it as forest ecosystems involved in the bioprospecting are far and few between.

Furthermore, the implementation of this soft-law approach to the forest conservation is subject to a number of challenges. *Inter alia*, the financial resources provided until now have proven themselves to be insufficient in solving the problem of deforestation.[1579] The lack of legally binding provisions dealing with financial availability and predictability while ensuring legal certainty and transparency results in the absence of stakeholders willing to invest in market-based forest projects. In some States with strong forest governance available, certain success was achieved; an example of this is Costa Rica.[1580] However, the contribution of market-based mechanisms to the conservation of forests has been generally modest until now.[1581]

1. *Contribution of Market-Based Soft Law Mechanisms to the Reduction of Deforestation Rates in Brazil*

The limited success in the reduction of deforestation rates in some regions can also be attributed to other factors. For instance, the reduction of deforestation in Brazil between 2005 and 2015 was also partly due to voluntary political commitments of the national and regional authorities, as well as the increased engagement of NGOs and other stakeholders. All these contributed to better forest law enforcement, reduction of illegal logging, and the establishment of new

---

Sttiftung, 11. January 2019. – Internet resource – https://www.boell.de/en/2019/01/11/redd-lost-decade-international-forest-conservation-0 [Last visit: 30.04.2023]

1578 *Wiersema* [n. 1576], pp. 5 ff.

1579 See Subsection D.IV.7 of Chapter 4.

1580 See *Albizu, Oreina Orrantia/Hidalgo, M. Mercedes Ortega/Madrigal, Olman Quirós/Arregui Javier Loidi*, Servicios Ambientales del Bosque: Ensayo en una Cuenca Atlántica Europea con Base en la Experiencia de Centroamérica, in: Revista de Biología Tropical, Volume 56, Issue 4, December 2008, pp. 287 ff.

1581 *Kill* [n. 1577], pp. 1 ff.

protected areas.[1582] The role of additional financial resources is difficult to estimate but it tends to be only complementary, as it will be shown below.

Importantly, Brazil had participated in REDD+-activities during this period. Among others, the Brazilian State of Acre explored REDD+ carbon credits selling opportunities within the Californian cap-and-trade market.[1583] Moreover, since the year 2012, Acre is participating in the REDD Early Movers (REM) Programme the government of the Federal Republic of Germany launched at the Rio+20 Conference in June 2012.[1584] In the course of this programme, Germany has provided more than USD 68 million as a pledge aimed to support Brazil's efforts in combating deforestation. The most important REDD+-initiative in which Brazil has participated is the Brazil-Norway bilateral agreement, in which USD 1.212 billion were disbursed as of 17 December 2018.[1585] Additionally, the major Brazilian oil company Petróleo Brasileiro S.A. also donated more than USD 7.713 million. However, it ought to be pointed out that all these pledges tend to underestimate the real socio-economic and environmental value of preserved forests, which far exceeds the USD 500 paid for each hectare of forest to be saved from degradation and deforestation.[1586] All grants were disbursed through the Amazon Fund, which serves to raise funds for non-reimbursable investments in the conservation and sustainable use of Amazonian rainforests.[1587]

1582 *Boucher, Doug*, How Brazil Has Dramatically Reduced Tropical Deforestation, in: Solutions, Volume 5, Issue 2, 2014, pp. 66–75; *Gebara, Maria Fernanda*, Can REDD+ help Brazil roll back rising deforestation rates? in: Forest News, 23 Jun 2017. – Internet resource – https://forestsnews.cifor.org/50288/can-redd-help-brazil-roll-back-rising-deforestation-rates?fnl=en [Last visit: 30.04.2023]; See also *Gebara, Maria Fernanda, Agrawan, Arul*, Beyond Rewards and Punishments in the Brazilian Amazon: Practical Implications of REDD+ Discourse, in: Forests, Volume 8, Issue 3, 2017, pp. 68 ff.

1583 *Boucher, Doug/Elias, Pipa/Faires, Jordan/Smith, Sharon*, Deforestation Success Stories: Tropical Nations Where Forest Protection and Reforestation Policies Have Worked, Union of Concerned Scientists, 2014, p. 12.

1584 Federal Ministry for Economic Cooperation and Development (BMZ), REDD+ in the State of Acre, Brazil: Rewarding a pioneer in forest protection and sustainable livelihood development, KfW Development Bank, 2017, p. 1. – Internet resource – https://www.kfw-entwicklungsbank.de/PDF/Entwicklungsfinanzierung/Themen-NEU/REDD-Early-Movers-Acre-Fact-Sheet.pdf [Last visit: 30.04.2023]

1585 See official site of the Amazon Fund – Internet resource – http://www.fundoamazonia.gov.br/en/donations/ [Last visit: 30.04.2023].

1586 *Boucher/Elias/Faires/Smith* [n. 1583], p. 12.

1587 See official site of the Amazon Fund – http://www.amazonfund.gov.br/en/home/ [Last visit: 30.04.2023]

The financial resources issued through these REDD+-programmes are nonetheless inadequate to compensate the loss of development opportunities resulting from the insistence on the sustainable use of forests and their resources.[1588] Therefore, the pledges are more a symbolic gesture of the German and Norwegian governments to demonstrate the commitment of the international community in assisting developing forest-rich States in their efforts to combat deforestation.[1589] That said, it can be argued that the political will of Brazil and some other forest-rich States have led to more stringent regulations and stricter forest law enforcement. The involvement of local residents serves to promote the conservation of tropical forests was also a decisive factor in the reduction of deforestation rates in certain regions.

Importantly, REDD+ projects in Brazil are not market-based ones, since the carbon credits are not sold at carbon certification markets and the financial resources for their implementation are derived from Norwegian and German public funds. Thus, it can be concluded that the additional financial aid played only a complementary role in the reduction of Brazil's deforestation rates and the purely market-based mechanisms have not been able to effect any significant change.

2. *Role of Market-Based Mechanisms in the Combating of Deforestation in Other Regions*

In a similar way, the reduction of deforestation rates in Central Africa was stimulated particularly by the shift to more sustainable forestry and agricultural practices, as well as by the establishment of new protected areas, whereby the financial aid played only a secondary role.[1590] For one, the growing urbanization of savannah areas has reduced the need for development of forests to accommodate the growing population.[1591] Moreover, oil and mineral mining has generated new and increased income.[1592] As a result, locals are able to spend more disposable income on imported food. This leads to a decrease in demand for

1588 See also *Boucher, Doug/Roquemore, Sarah/Fitzhugh, Estrellita*, Brazil's Success in Reducing Deforestation, in: Tropical Conservation Science, Special Issue, Volume 6, Issue 3, 2013, p. 439; *Boucher/Elias/Faires/Smith* [n. 1583], p. 12.

1589 *Boucher/Elias/Faires/Smith* [n. 31], p. 12; *Boucher/Roquemore/Fitzhugh* [n. 36], p. 439.

1590 *Boucher/Elias/Faires/Smith* [n. 31], p. 36.

1591 Ibid, p. 37.

1592 Ibid, p. 37.

local agricultural produce. In addition, since most natural rainforests in Central Africa are situated far away from major urban, there is no immediate need to exploit them for fuel, construction materials and land for industrial and agricultural activities.[1593] Consequently, the human pressure on natural rainforests was reduced.[1594] Particularly noteworthy are the number of internationally sponsored activities that were launched in the Congo Basin in the past years. Since 2002, the Congo Basin Forest Partnership (CBFP) has been active in the region; it is currently facilitated by the European Union.[1595]

Another example is the situation in Papua New Guinea. There, the proposal for the development of REDD+, which is presently the mechanism with the highest potential to secure fundraising for the protection of forests, received widespread support since 2005. However, deforestation in Papua New Guinea continues at alarming rates.[1596] Illegal logging represents one of the primary causes of deforestation in this tropical country.[1597] According to statements and news reports, the income from illegal logging does not contribute to the development of local economy and the improvement of living conditions; instead, the resultant environmental degradation adversely impacts local residents.[1598] With REDD+ being included in the Paris Agreement, Papua New Guinea has been moving – slowly but steadily – towards the full implementation of its mechanisms with the entire international community. That said, the Papua New Guinean government still fails to enforce its own laws in the fight against illegal logging, while other States do not hesitate to import timber that is harvested illegally and unsustainably.

Currently implemented REDD+-projects have clearly demonstrated their insignificant contributions to environmental conservation. It is estimated that the majority of these projects has only either low or insignificant impact on

---

1593 Ibid, p. 38.

1594 Ibid, p. 38.

1595 Ibid, p. 38.

1596 *Gabatiss, Josh*, Alarming photos reveal devastating scale of rainforest destruction in Papua New Guinea, in: Independent, 21 March 2018. – Internet resource – https://www.independent.co.uk/environment/papua-new-guinea-rainforest-destruction-photos-deforestation-global-witness-illegal-logging-a8265451.html [Last visit: [30.04.2021]; *Cannon, John C.*, Stop importing illegal timber, PNG activists tell China at APEC summit, in: Mongabay, 26 November 2018. – Internet resource – https://www.eco-business.com/news/china-should-stop-importing-illegal-timberactivists/ [Last visit: 30.04.2023]

1597 *Cannon* [n. 1596].

1598 Ibid.

the promotion of forest conservation;[1599] some of them even have a negative impact.[1600]

On the other hand, one should not underestimate the potential of additional finance and market-based approaches in promoting the conservation of forests. Certainly, without additional financial resources, progress would have been even more limited.[1601] For instance, it was discovered that the activities related to the preparations for the implementation of REDD+ have led to the reform of forest governance in Indonesia.[1602] For one, they stimulated the introduction of the moratorium on forestland conversions and accelerated the reform of land tenure legislation by drawing a clear distinction between the state-owned and indigenous lands.[1603] Moreover, Indonesian authorities achieved a significant reduction of deforestation rates between 2017 and 2018. However, it remains unclear to what extent REDD+ preparatory activities contributed to these legislative breakthroughs in Indonesia, since the positive impact of these mechanisms is likely to be insignificant due to the uncertainties associated with the financing of forest-related climate change mitigation mechanisms. Nevertheless, Indonesia can still count on more predictable and sustainable REDD+-finance, thanks to the involvement and engagement of the Norwegian government.[1604]

However, according to some estimations based on the FAO analysis, the reduction of deforestation rates since 2000 appears to be simply because the most profitable forest areas have already been cleared.[1605] If this assumption is true, it

---

1599 *Duchelle, Amy E./Simonet, Gabriela/Sunderlin, William D./Wunder, Sven*, What Is REDD+ Achieving on the Ground?, in: Current Opinion in Environmental Sustainability, Volume 32, 2018, pp. 134 f.; See also *Milne, Sarah/Mahanty, Sangoty/To, Phuc/Dressler, Wolfram/Kanowski, Peter/Thavat, Maylee*, Learning From "Actually Existing" REDD+: A Synthesis of Ethnographic Findings, in: Conservation & Society, Volume 17, Issue 1, 2019, pp. 92 f.

1600 See also *Milne/Mahanty/To/Dressler/Kanowski/Thavat* [n. 1599], pp. 93 f.

1601 *Sills, Erin O./Atmadja, Stibniati S./de Sassi, Claudio/Duchelle, Amy E./Kweka, Demetrius L./Resosudarmo, Ida Aju Pradnja/Sunderlin, William D.*, REDD+ on the Ground, A Case Book of Subnational Initiatives Across the Globe, CIFOR, Boor, Indonesia, 2014, p. 421.

1602 *Agung, Putra/Galudra, Gamma/Van Noordwijk, Meine/Maryani, Retno*, Reform or reversal: the impact of REDD+ readiness on forest governance in Indonesia, in: Climate Policy, Volume 14, Issue 6, 2014, pp. 760 ff.

1603 *Agung/Galudra/Van Noordwijk/Maryani* [n. 1602], pp. 761 f.

1604 Ibid.

1605 *MacKenzie, Catherine P.*, Lessons from Forestry for International Environmental Law, in: RECIEL, Volume 21, Issue 2, 2012, p. 115; *FAO*, State of the World's Forests 2011, FAO, United Nations, 2011, pp. 1–28.

means that the rates of deforestation would likely accelerate again as soon as new technologies are made available, or when rising market prices for forest goods and agricultural produce make forest clearing a worthwhile enterprise. However, it is important to note that the decreased deforestation rates can be attributed not also to market-based mechanisms, but also to the expansion of protected areas, other conservational measures, and the introduction of more sustainable forest management practices.[1606]

Overall, the influence of financial resources on forest conservation, including those derived from market mechanisms, requires further analysis that falls beyond the scope of this present research. Nevertheless, it can be argued that the impact of additional finances, which was widely recognized as the most important factor in promoting environmental protection since the early 2000s, was insufficient to halt or reverse deforestation and forest degradation throughout the world, particularly in the tropics. Moreover, the reluctance of donor States to give more means that development aids are unlikely to increase in the near future. Even if they were willing to offer more financial assistance, their domestic resources are also limited, especially in times of recession when the books are in the red. The steadily increasing domestic resources in the Third World should also be considered; a portion of this could be redirected to forest conservation and the improvement of local livelihoods in developing States. In other words, there is a need for more South-South cooperation. Even though States are slowly implementing market-based mechanisms like REDD+, there is no absolute certainty that these mechanisms will yield significant financial resources.

3. *Role of Market-Based Mechanisms in the Conservation of Temperate and Boreal Forests*

Despite the overall increase in forest areas in most of the developed States, high-value natural forests are still being destroyed and degraded in the northern latitudes. For example, on the one hand, the Canadian forest authorities might argue that its national forests are properly managed, citing the fact that the damage caused by logging is only temporary and that the cleared forests will regenerate.[1607] On the other, the statement in question does not consider the

1606 *FAO* [n. 1605], pp. 1–28.

1607 *Natural Resources Canada*, Deforestation in Canada – What Are the Facts? Canadian Forest Service Science-Policy Notes, May 2008. – Internet resource – https://cfs.nrcan.gc.ca/pubwarehouse/pdfs/28159.pdf [Last visit: 30.04.2023]

differences between mature and regenerating forests, as well as the environmental damage incurred at the moment of forest clearance.

The situation in Russia deserves particular attention. Although the forest-rich Russia is officially a developing State, it was never regarded as a major beneficiary of financial aids for forest protection from international donors. Deforestation and forest degradation are a source of preoccupation in Russia: 365,000 km^2 of forest areas were lost between 2000 and 2012, while even larger areas were cleared temporarily.[1608] In the 2012 review of forest condition, the Russian federal government estimated a greatly reduced average ages of trees in the State's forests. This is a clear evidence that regenerating forests and monoculture tree plantations with lower ecological values were widely replacing old-growing forests.[1609]

Considering the role of boreal and temperate forests in sustaining the global environment, conservation of forests in the northern hemisphere should be considered as a major international concern. Forest degradation in the northern latitudes cannot be alleviated through development assistance, since many States in the northern hemisphere are developed States and hence not meet the requirements for receiving financial aids. At the same time, by investigating the economic value of the forests' role in ecology, the developed States might be propelled to put in more efforts in the conservation of temperate and boreal forests. For instance, the multiple functions of these forests ought to be examined and recognized, including their role as sinks and reservoirs of greenhouse gases. However, to step up forest conservation efforts in the northern latitudes, the ways they manage and exploit forest resources must be revisited and amended. This can be achieved through the establishment of clear forest-specific regulations on the international level. For this reason, promoting forest protection solely by relying on financial resources while turning a blind eye on the extending protected areas and introducing gentle forest use practices is unsustainable and cannot be seen as a way to conserve boreal and temperate forests. Therefore, it can be argued that stronger international regulations could help consolidate and strengthen policies and procedures that would allow the conditions of forests in developed States to improve.

1608 *Dobrovidova, Olga*, Russia tops 2000–2012 deforestation charts, in: Climate Home News, 15.11.2013. – Internet resource – https://www.climatechangenews.com/2013/11/15/russia-lost-more-forest-cover-in-2012-than-any-other-country/ [Last visit: 30.04.2023]

1609 Ibid.

II. Estimating the Overall Role of Market-Based Mechanisms in the Conservation of Forests

Nevertheless, development assistance and market-based mechanisms should still be regarded as a major factor in the promotion of forest conservation worldwide but should not be seen as a panacea for deforestation and forest degradation. It is also noteworthy that the lacking enhancement of the available assistance funds in current situation compromises the future implementation of the REDD+-mechanism, as much of the funding has been provided by the governments of developed States until now, following the multilateral and bilateral agreements that were concluded between donor and recipient States.[1610] Voluntary markets have played only a secondary role in this process.[1611] The major forest-rich States that have mentioned the REDD+-mechanism in their INDCs have also overlooked the role of financial help provided through multilateral or bilateral cooperation programmes.[1612] Major donor States did not include the assistance to forest-rich States in implementing REDD+ as a climate change mitigation initiative in their INDCs. This means that they prefer to consider their participation in the implementation of the REDD+-mechanism as a purely voluntary commitment beyond their duties and obligations determined by the UNFCCC.[1613] The developed States have been actively assisting the developing ones in the elaboration of their INDCs. The very inclusion of REDD+ in the INDCs of forest-rich States indicates a certain level of confidence in the availability of financial aids in the future. Importantly, a number of forest-rich States have received much international aid for biodiversity conservation in the recent decades.[1614] The main recipients of this kind of aids are the States who included REDD+ in their INDCs.[1615] This suggests that the provision of financial

1610 *Hein, Jonas/Guarin, Alejandro/Frommé, Ezra/Pauw, Pieter*, Deforestation and the Paris Climate Agreement: An Assessment of REDD+ in the National Climate Action Plans, in: Forest Policy and Economics, Volume 90, 2018, p. 10.

1611 *Hein/Guarin/Frommé/Pauw* [n. 1610], p. 10.

1612 See, for example: *Federative Republic of Brazil*, Intended Nationally Determined Contribution, Towards Achieving the Objective of the United Nations Framework Convention on Climate Change, 2015, p. 4. – Internet resource – https://www4.unfccc.int/sites/ndcstaging/PublishedDocuments/Brazil%20First/BRAZIL%20iNDC%20english%20FINAL.pdf [Last visit: 30.04.2023]

1613 *Hein/Guarin/Frommé/Pauw* [n. 1610], p. 10.

1614 *Miller, Daniel C./Agrawal, Arun/Roberts J. Timmons*, Biodiversity, Governance, and the Allocation of International Aid for Conservation, in: Conservation Letters, Volume 6, Issue 1, 2013, p. 14.

1615 *Hein/Guarin/Frommé/Pauw* [n. 1610], p. 10.

resources for biodiversity conservation could have been interpreted as a signal for the availability of REDD+ financial assistance in the future.

However, the predictability and availability of forest finances should not depend on the indirect signals from the donor States. For the purpose of the consistent and effective implementation of REDD+, an international financial mechanism should be established. This requires precise legally binding international regulations, which ideally must lay down the obligations of different States concerning the provision of adequate funds for REDD+ implementation.[1616] However, the hope to elaborate and adopt these legal regulations in relation to REDD+ is vanishing: instead of extending stronger legal regulations, which are common for the international legal regime for combatting climate change, the provisions governing the development and implementation of these mechanisms are vague, superficial and ambiguous – just like those of the international legal regime for forests[1617]. Against this backdrop, it must not be expected that the deficiencies of the international legal regime for forests would be resolved only through the development of REDD+ and similar mechanisms. For this purpose, it is necessary to tackle the heart of the problem, viz. the lack of strong legally binding provisions targeted specifically at forests. If the obstacles to the elaboration of a legally binding forest-specific treaty are not resolved and the lack of consensus on major forest issues is not tackled, any new mechanism or regulation aimed at resolving the environmental or socio-economic problems linked to forests is destined to acquire the character of the international legal regime for forests – voluntary, vague, and ineffective.

While strong legally binding regulations for REDD+ implementation are absent, there is strong likelihood that this mechanism will only achieve a partial success in combating deforestation.[1618] Obviously, when properly implemented, REDD+ can have positive outcomes for the forest ecosystems and biodiversity, but these mechanisms are unlikely to halt and reverse forest loss worldwide.

1616 See also: *Lang, Chris*, REDDheads: The people behind REDD and the climate scam in Southeast Asia, Stiftung Asienhaus, 2016, pp. 1 ff. – Internet resource – https://www.asienhaus.de/archiv/asienhaus/kattermann-stiftung/user_upload/REDDheads.pdf [Last visit: 30.04.2023]

1617 *Wiersema* [n. 1576], p. 66; *Gupta, Joyeeta/Sanchez, Nadia*, Global Green Governance: Embedding the Green Economy in a Global Green and Equitable Rule of Law Polity, in: RECIEL, Volume 21, Issue 1, 2012, pp. 14 f.

1618 *Cadman, Timothy/Maraseni, Tek/Ma, Hwan Ok/Lopez-Casero, Federico*, Five Years of REDD+ Governance: The Use of Market Mechanisms as a Response to Anthropogenic Climate Change, in: Forest Policy and Economics, Volume 79, 2017, pp. 12 ff.

It is unlikely that the soft law market-based mechanisms will help to solve the problem of deforestation and forest degradation without the need to recourse to stronger legal provisions. Indeed, a number of initiatives like FLEGT, Participatory Forest Management and the already mentioned REDD+ often do not meet the initial expectations.[1619] In the course of development and implementation, they run the risk of evolving into some kind of conservationist "fads" that attract additional financial and administrative resources, but yet fail to considerably change the situation on the ground.[1620] When clear regulations defining the desirable outcomes of the REDD+-mechanism, such as reducing deforestation rates, and the ways to achieve these goals are absent, the initial expectations are rarely fulfilled during project implementation. This sometimes results in the relevant stakeholders rapidly losing their interest to collaborate on the project.[1621] In certain cases, the authoritative regulations impeding or restricting forest clearance have proven themselves to be more efficient in terms of reducing deforestation rates compared to the REDD+-mechanism,[1622] However, it is unclear if such authoritative forest clearance restrictions were inspired by REDD+.[1623]

Consequently, the need for enhancing financial assistance and for promoting market-based mechanisms should not be denied. However, while recognizing their contribution to the preservation of forest ecosystems worldwide, it also has to be acknowledged that precise, detailed, and legally binding regulations are required to improve the effectiveness of international legal regime for forests and to highlight its contribution to the conservation of global forest ecosystems. As

---

1619 *Lund, Jens Friis/Sungusia, Eliezeri/Mabele, Mathew, Bukhi/Scheba, Andreas*, Promising Change, Delivering Continuity: REDD+ as Conservation Fad, in: World Development, Volume 89, January 2017, pp. 124 ff.; *Rutt, Rebecca L./Myers, Rodd/Ramcilovic-Suominen, Sabaheta/McDermott, Constance*, FLEGT: Another "Forestry Fad"? in: Environmental Science and Policy, Volume 89, 2018, pp. 266 ff.

1620 *Rutt/Myers/Ramcilovic-Suominen/McDermott* [n. 1619], pp. 266 f.

1621 *Massarella, Kate/Sallu, Susannah M./Ensor, Jonathan E./Marchant, Rob*, REDD+, Hype, Hope and Disappointment: The Dynamics of Expectations in Conservation and Development Pilot projects, in: World Development, Volume 109, September 2018, pp. 376 ff.; *Chomba, Susan/Kariuki, Juliet/Lunda, Jens Friis/Sinclair, Fergus*, Roots of Inequity: How the Implementation of REDD+ Reinforces Past Injustices, in: Land Use Policy, Volume 50, 2016, pp. 210 ff.

1622 *Duchelle, Amy E/de Sassi, Claudio/Jagger, Pamela/Cromberg, Marina/Larson, Anne M./Sunderlin, William M./Atmadja, Stimiati S./Resosudarmo, Ida Aju Pradnja/Pratama, Christy Desta*, Balancing Carrots and Sticks in REDD+: Implications for Social Safeguards, in: Ecology and Society, Volume 22, Issue 3, 2017, p. 2.

1623 *Agung/Galudra/Van Noordwijk/Maryani* [n. 1602], p. 760.

in the case of REDD+, a lack of strong provisions diminishes the potential of additional financial resources in contributing to forest protection. Ignoring the weak environmental and social safeguards for now, the uncertainties in determining forest reference emission levels to serve as a baseline for the estimation of progress achieved in REDD+ implementation demonstrate how the lack of clearly defined rules results in a situation when the measured and verified emission reductions do not correspond to the real conditions of tropical forests.[1624] In fact, there are significant discrepancies between the achieved emission reductions as stated in the correspondent reports submitted by tropical States and the actual extent of forest cover determined by analyzing remote sensing data.[1625] This is why stricter and more precise regulations are needed to ensure a consistent and environmentally friendly implementation of projects that are aimed at promoting forest conservation through the enhancement of financial help for forest-rich States. Concerning the FREL baseline for REDD+-projects, the Governing Bodies of the UNFCCC are preliminary responsible to develop precise rules for its application, which would ensure the adequate monitoring of the conditions of forests. However, considering that there is a multitude of different initiatives that are aimed at the monetary evaluation of multiple functions of forests, there is a need for a definitive set of precise regulations. These regulations are most likely to be adopted through an agreement dedicated to forests specifically.

The lack of stronger provisions for REDD+ implementation on the international and national levels, particularly in relation to establishing clear tenure rights and strengthening forest governance, represents another risk for the promotion of forest protection in developing States.[1626] It can impede the progress not only in the course of REDD+ implementation, but also in the process of applying other forest-related policies that have relatively stronger provisions concerning tenure issues and the promotion of forest governance, such as FLEGT.[1627]

1624 See Subsection D.IV.2 of Chapter 4.

1625 *Sax, Sarah*, Brazil to receive first-ever results-based REDD+ payments, but concerns remain, in: Mongabay, 1 March 2019. – Internet resource – https://news.mongabay.com/2019/03/brazil-to-receive-first-ever-results-based-redd-payment-but-concerns-remain/ [Last visit: 30.04.2023]

1626 See *Dooley, Kate/Ozinga, Saskia*, Building on Forest Governance Reforms Through FLEGT: The Best Way of Controlling Forests' Contribution to Climate Change?, in: RECIEL, Volume 20, Issue 2, 2011, p. 170.

1627 Ibid.

Thus, the very fact that the provision of additional financial assistance, including through market-based mechanisms, is likely to play a crucial role in a hypothetical forest convention or any other legally binding forest-specific document does not make their adoption obsolete. The development of international legal regime for forests should not be restricted only to one option. The promotion of market-based mechanisms, for instance, can and should coexist with precise and strong provisions specifically dedicated to forests. In the case of REDD+, the negligence of legal regulations leads to the prevalence of purely technical and pragmatic approaches to the implementation of projects and distribution of benefits in many cases.[1628] This results in the worsening of social injustice and the disregard for environmental issues. Moreover, ILCs – and even local authorities – remain underrepresented and disadvantaged in the course of benefit sharing.[1629] Considering the great variety of forest-related projects, it becomes all the more important to unify and standardize policies and procedures to ensure adequate representation of the most vulnerable stakeholders as well as to guarantee fair and equal benefit sharing. These could be established by an international forest convention.

One may either follow the approach in promoting forest conservation based on better implementation of existing forest-related agreements, or advocate drafting new legally binding forest-specific documents. In both cases, strong provisions ought to be adopted and the compliance must be enforced.[1630] The examples above reveal that the promotion of forest protection based on existing international agreements, which are not dedicated specifically to forests but tend to regulate forest issues as a side issue, has been ineffective until now. Their contribution to combating deforestation and forest degradation was not sufficient to halt and reverse trends of forest loss. At the same time, the Governing Bodies of pertinent conventions have failed to elaborate and implement strong legal provisions to deal with forests within the mandate of relevant treaties. Instead, the "dilution" of the international legal regime for forests has taken place. It is manifested in the fact that the most of resources and efforts has been directed to the promotion of "peripheral" issues, such as the promotion of reduction of

---

1628 *Myers, Rodd/Larson, Anne M./Ravikumar, Ashwin/Kowler, Laura F./Yang, Anastasia*, Messiness of Forest Governance: How Technical Approaches Suppress Politics in REDD+ and Conservation Projects, Global Environmental Change, Volume 50, May 2018, p. 322.

1629 Ibid.

1630 *MacKenzie, Lessons from Forestry* [n. 1605] pp. 125 f.

emissions from forests.[1631] Admittedly, the existent forest-related provisions and mechanisms have not yet been completely implemented; in other words, their potential in contributing to the regulation of forest issues has not been fully achieved. However, due to the lack of strong legal provisions and mechanisms to ensure compliance, they are unlikely to be fully implemented in the near future. If the international community endeavours to scale up their efforts to preserve forests worldwide, the adoption of legally binding documents specifically dedicated to forests and also the strict enforcement of their implementation must return on the agenda of international meetings on forest issues.

C. Forest-Related Issues on the Agenda of International Negotiations

Forest-related conventions and soft law documents dealing with forests, which were analyzed in the previous chapters, give insight into the main issues that currently constitute the core of international forest-related policies. These main issues provide an indicative framework for the further development of the international legal regime for forests. However, with the controversial and fragmented character of the international legal regime for forests in mind, one can assume that this indicative framework is not free from defects or shortcomings. The main forest-related topics that are incorporated in the existing international documents dealing with forests are briefly analyzed below.

I. Sustainable Forest Management and the Quest for Equilibrium Between the Conservation and Utilization of Forest Ecosystems

Among other things, it is argued that the predominant approach to forest conservation consists in regulating both the conservation and utilization aspects of forest governance in an equal and balanced manner.[1632] However, the very

---

1631 See also: *McDermot, Constance L./Levin, Kelly/Cashore, Benjamin*, Building the Forest-Climate Bandwagon: REDD+ and the Logic of Problem Amelioration, in: Global Environmental Politics, Volume 11, Issue 3, August 2011, pp. 85 ff.; *Singer, Benjamin/Giessen, Lucas*, Towards a Donut Regime? Domestic Actors, Climatization, and the Hollowing-out of the International Forests Regime in the Anthropocene, in: Forest Policy and Economics, Volume 79, 2017, pp. 69 ff.

1632 *Eikermann* [n. 1574], p. 136; *Schrijver, Nico*, Sovereignty over natural resources: balancing rights and duties, 2008, pp. 292 ff.

concept of treating the conservation and utilization issues equally is vague and unclear. The utilization of forests can include a wide range of activities. Some of these activities, such as the promotion of ecotourism, the gathering of samples of forest biodiversity for bioprospecting, and the promotion of the multiple ecological functions of natural forests, can have positive impacts on forest biodiversity. In the last case, the conservation and utilization of forests is basically incorporated in one activity, as the utilization of the ecological benefits of forests presupposes the preservation and restoration of forest ecosystems. Other activities, including forest clearance, conversion of forestlands, and the establishment of monoculture plantations certainly have detrimental effects on forest ecosystems.

Consequently, the concept of forest governance that seeks to achieve the balance between the conservation and utilization of forests has to be further elaborated in order to become operational. International organisations and national authorities should further clarify which activities should be applied in different forests in order to achieve the balance between the conservation and utilization of forests. For this purpose, provisions that would consider different characteristics of different types of forests should be adopted.

The concept of the balance between the utilization and conservation of natural resources is also enshrined in Principle 21 of the Stockholm Declaration. This principle indicates that States have the right to exploit their natural resources but, at the same time, they bear responsibility to ensure that relevant activities do not cause damage to the environment of the areas beyond their national jurisdiction. The Stockholm Declaration is a non-legally binding document; however, given that the substance of Principle 21 was applied also in other international documents and court decisions, including the Trail Smelter Arbitration,[1633] it is widely acknowledged that the principle is already incorporated in international environmental law as a customary norm.[1634] Noteworthy, Principle 21 does not directly refer to the conservation of natural resources but only to the prevention

---

1633 Trail Smelter Arbitration (United States v. Canada), Arbitral Judgements of 16.04.1938 and 11.03.1941, Rewards of International Arbitral Awards, Volume III, pp. 1905–1982.

1634 Legality of the Threat or Use of Nuclear Weapons, Advisory Opinion, ICJ Reports 1996, pp. 241–242, Paragraph 29; See also *Altea, Maria Pia M.*, Principle 21 of the Stockholm Declaration: A Customary Norm of International Environmental Law, in: Ateneo Law Journal, Volume 41, Number 2, 1997, pp. 429–443; *Birnie, Patricia/ Boyle, Alan/Redgewell, Catherine*, International Law and the Environment, 2009, pp. 158 ff.; *Proelß, Alexander*, Prinzipien des Internationales Umweltrechts in: *Proelß, Alexander* (ed.), Internationales Umweltrecht, 2017, pp. 75 f.;

of environmental damage to the areas beyond their national jurisdiction. Moreover, in this principle, the duty to preserve natural resources is vaguer than the right to exploit them.[1635] Specifically in the case of forests, it can be argued that the wording of the principle permits forest owners and users to prioritize exploitation issues over the conservation of forests.[1636]

The uncertainties in relation to the concept of the equilibrium between conservation and utilization aspects are directly related to the ambiguity of the concept of "sustainable forest management", whose legal content remains unclarified.[1637] Although this concept has been incorporated into numerous international treaties and soft-law documents dealing with forests, there is a clear lack of common understanding on how it ought to be operationalized.[1638] Empirical evidence suggests that this concept, at least as it is currently interpreted and implemented by forest-rich States and other actors, is not sufficiently robust in order to ensure the conservation of forest biodiversity and the promotion of multiple forest functions.[1639] After several agencies, including the FAO and the ITTO developed criteria and indicators for the implementation of SFM, this concept has finally become suitable for practical application.[1640] It should be noted, however,

1635 *Brunnée, Jutta*, A conceptual framework for an international forest's convention: customary law and emerging principles', in: Canadian Council on International Law (ed.), Global Forests and International Environmental Law, 1996, pp. 41 ff.; *Humphreys, David,* The Elusive Quest for a Global Forests Convention, in: RECIEL, Volume 14, Issue 1, 2005, p. 2.

1636 See Section B of Chapter 2.

1637 *Eikermann* [n. 1574], p. 139; *Maguire, Rowena*, The international regulation of sustainable forest management: doctrinal concepts, governing institutions and implementation, PhD thesis, Queensland University of Technology, 2010, p. 54.

1638 *Maguire* [n. 1637], p. 54.

1639 *Rice, Richard E./Gullison, Raymond E./Reid, John W.*, Can Sustainable Management Save Tropical Forests?, in: Scientific American, Volume 276, Number 4, 1997, pp. 44–49; *Pearce, David/Putz, Francis E./Vanclay, Jerome K.*, Sustainable Forestry in the Tropics: Panacea or Folly?, in: Forest Ecology and Management, Volume 172, Issues 2–3, 2003, pp. 230 ff.; *Zimmerman, Barbara L./Kormos, Cyril F.*, Prospects for Sustainable Logging in Tropical Forests, in: BioScience, Volume 62, Number 5, May 2012, pp. 479 ff.; *Nasi, Robert/Frost, Peter, G. H.*, Sustainable Forest Management in the Tropics: Is Everything in Order But the Patient Still Dying?, in: Ecology and Society, Volume 14, Issue 2, 2009. –Internet resource – https://www.ecologyandsociety.org/vol14/iss2/art40/ [Last visit: 30.04.2023]

1640 *Wilkie, Mette Løyche/Holmgren, Peter/Castañeda, Froylan*, Sustainable Forest Management and the Ecosystem Approach: Two Concepts, One Goal, 2003, pp. 4 f.;

that, in spite of the fact that certain progress has been achieved in relation to the development of qualitative indicators for SFM,[1641] quantitative indicators continue to play an important role. The promotion of qualitative indicators is of crucial importance. This is because the prevailing understanding of SFM as a process, which foresees that the same amount of trees is planted as it is cut and removed from forests, fails to take into account multiple forest functions.[1642] Such an approach may, therefore, prove incompatible with the natural life cycles of forest ecosystems, as it impedes to achieve the natural level of biodiversity and maturity of forest ecosystems.[1643] Such indicators as forest area and canopy cover may remain stable or even increase with time, whereas forest biodiversity, regeneration capability, and vitality may dramatically decline.[1644] This is why, for the purpose of the enhancement of the effectiveness of international forest law, it should be recognized that the concept of SFM as it stands now should not be applied at least to those forest ecosystems where the accent should be on the maintenance and enhancement of forest biodiversity, forest vitality and regeneration capability. The aim of forest management in these forest ecosystems should be to achieve natural levels of biodiversity, vitality, and maturity. Otherwise, the concept of SFM should be revised in a way that would permit to ensure the protection of forests and forest biodiversity. For example, the establishment of monoculture tree plantations could be perceived as an activity eligible under the scope of the provisions regulating the application of this concept as long as their establishment does not concern converted forest areas and the utilization of these plantations contributes to the conservation and restoration of natural and near-natural forests with high levels of biodiversity.

It should be noted, however, that the latest developments in international environmental law resulted in that a stronger emphasis was put on the conservation-oriented aspects of forest management. Inter alia, the Aichi Biodiversity Target and the Millennium Development Goals[1645] that were replaced by

---

ITTO, Criteria and indicators for the sustainable management of tropical forests, 2016, p. 1 f.; See also *Maguire* [n. 1637], p. 54.

1641 See *Wilkie/Holmgren/Castañeda* [n. 1640], pp. 4 ff.

1642 *Zimmerman/Kormos* [n. 1639], p. 485.

1643 Ibid.

1644 Ibid.

1645 United Nations A/RES/55/2, United Nations Millennium Declaration, General Assembly Resolution, 18 September 2000. – Internet resource – https://undocs.org/A/RES/55/2 [Last visit: 30.04.2023]

the Sustainable Development Goals[1646] should be mentioned in this regard.[1647] However, the destruction of particularly valuable natural forest ecosystems continues.[1648] It is also problematic that the two major forest-related instruments of the UNFCCC regime, namely, REDD+ and LULUCF, do not pay sufficient attention to the conservation of natural forests.[1649] The legal framework for their implementation does not contain the definition of a natural forest and does not establish a duty to conserve these forests, while the most pivotal provisions are deprived of any legally binding force.[1650]

Thus, current attempts to achieve the equilibrium between the conservation and utilization of forests and to promote SFM have resulted in the lack of attention to the conservation of natural forests. Pertinent provisions are not sufficiently robust and effective. In spite of the fact that a great number of conventions and soft law documents were adopted, the relevant issues remain particularly underrepresented within the realm of international environmental law.[1651] It can be suggested that if a legally binding forest convention was adopted, it could clarify the issue of the equilibrium between the conservation and utilization of forests. This document could prioritize the conservation of natural forests and near-natural afforestation and establish the obligation to protect natural forests. In order to promote the implementation of relevant provisions, a hypothetical forest convention could establish an effective financial mechanism and other monetary and non-monetary mechanisms could be elaborated.

1646 United Nations A/RES/70/1, Resolution adopted by the General Assembly on 25 September 2015: Transforming our world: the 2030 Agenda for Sustainable Development, 2015. – Internet resource – https://undocs.org/en/A/RES/70/1 [Last visit: 30.04.2023]

1647 *Morales-Hidalgo, David/Oswalt, Sonja N./Somanathan, E.*, Status and Trends in Global Primary Forest, Protected Areas, and Areas Designated for Conservation of Biodiversity from the Global Forest Resources Assessment 2015, in: Forest Ecology and Management, Volume 352, 7 September 2015, pp. 68 ff.

1648 *Mackey, Brendan/DellaSala, Dominick A./Kormos, Cyril/Lindenmayer, David/Kumpel, Noelle/Zimmerman, Barbara/Hugh, Sonia/Young, Virginia/Foley, Sean/Arsenis, Kriton/Watson, James E.M.*, Policy Options for the World's Primary Forests in Multilateral Environmental Agreements, in: Conservation Letters, Volume 8, Issue 2, 2015, p. 140.

1649 Ibid, p. 144.

1650 See Subsections C.II and D.V of Chapter 4.

1651 *Mackey/DellaSala/Kormos/Lindermayer/Kumpel/Zimmerman/Hugh/Young/Foley/Arsenis/Watson* [n. 1648], p. 145.

II. Duty to Conserve and Protect Forest Ecosystems

General due diligence obligations of public international law are too vague and superficial to have significant consequences for the conservation of forests. The obligation to protect forests cannot be derived from Principle 21 of Stockholm Declaration and Principle 2 of Rio Declaration.[1652] For this reason, it is necessary to analyze whether other due diligence duties can be derived from the international legal regime for forests. In spite of the fact that forest-relevant issues have been on the agenda of international negotiations for decades, the international legal regime for forests does not contain any provision that would establish a general duty to protect and preserve forest ecosystems. Principle 1 of the Rio Forest Principles contains precisely the same wording as the above-mentioned text of Principle 21 of the Stockholm Declaration. Thus, it does not sufficiently contribute to the development of forest-specific conservationist duties. The NLBI only stipulates that States are primarily responsible for the implementation of SFM,[1653] but it neither establishes any legal duties nor gives rise to strong political commitments. As it was stated above, the Rio Forest Principles and the NLBI are non-legally binding. Moreover, unlike the Rio and Stockholm Declarations, their respective provisions have not yet acquired the character of customary law, as the relevant criteria is not fulfilled. These criteria include the duration, consistency, and generality of State's practice, as well as the recognition of an international provision as *opinio juris*.[1654] Even if the provisions of the Rio Forest Principles and the NLBI had already represented international customary law, they would not have established a duty to protect world forest ecosystems. The reason for it is that their provisions dealing with SFM only stipulate the responsibility of States to manage their forests sustainably.

Major forest-related conventions do not establish any legal obligation to conserve forests. In this regard, Article 3 of the CBD once again reiterates Principle 21 of the Stockholm Declaration without establishing any specific conservationist duties in relation to forests and their biodiversity. According to Article 4.1(d) of the UNFCCC, States shall only promote sinks and reservoirs of carbon

---

1652 See Subsections B.I and D.III of Chapter 2.

1653 United Nations GA/RES/62/98, Non-Legally Binding Instrument on All Types of Forests, Sixty second Session, 17 December 2007, Paragraph 2b. – Internet resource – https://documents-dds-ny.un.org/doc/UNDOC/GEN/N07/469/65/PDF/N0746965.pdf?OpenElement [Last visit: 30.04.2023]

1654 *Fon, Vincy/Parisi, Francesco*, Stability and Change in International Customary Law, in: Supreme Court Economic Review, Volume 17, 2009, pp. 282 ff.

dioxide, including forests, but they are not obliged to preserve them. It is submitted that the UNFCCC as a non-conservation-oriented treaty cannot deliver a provision that could give rise to the legal duty to protect world forests. What concerns the ITTA, its main objective consists in the promotion of international trade with legally harvested tropical timber and of SFM without establishing legal duties to protect forests.

This is why there is a need for strong provisions that would establish the duty to protect and preserve existing forest ecosystems similar to the duties established by Article 192 of the United Nations Convention on the Law of the Sea in relation to the Ocean.[1655] Such a duty could include the obligation to act with due diligence in order to prevent a threat or damage to forest ecosystems. Moreover, from the background, that it always costs less to avoid environmental harm than to repair it, this obligation could incorporate the precautionary principle. According to this principle, States would be obliged to take measures in order to avoid serious damage to forest ecosystems, even if there are scientific uncertainties concerning the impacts of human activities on forests.[1656] Due to their restricted legal mandate, the existing forest-related conventions cannot establish any specific and robust obligations to conserve forests. However, this task could be fulfilled by a legally binding convention specific to forests.

III. Maintaining and Enhancing the Existing Forest Cover

Another issue that is currently underestimated in the international legal debate on forests is the need for maintaining at least the existing forest cover on a global and regional level.[1657] In spite of the fact that States, according to the New York Declaration on Forests, committed themselves to make efforts in order to completely halt the loss of natural forests by the year 2030, this commitment is neither legally binding nor enforceable.[1658] It can be evaluated positively that the Global Forest Goal 1 contained in the United Nations strategic plan for forests 2017–2030 foresees the enhancement of the global forest area by 3% until 2030.[1659] However, the Strategic Plan is a legally non-binding document and

1655 United Nations Convention on the Law of the Sea, adopted on 10 December 1982 and entered into force on 16 November 1994, in: 1833 UNTS 3.

1656 *Read, Ruppert/O'Riordan, Tim*, The Precautionary Principle Under Fire, in: Environment: Science and Policy for Sustainable Development, Volume 59, Issue 5, 2007, pp. 4 ff.

1657 *Eikermann* [n. 1574], p. 140.

1658 See Subsection D.IV of Chapter 2.

1659 See Subsection D.IV of Chapter 2.

does not clarify how its goals can be achieved. What concerns forest-related international conventions, no one of these documents contains any provision that would literally stipulate the duty to maintain and enhance the global forest area, although the UNFCCC, the CBD, the UNCCD, and other treaties refer to the importance of maintaining and promoting multiple forest functions.[1660] These functions cannot be promoted without the preservation and enhancement of global forest ecosystems.[1661] However, these legally binding international environmental treaties do not contain any provisions that would go beyond vague statements underlining the forest's environmental role, and generalized calls for action. The provisions in question can merely give rise to vague commitments to undertake actions for the maintenance of the current global forest area. Consequently, provisions that would establish legally-binding commitments to undertake urgent measures in order to maintain and enhance the global forest area are still absent. If such regulations were adopted, they would still be insufficient in order to promote the comprehensive protection of forests. For this reason, in order to be effective, they would need to stipulate specific and robust obligations to conserve natural forests and to promote multiple forest functions. This is why such hypothetical regulations would be consistent with the preservation of natural forests only if they foresaw the enhancement of forest areas under conservation. For this purpose, these provisions would also need to elaborate further the concept of SFM. It can be argued that it is more likely that such comprehensive regulations will be included in a legally binding forest convention rather than incorporated in any existing environmental treaty.

IV. Financing the Conservation and Sustainable Use of Forests

There is an international consensus that additional financial and technical resources are needed in order to sustain the efforts of forest-rich States that endeavor to conserve their forest ecosystems.[1662] Despite the fact that different forest-related financial mechanisms, including REDD+, have been developed, the actual availability of financial resources is still uncertain.[1663] Obviously, developed States and private investors expect that they will retrieve certain benefits from the implementation of REDD+. Moreover, they expect that these benefits

1660 *Mackey/DellaSala/Kormos/Lindermayer/Kumpel/Zimmerman/Hugh/Young/Foley/Arsenis/Watson* [n. 1648], p. 144.

1661 Ibid, p. 144.

1662 *Eikermann* [n. 1574], p. 142.

1663 Ibid.

would not only be of ecological, but also of economic value. Such complex state-guided programmes that are developed without the involvement of market-based mechanisms do not permit to identify and evaluate direct economic benefits for the investors.[1664] For this reason, direct payments from the public funds of developed States were supposed to play only a secondary role in the international efforts dedicated to forest conservation. From this background, attempts have been undertaken to develop market-based approaches for the evaluation of the multiple ecological functions of forests, including the market-based approach to the implementation of REDD+ and to the promotion of ABS. However, markets currently fail to raise enough funds for forest protection, especially in the case of REDD+.[1665] In spite of the fact that forest carbon offset markets are growing, these mechanisms are still unattractive for private investors.[1666] Despite the fact that one of the reasons for the failure of the previous negotiations of a forest convention consisted exactly in the lack of consensus on the issue of enhancing forest-related finance, the option of direct payments from public funds has been revitalized again by States, as it is the case with the JMA mechanism.[1667] This is why, it can be assumed that the provision of financial resources from the public funds of developed States and its possible combination with market-based mechanisms should receive more attention in the activities of the competent international bodies.

Against this backdrop, it can be argued that environmental benefits alone are not sufficient in order to persuade developed States and private actors to provide the additional finance for forest conservation. The combination of ecological and economic benefits is supposed to be achieved through REDD+ and other PES-mechanisms, however, the legal framework for their implementation is still too vague to ensure their environmentally-friendly implementation.[1668] A financial scheme that would consider all the important ecological functions of forests and would incorporate a holistic approach to their conservation is still absent. Moreover, even in the cases where sufficient financial resources are available, the environmental feedback of these initiatives and projects is low due to a lacking will of forest owners to invest sustainably into forest conservation, because pure

1664 See ibid.

1665 *Norman, Marigold/Nakhooda, Smita*, The State of REDD+ Finance, CGD Climate and Forest Paper Series Number 5, Working Paper 378, 2015, pp. 13 ff., 34 ff.

1666 Ibid.

1667 See Subsection D.IV.6 of Chapter 4.

1668 See Subsection D.IV.7 of Chapter 4.

economic interests prevail on the side of the beneficiaries of payments.[1669] Apart from the poor governance on a local level, this situation is exacerbated by the lack of strong, detailed, and precise international legal provisions that would regulate forest finance. It can be argued that, instead of a set of different mechanisms, whose implementation is guided by vague and inconsistent regulations, a unique international framework for forest finance should be developed. Such an international legal framework should ensure the conservation of natural forests and promote the sustainable utilization of forests and near-natural afforestation. In order to promote the efficient utilization of provided financial resources in forest-rich States, the predictability and availability of forest finance should be ensured, and compliance and monitoring mechanisms should be established. The transfer of financial resources should be subject to the compliance with conservation-oriented provisions. At the same time, it is necessary to ensure that the financial mechanism in question would be acceptable for the providers of financial resources. Moreover, this financial mechanism would have to create a powerful stimulus for the intensification of efforts dedicated to the conservation of forests, while contributing to the improvement of local livelihoods. Consequently, it can be argued that there is a strong need for more detailed and legally binding international legal provisions dealing with forest finance.

V. Combating Illegal Logging and Trade in Illegally Harvested Timber

One of the major issues that have been confronted by international forest-related treaties is the combating of illegal logging and of the trade in illegally harvested timber. In particular, the ITTA deals therewith. The ITTA mentions the issue on several places, by indicating explicitly the need for the enhancement of the capacity of its Member States to combat this malign phenomenon, which represents one of the driving forces of deforestation and has negative impacts on the socio-economic development in forest-rich States.[1670] Moreover, the Council of the ITTA has to include information, provided by Member States in relation to the established control systems designed in order to monitor and to combat the illegal harvesting of and the illegal trade in tropical timber, in its biannual reviews of the international tropical timber situation.[1671] However, the Council

1669 *Eikermann* [n. 1574], p. 142.

1670 International Tropical Timber Agreement, adopted on 27 January 2006 and entered into force on 7 December 2011, in: 2797 UNTS 75, Article 1(n).

1671 Article 28(e).

is not required to pronounce its judgement on the information provided by Member States and no measures are foreseen in order to assist forest-rich States in the implementation of these control systems.

It can be assumed that only scarce international legal provisions endeavored to tackle illegal logging and the trade in illegally harvested timber are available. Additionally, those cases where the trade in illegally harvested timber occurs through a third state are not tackled. For example, the timber harvested in Papua New Guinea and processed in China is sold all around the world.[1672] Consequently, a legally binding international instrument that would specifically deal with the problem of illegal logging is needed. Thereby, an efficient international controlling system should be established in order to monitor the origins of timber and of forest products traded internationally. It can be argued that this system would not violate the provisions of the GATT as long as same regulations would be applied in relation to the trade in illegally harvested timber on the national and international levels.[1673]

VI. Promoting Human Rights and Equity

The issue of equity plays a major role in international forest-related documents. Particularly, it is related to the equal participation of all stakeholders, especially of the most vulnerable ones, in the processes of decision-making and of the implementation of forest-related policies, procedures, and mechanisms.[1674] Moreover, it incorporates the equal and fair sharing of benefits derived from the utilization of forest resources, from the implementation of REDD+ and from other similar mechanisms and projects.[1675] The idea of fair benefit sharing is, inter alia, incorporated in ABS. Finally, the issue of equity in the context of forest governance foresees the provision an adequate compensation to stakeholders

---

1672 See Subsections B.IV and B.V of Chapter 2.

1673 See *Eikermann* [n. 1574], p. 144.

1674 *Eikermann* [n. 1574], p. 144; *Colchester, Marcus/Boscolo, Marco/Contreras-Hermosilla, Arnoldo/Del Gatto, Filippo/Dempsey, Jessica/Lescuyer, Guillaume/Obidzinski, Krystof/Pommier, Denis/Richards, Michael/Sembiring, Sulaiman N./Tacconi, Luca/Rios, Maria Teresa Vargas/Wells, Adrian,* Justice in the forest: Rural livelihoods and forest law enforcement, 2006, pp. 2 ff.; *Paladino, Stephanie/Fisket, Shirley J.*, The carbon fix: forest carbon, social justice, and environmental governance, 2017, pp. 140 ff.

1675 *Eikermann* [n. 1574], p. 144; *Colchester/Boscolo/Contreras-Hermosilla/Del Gatto/Dempsey/Lescuyer,/Obidzinski/Pommier/Richards/Sembiring/Tacconi/Rios/Wells*, [n. 1674], pp. 44, 57; *Paladino, Stephanie/Fisket, Shirley J.* [n. 1674], pp. 141 f.

that must abandon industrial and logging activities in forests in order to ensure their conservation.[1676] However, in the last case, temporary economic disadvantages should be weighed up against the benefits for forest owners resulting from the promotion of multiple forest functions.

The promotion of equity in the process of forest governance depends on two issues. Firstly, the forest-related owner and user rights of all relevant stakeholders, including ILCs, should be recognized and assured on the national and international levels.[1677] This presupposes the establishment of secure land tenure patterns, combined with the decentralization of forest governance and with the promotion of community-based ownership. It is important to note that local communities are more likely to manage forests sustainably if their owner and user rights over forestlands are ensured. Secondly, the rights of ILCs and other vulnerable groups should be recognized and promoted on the international level and their representatives should be directly involved into the process of decision-making.[1678] The international legal framework for the implementation of forest-related policies and mechanisms should provide for that no project is implemented until the PIC of ILCs is acquired, and the effective control from their side over the aspects of project implementation that regard their rights and interests is assured. The requirement of PIC and MAT should not be limited to ABS, but should be extended to REDD+ and other forest-related mechanisms. Moreover, in cases where the rights of ILCs are not adequately promoted at the national level, international law could partially fill this gap by providing the representatives of ILCs with comprehensive and direct membership and participation rights in relation to the activities of forest-related organisations.

As for the moment, a notable progress in this regard has been achieved after the Nagoya Protocol entered into force. However, its provisions dealing with the protection of the rights of ILCs are still weaker than its regulations endeavored to strengthen the transparency and legal certainty for foreign investors.[1679] In view of the fact that the scope of the Nagoya Protocol is limited to the regulation of the utilization of genetic resources, it cannot be applied to all forest-related activities. Therefore, it is necessary to strengthen the pertinent provisions of the

1676 *Eikermann* [n. 1574], p. 144; *Colchester/Boscolo/Contreras-Hermosilla/Del Gatto/Dempsey/Lescuyer,/Obidzinski/Pommier/Richards/Sembiring/Tacconi/Rios/Wells*, [n. 1674], pp. 44, 57.

1677 *Eikermann* [n. 1574], p. 145.

1678 *Eikermann* [n. 1574], p. 144.

1679 See Subsection D.II.5 of Chapter 3.

Nagoya Protocol dealing with the issues of equity and human rights, and include similar provisions in another legal regimes dealing with forests. Another solution may consist in the elaboration of specific regulations for the promotion of equity in the context of forest-related policies. Considering the obstacles on the way towards the further development of international law specific to forests, the first solution seems to be a more plausible one at first glance. However, the problem of the promotion of equity and of human rights in the course of the implementation of forest-related activities is so complex that it arguably requires robust and specific regulations that could be delivered only by legal documents specific to forests.

VII. Bottom-Up Approaches and the Decentralization of Forest Governance

The development of the international legal regime for forests is currently underpinned by the promotion of a bottom-up approach and the enhanced decentralization of forest governance.[1680] The praxis has shown that these processes can have a positive impact on forest conservation.[1681] It is noteworthy that a number of conservational initiatives have originated from NGOs, other civil society groups, and from the groups of forest dependent people.[1682] In general, bottom-up approaches foresee the promotion of the participation of the stakeholders that are directly engaged in forest management and/or dependent on forests in the development and implementation of forest-related policies.[1683] It is submitted that bottom-up approaches permit one to identify the most suitable and effective practices, procedures, and initiatives originating on the lowest levels of forest governance. Moreover, it permits to promote them more effectively,

---

1680 *Moeliono, Moira/Wollenberg, Eva/Limberg, Godwin*, The Decentralization of Forest Governance: Politics, Economics and the Fight for Control of Forests in Indonesian Borneo, 2009, pp. 4–7; *Andersson, Kriter*, Understanding Decentralized Forest Governance: An Application of the Institutional Analysis and Development Framework, in: Sustainability: Science, Practice and Policy, Volume 2, Issue 1, 2006, pp. 25 ff.

1681 *Andersson* [n. 128], pp. 26 f.

1682 *Martin, Pamela L.*, Global Governance from the Amazon: Leaving Oil Underground in Yasuní National Park, Ecuador, in: Global Environmental Politics, Volume 11, Issue 4, 2011, pp. 38 ff.

1683 Cf. *Rodriguez-Ward, Dawn/Larson, Anne M./Ruesta, Gordilla Harold*, Top-down, Bottom-up and Sideways: The Multilayered Complexities of Multi-level Actors Shaping Forest Governance and REDD+ Arrangements in Madre de Dios, Peru, in: Environmental Management, Volume 62, Issue 1, 2018, pp. 99 ff.

including through the development of a specific legal framework. Additionally, it permits to facilitate the adjustment of international regulations in accordance with the needs and expectations of local stakeholders.

The importance of the development and promotion of bottom-up approaches is signalized also by the fact that the effectiveness of any forest-related policies depends on the support for their implementation from the side of key local stakeholders.[1684] If the relevant policies are not acceptable for the domestic actors in forest-rich States, their implementation may become practically unenforceable. In order to ensure a positive outcome of forest policies, it is necessary to ensure that the local circumstances of forest-rich States are fully considered.[1685]

In this regard, the experience with the development of initiatives for the protection of biodiversity hotspots in the Yasuní National Park, Ecuador can be mentioned. After significant oil stocks had been discovered in this rainforest region, the Ecuadorian government fostered its efforts in order to convince the international community to provide additional financial resources as a kind of compensation for abandoned drilling projects, which allowed to develop the national economy and to improve local livelihoods, while conserving tropical rainforests.[1686] The role of local actors, of indigenous communities, and of civil society in the development of the relevant policies is particularly appreciated.[1687]

The promotion of bottom-up approaches is deeply related to the decentralization of forest governance, which, on its turn, is linked to the establishment of a cross-sectoral approach. The decentralization of forest governance has proven to be beneficial for the protection of forests as deforestation rates have reduced in those tropical regions of Latin America where forest governance was decentralized. However, it should be noted that these benefits have only taken place in the areas where local forest users and owners were involved into the cooperation with local forest authorities.[1688]

1684 *Singer/Giessen* [n. 1631], p. 76.

1685 *Bull, Garry Q./Boedhihartono, Agni K./Bueno, G./Cashore, B./Elliot, C./Langston, James D./Riggs, R. A./Sayer, J.*, Global Forest Discourses Must Connect with Local Forest Realities, in: International Forestry Review, Volume 20, Issue 2, 2018, pp. 163 ff.

1686 *Martin* [n. 1682], pp. 163 ff.

1687 Ibid, p. 164.

1688 *Wright, Glenn D./Andersson, Krister P./Gibson, Clark C./Evans, Tom P.*, Decentralization Can Help Reduce Deforestation When User Groups Engage with Local Government, in: Proceedings of the National Academy of Sciences of the United States of America, Volume 113, Number 52, 2016, p. 14958.

The importance of promoting the participation of non-state actors and of the cross-sectoral approach was recognized also in the course of the implementation of REDD+-related activities. It was found out that the REDD+-mechanism contributes to the wider engagement of different social groups in the process of forest management. It also contributes to the promotion of cross-sectoral approach.[1689]

However, a bottom-up approach and the decentralization of forest governance can also have negative impacts in certain cases. The growing influence of domestic actors and policies on the international legal regime for forests has likely contributed to its weakening and to the shift of its focus on the periphery of this regime, where there are links to other international legal regimes, which occurs on the cost of the development of forest-specific regulations.[1690] Moreover, in some cases, the promotion of the participation of local farmers in forest management activities resulted in the enhancement of deforestation rates, as it happened in the Peruvian Amazon, where local farmers were largely involved in the production of palm oil.[1691] Although the main scope of Company-Community Partnerships (CCPs) established in this region consisted in the enhancement of palm oil production through higher productivity and the more effective management of rural land areas, they were also supposed to contribute to the social and environmental sustainability by diversifying income sources for local farmers, and, consequently, by reducing the anthropogenic pressure on local forests.[1692]

However, in reality, CCPs have increased social tensions and the environmental harm associated with the production of palm oil.[1693] One of the reasons for this consists in the lack of the effective monitoring and supervision from the side of national authorities and international organizations, whereas regulatory and implementation activities are merely fulfilled by private companies.[1694]

1689 *Fujisaki, Taiji/Hyakumura, Kimihiko/Scheyvens, Henry/Cadman, Tim*, Does REDD+ Ensure Sectoral Coordination and Stakeholder Participation? A Comparative Analysis of REDD+ National Governance Structures in Countries of Asia-Pacific Region, in: Forests, Volume 7, 2018, Article Number 195.

1690 *Benjamin/Giessen* [n. 1631], p. 76.

1691 *Bennet, Aoife/Ravikumar, Ashwir/Paltán, Homero*, The Political Ecology of Oil Palm Company-Community Partnerships in the Peruvian Amazon: Deforestation Consequences of the Privatization of Rural Development, in: World Development, Volume 109, 2018, p. 38.

1692 Ibid.

1693 Ibid, pp. 36 ff.

1694 Ibid, p. 40.

Therefore, the initiatives endeavored to be implemented in forest-rich States should deliver the incentives for local stakeholders to promote forest protection. At the same time, it is necessary to promote the national and international monitoring of the social and environmental impacts of implemented activities.

Thus, the promotion of bottom-up approaches to forest governance combined with its decentralization does not imply that the international community and national authorities should not participate actively in the development and implementation of forest-related policies. It should rather be regarded as an opportunity for a wider involvement of other relevant stakeholders into the process of forest governance. Bottom-up approaches should be properly integrated in the international legal regime for forests. At the same time, the role of international institutions and national authorities should be strengthened in order to ensure the comprehensive and holistic implementation of forest-related policies.

VIII. Promotion of Monitoring, Assessment, and Reporting

The establishment and promotion of systems for monitoring and assessing the conditions of forest ecosystems is another important issue to be considered in the future development of the international legal regime for forests. Such systems must also serve in order to ensure the consistent and comprehensive implementation of forest-related policies. According to the NLBI, States are expected to monitor and evaluate the progress towards the implementation of forest-related policies.[1695] However, the significance of this provision, which is a legally non-binding one, is further relativized by its weak wording, which stipulates that States "should" only comply with it. The next provision, according to which national progress reports on the activities undertaken by States should be submitted to the UNFF, is also formulated in a weak and vague manner.[1696] It indicates clearly that States are only supposed to submit these progress report on a voluntarily basis. Moreover, the submission of reports is subject to the availability of financial and technological resources. Indeed, States are supposed to report on their progress in the achievement of the main purpose of the NLBI, which is to strengthen the political commitment to the implementation of SFM and to the achievement of the Global Objectives on Forests.[1697] However, it is not explicitly stipulated that the conditions of forest ecosystems must be described in these reports. Moreover, the UNFF is not currently empowered to promote the

1695 NLBI, Articles 8.
1696 NLBI, Articles 9.
1697 Ibid.

compliance with its forest-related documents in cases where any inconsistencies in the States' forest-related activities are revealed. The provisions dealing with monitoring, assessing and reporting were included also into the legal framework for the implementation of REDD+. However, as it was shown in the previous chapters, the relevant provisions are vague and unclear, and a robust compliance system endeavored to promote the conservation of natural forests under REDD+ is still needed.

Monitoring, assessing and reporting of forest conditions is further complicated by the absence of an internationally recognized, legally binding, and comprehensive definition of forests. As a result, international organizations apply different forest definitions. By choosing a pertinent forest definition, different stakeholders are guided by their vision of forest management.[1698] The common feature is that those definitions that are currently applied are usually based on the definition of the FAO, which is characterized by the accentuation of technical quantitative criteria, such as the minimal tree crown cover (not less than 10%), the height of trees (must exceed 5 meters), and the minimal land area requirement (not less than 0.5 ha). According to the definition in question, monoculture tree plantations are also treated as forests.[1699] Consequently, the definition used by the FAO is insufficient for the promotion of forest conservation, as it does not take into account the differences between different types of forests and between, on one side, those trees that are planted for the purpose of near-natural re/afforestation and, on the other, monoculture tree plantations.[1700] The evaluation and assessment of the global forest cover based on such a forest definition that operates only with quantitative indicators can be misleading. It can create an illusion that the rates of deforestation are reducing or even the global forest area is increasing, whereas in reality the conditions of forests are worsening due to the fragmentation of forest area, unsustainable harvesting practices and the conversion of natural forests. For example, the FAO Forest Resources Assessment demonstrates that national authorities and international organizations can change the estimation of forest land area by modifying the relevant quantitative

1698 *Chazdon, Robin L./Brancallion, Pedro H. S./Laestadius Lars, Bennett-Curry Aoife/Buckingham, Kathleen/Kumar Chetan/Moll-Rocek, Julian/Vieira, Célia Guimarães Ima/Wilson, Sarah Jane*, When Is a Forest a Forest? Forest Concepts and Definitions in the Era of Forest and Landscape Restoration, in: Ambio, 2016, Volume 45, Issue 5, 2016, p. 539.

1699 Ibid, p. 542.

1700 Ibid, p. 548.

indicators.[1701] For example, the requirement to the minimal size of forest land area can be reduced from 1 ha to 0,5 ha; whereby the issue of forest fragmentation is not taken into consideration.[1702] In spite of this fact, credit should be given to the FAO for extending the share of land areas listed as forest area. Such an approach permits to apply forest conservation policies to land areas that were not previously considered as forests. In spite of the fact that such a practice can deliver certain benefits, a robust definition of forests that would permit to differentiate between different types of forest ecosystems is needed in order to ensure the conservation of natural forests and of forest biodiversity.

Consequently, an adequate international system for the monitoring and assessing of the conditions of forests is still absent, as well as a comprehensive and all-embracing definition thereof.

IX. Promoting Compliance and Law Enforcement

Finally, it should be underlined that the international legal regime for forests does not provide enough regulations and mechanisms for the promotion of the implementation of its forest-related legally binding provisions and legally non-binding positive incentives on the national level. It does not ensure the compliance with the provisions of international forest-related documents and fails to promote the enforcement thereof.[1703] The issue of promoting compliance on a national level goes beyond the provision of financial and technical assistance. The enhanced implementation of the provisions in question can be promoted through the adequate monetary evaluation of the ecosystem services of forests. This means that the implementation of SFM can generate a bulk of financial resources for the conservation of forests. Moreover, the non-monetary benefits that forest-rich States receive from the maintenance and promotion of forest functions, can pursue them to promote compliance. Such non-monetary benefits foresee the recognition of efforts of the States in question, and the improvement of their position on the international level. Additionally, an adequate mechanism for ensuring compliance, which should also foresee sanctions for the breech of forest-related treaties, is needed. In this regard, it can be seen as negative that despite the years of the development of REDD+ and of the ABS regime,

---

1701 Ibid, p. 543.

1702 Ibid.

1703 For the importance of compliance in international environmental law, see: *Kiss, Alexandre/Shelton, Dinah*, International environmental law, Third Edition, 2004, pp. 275 ff.

a comprehensive instrument for promoting the monetary benefits of forest conservation is still absent. It is also problematic that with the exception of the CITES, no effective sanction mechanism has ever been established by any treaty.

It can be argued that sanctions should be developed in order to be applied against those States that continue to promote unsustainable forest management practices, destroy natural forests, and tolerate or promote the illegal harvesting of and the illegal trade in timber and other forest products. As the example of the CITES has demonstrated, these sanctions can include the exclusion of certain States from those profitable markets of the ecosystem services of forests that are established according to the provisions of the Conventions concerned. Furthermore, they can foresee the restrictions of access to the conventional markets of forest timber and other forest products. It can be argued that these sanctions would be compatible with the provisions of international trade law as long as national stakeholders would be treated in the same way as the foreign ones, in a case of the infringement of forest-related provisions.[1704]

Overall, the promotion of compliance is further hampered by a lack of a legally binding instrument specific to forests. Indeed, it is stated that only a legally binding treaty can establish an effective compliance mechanism.[1705]

Thus, the existing international legal regime for forests permits to draw a conclusion on main issues that should be included in any international legal document that is specific to forests. However, all of them are not sufficiently covered by international law. As the next step, the main forest-related issues that have not yet found their way into the existing forest-related documents should be analyzed.

IX. Creating an International Agency on Forest Issues

Although a number of international institutions on forest issues have been established, no one of them deals with all the major forest-related issues in a holistic manner. Instead, there is a bulk of institutions with conflicting authorities that deal with different aspects of forest governance. The need for one unique international agency that would deal with forest-related issues in a holistic manner has never been articulated by forest-related documents. Such an institution would promote the compliance and facilitate the monitoring, assessment and

1704 See Subsection B.IV.2 of Chapter 2.

1705 *Davenport, Deborah. S.*, An Alternative Explanation for the Failure of the UNCED Forest Negotiations, in: Global Environmental Politics, Volume 5, Issue 1, 2005, p. 105.

evaluation of the conditions of world forest ecosystems. Instead, a number of different forest-related institutions are currently dealing with forest-related issues, with the UNFF and the CPF serving for the coordination of their activities.

The creation of an international agency on forest issues can potentially foster the conservation of forests and promote fundraising and law enforcement. In order to fulfil these tasks, such an agency has to be adequately empowered. This could be only achieved if such an agency were created in accordance with the provisions of a legally binding forest treaty. As long as such a treaty is absent, the establishment of a unique international agency on forests is highly unlikely. The UNFF, the CPF, and the FAO are unlikely to fulfil this role, because they are not adequately empowered by any international document. For this purpose, it can be proposed to create a new agency on forests or to reform the UNFF. It is submitted that both of these tasks are more likely to be accomplished by a legally binding treaty on forests.

X. Promotion of Non-Climate and Non-Biodiversity-Related Forest Functions and the Role of Degraded Lands Restoration

It should be pointed out that, in spite of the fact that the scientific evidence for the role of forests in the formation and distribution of atmospheric precipitations and their hydrological function is growing, the relevant issues have not yet found their way into international forest-related documents. Although the UNCCD underlines that the loss of vegetation cover represents one of the major driving forces of desertification,[1706] it does not specify how the relevant function of forests can be maintained and promoted.

Moreover, the international legal regime for forests fails to consider the role that sustainable activities outside of the forest sector can play in the conservation of forests. REDD+ and other international mechanisms suffer from the same shortcoming, namely, from their inability to consider the potential for reducing the human pressure on forests by restoring degraded agricultural lands. Bearing in mind that the strong need for the enhancement of agricultural produce is the main driver for forestland conversion, it is necessary to restore the degraded land areas that are currently deprived of forests and not deemed for afforestation and can be reintegrated in agricultural productive cycles. Instead of converting forest lands for agricultural purposes, States should undertake efforts in order

1706 UNCCD, Article 1(f).

to restore at least 2 billion hectares of degraded lands worldwide.[1707] Some areas could be restored for agricultural purposes, whereas other areas could be afforested or reforested.[1708] In order to ensure the sustainability of the restoration of degraded lands, some part of agricultural areas should also be covered with forests and the forest vegetation.[1709] The restoration for agricultural purposes should contribute to the reduction of the anthropogenic pressure on forest ecosystems, while the afforestation and reforestation activities should contribute to the enhancement of the global forest area. For this purpose, the near-natural reforestation and afforestation practices should preferably be applied.

A number of international documents refer to the restoration of degraded lands. For example, in the document "The Future we Want", adopted at the Earth Summit 2012 in Johannesburg, States committed themselves to enhance their efforts in order to achieve a "land degradation neutrality on the global scale".[1710] However, the document in question does not clarify further how the restoration of degraded lands should be implemented. The Bonn Challenge, launched by the Government of the Federal Republic of Germany and the IUCN in 2011, should be cited in this regard.[1711] However, there is still a lack of clearness on the role that the international community should play in the restoration of degraded lands. It can be argued that a comprehensive international program of action is required. The restoration of degraded lands can contribute to the combating of deforestation and forest degradation. It is also noteworthy that a significant portion of degraded lands is still classified as or used to be forestlands.[1712] Therefore, their restoration should include afforestation and reforestation activities. Consequently, an international program of action on the restoration of degraded lands ought to be linked directly to international forest governance. Eventually,

---

1707 The data from the official Internet site of the Bonn Challenge. – Internet resource– https://www.bonnchallenge.org/what-our-global-restoration-opportunity [Last visit: 30.04.2023]

1708 Ibid.

1709 Ibid.

1710 United Nations A/RES/66/288, Resolution adopted by the General Assembly on 27 July 2012: The Future We Want, Outcome document of the United Nations Conference on Sustainable Development, United Nations, New York, 2012, Paragraphs 205 ff. – Internet resource – https://undocs.org/A/RES/66/28 [Last visit: 30.04.2023]

1711 See the official Internet site of the Bonn Challenge. – Internet resource – https://www.bonnchallenge.org/what-our-global-restoration-opportunity [Last visit: 30.04.2023]

1712 Ibid.

it can be adopted as a part of any international document specifically dealing with forests.

D. Adoption of a Forest Convention Versus Other Approaches

I. Revisiting the Option of the Adoption of a Legally Binding Forest Convention

In order to analyze the probability of the adoption of a legally binding forest convention, it is necessary to reconsider briefly the main reasons why its negotiations have failed in the past.

The concerns of developing States about their sovereign rights are often cited as a primary reason for the lack of consensus in relation to the conclusion of a legally binding forest convention.[1713] Following this logic, the adoption of such a convention is considered to be redundant and counterproductive.[1714] Other researchers suggest that the very fact that the international legal regime for forests has been inefficient in tackling the problem of deforestation is largely determined by the past failures of the negotiations of a forest convention.[1715]

It is argued that the failure of the negotiations in question can be attributed to the fact that developing States were unwilling to give up even the smallest share of their sovereign rights.[1716] According to another point of view, the absence of a forest convention can be explained by the lack of will on the side of key States, including the USA, to guide the negotiations in the direction that would make the adoption of such a treaty acceptable for developing States.[1717] Following the logic that cooperation always permits to achieve a better outcome for all the members of a negotiation process in comparison to non-cooperation, it can be assumed that it could still have been possible to achieve a consensus on forest issues.[1718]

1713 *Gupta, Joyeeta/van der Grijp, Nicolien/Kuik, Onno*, Climate Change, Forests and REDD+: Lessons for Institutional Design, 2013, p. 55.

1714 Ibid.

1715 *Davenport, Deborah. S.*, An Alternative Explanation for the Failure of the UNCED Forest Negotiations, in: Global Environmental Politics, Volume 5, Issue 1, 2005, pp. 110 ff.; *Humphreys, David*, Logjam: Deforestation and the crisis of global governance, 2006, p. 150; *Neeff, Till/Göhler, Daniela/Ascui, Francisco*, Finding a Path for REDD+ between ODA and the CDM, in: Climate Policy, Volume 15, Issue 2., 2014, pp. 151 ff.

1716 *Davenport* [n. 1715], p. 115.

1717 Ibid.

1718 Ibid, p. 116.

It is argued that this assumption tends to be true both in the case where all the members of a negotiation process are interested in the adoption of a legally binding document, and in the situation when one of negotiators or a group of negotiators decisively rejects the very idea of any such document.[1719] The second situation described as an "asymmetric deadlock" has supposedly emerged in the course of the forest-related negotiations prior to and the Rio Summit in 1992.[1720]

However, it can be argued that the "asymmetric deadlock" did not exist from the very beginning of the negotiations as the very readiness of States to enter the negotiation process indicates that they regarded the adoption of a legally binding document as a plausible solution. The very fact that forest issues were included on the official agenda of the preparatory negotiations for the Rio Summit indicates that States have agreed on the necessity to cooperate on the elaboration of global forest-related policies.[1721] A number of facts justify this assumption. In this regard, one can cite the Malaysian list of options for enhancing the transfer of technologies and financial resources endeavored to compensate the opportunity losses associated with the reduction of human pressure on forest ecosystems.[1722] This list was presented by Malaysia at the Prep Com 2, although officially the delegation of this State had already previously rejected the option of a forest convention.[1723] Moreover, in the aftermath of the Prep Com 2, the Malaysian Delegation highlighted its statement according to which the studies on socio-economic functions of forests should preclude the elaboration and adoption of a forest convention.[1724] The same conclusions can be drawn from the speech of Malaysian Prime Minister at the UNCED in 1992.[1725] In fact, in his

1719 Ibid.

1720 Ibid.

1721 *Steiner, Melanie*, The Journey from Rio to Johannesburg: Ten Years of Forest Negotiations, Ten Years of Successes and Failures, in: Golden Gate University Law Review, Volume 32, Issue 4: Rio's decade: Reassessing the 1992 Earth Summit, January 2002, pp. 630, 635 ff.

1722 *Grubb, Michael/Koch, Matthias/Thomson, Koy/Munson, Abby/Sullivan, Francis*, The "Earth Summit" Agreements: A Guide and Assessment, an Analysis of the Rio'92 UN Conference on Environment and Development, 1993, p. 36; *Taib, Fauziah Mohd*, Malaysia and UNCED: An Analysis of a Diplomatic Process: 1989–1992, 1997, p. 58.

1723 *Taib* [n. 1722], p. 59.

1724 *Environmental and Energy Study Institute*, Earth Summit Update, Washington D.C., July 1991, p. 1.

1725 United Nations, Statement by his H. E. Dato' Seri Dr. Mahathir bin Mohamad, Prime Minister of Malaysia, in: Report of the United Nations Conference on Environment

statement, the Head of Malaysian government particularly underlined the issue of equity and put the blame for the failure of negotiations on developed States. According to him, these States were unwilling to cooperate in order to develop and implement effective and comprehensive rules and policies aimed at the resolution of environmental problems.[1726]

At any time, despite the fact that the "asymmetric deadlock" had emerged during the negotiations in question, key States might still have been able to pursue their partners to change their attitude to the adoption of a legally binding forest convention.[1727] This could have been done by enhancing the value of potential benefits that could have been delivered to forest-rich States through the adoption of a convention. At the same time, these States should have been aware of the negative impacts resulting from the failure of negotiations.[1728] Consequently, the developed States led by the USA were able to influence the position of developing States by promoting the incentives for the adoption of a treaty.[1729] Theoretically, the ability of one party to change the position of another party is limited only by the available resources, which the first party can apply for this purpose. However, it is also limited by the rules of international law, which largely impede the application of force, such as a punishment or the threat of a punishment, in order to influence the opinion of the counterpart.[1730] In this situation, developed States still had the ability to apply incentives in order to persuade developing ones led by Malaysia to change their mind. It is submitted that the USA had at that moment the necessary resources and powers to influence the position of this group of States by providing them with further rewards.[1731] However, the USA were unwilling to do so as the potential benefits associated with the adoption of a forest convention could not outweigh for them the costs of leading negotiations to a successful outcome.[1732]

It can be argued that the decision of Malaysia to oppose the adoption of a convention was not final and that it was rather motivated by the considerations of equity than by sovereignty concerns.[1733] Indeed, any scenario in which

---

and Development, Volume 3, Statements Made by Heads of State or Government at the Summit Segment of the Conference, Rio De Janeiro, 3–14 June 1992, pp. 230 ff.

1726 Ibid.

1727 *Davenport* [n. 1715], p. 112.

1728 Ibid.

1729 Ibid.

1730 Ibid.

1731 Ibid.

1732 Ibid.

1733 *Grubb/Koch/Thomson/Munson/Sullivan* [n. 1722], p. 36.

developing States would have had to bear the main costs of forest conservation was unacceptable for them. Based on all these statements, it is submitted that the position of States still could have been influenced in such a way that the adoption of a legally binding forest convention would have become feasible.

Against this backdrop, it can be argued that it was not only the sovereignty concerns of Malaysia, but also the lack of leadership on behalf of developed States that played a crucial role in the failure of the negotiations of a legally binding forest convention in Rio.[1734]

It is noteworthy that despite their opposition to the adoption of a forest convention during the Rio Summit in 1992, Malaysia and India, supported by the United Kingdom and Canada, launched in 1994 the Intergovernmental Working Group on Forests (IWGF) and the Workshop "Towards sustainable Forestry: Preparing for CSD 1995" that both contributed to a certain extent to the development of common policies on forests.[1735] They were deemed to support the organization of the third meeting of the Commission on Sustainable Development held in April 1995, which was specifically dedicated to forests.[1736] In reality, the contribution of these initiatives to the development of common policies was insignificant as elaborated documents were too vague and superficial for the implementation on the ground.[1737] However, the collaboration of developing and developed States, as well as the increased readiness to cooperate demonstrated by forest-rich States during the Conferences in Bandung and New Delhi in 1993, can be interpreted as a signal that the elaboration of a forest convention was still possible.[1738] It is submitted that the awareness of possible benefits that could be delivered for them through the adoption of a convention was growing among developing States.[1739] Additionally, in the course of the activities under the Intergovernmental Panel on Forests, a number of States, including Malaysia, Papua New Guinea, Costa Rica and Philippines advocated the creation of a time-bounded negotiating committee on the elaboration of the "legally

---

1734 *Hajost, Scott A.*, The role of the United States, in: *Campidoglio, Luigi/Pineschi, Laura/Siniscalco, Domenico/Treves, Tullio* (eds.), The Environment after Rio, International Law and Economics, International Environmental Law and Policy Series, 1994, pp. 16 ff.

1735 *Schulte zu Sodingen, Beate,* Der völkerrechtliche Schutz der Wälder, 2002, pp. 227 f.

1736 Ibid.

1737 Ibid, p. 228.

1738 *Schulte zu Sodingen* [n. 1735], p. 228; *Hönerbach, Frank*, Verhandlung einer Waldkonvention – Ihr Ansatz und Scheitern, 1996, p. 38.

1739 *Schulte zu Sodingen* [n. 1735], p. 228.

binding instrument on all types of forests".[1740] However, it is argued that it were not only environmental concerns, but also economic considerations that promoted the shift of the position of this group of States.[1741]

Consequently, after the Rio Summit 1992, the differing positions of States in relation to the elaboration and adoption of a legally binding convention were no longer determined by the North-South differences.[1742] Supporters and opponents of a legally-binding forest convention were present on both sides, which resulted in the formation of unexpected alliances between developing and developed States. The steadily changing positions of States signified that their attitude to the adoption of a convention was rather determined by their short-term economic and political interests than by environmental concerns.

Since its establishment in 2000, the UNFF has been empowered to analyze possible options for strengthening the international legal regime for forests by considering, *inter alia*, the option of elaboration of a legally binding treaty.[1743] However, by that time the general support for a forest convention had decreased as more States were pleading for the elaboration and implementation of market-based approaches to global environmental governance.[1744] The application of such market-based approaches does not necessarily require the adoption of legally binding documents. The commitment to market-based approaches has largely influenced the negotiations of a forest convention after the year 2000.[1745] This fact made the conditions even more unfavorable for the adoption of such a convention. Additionally, some civil society groups opposed the adoption of a legally binding convention, while they feared that States were able to negotiate only a treaty regulating economic issues with weak environmental commitments.[1746] The impact on forest protection of such a convention was supposed to be either neutral or negative.[1747]

1740 Ibid, p. 229.

1741 *Humphreys*, the Elusive Quest [n. 1635], pp. 3 f.

1742 *Davenport, Deborah S./Wood, Peter*, Finding the Way Forward for the International Arrangement on Forests: UNFF-5, -6 and -7, in: RECIEL, Volume 15, Issue 3, 2006, p. 318.

1743 ECOSOC Resolution 2000/35, Report on the forth Session of the Intergovernmental Forum on Forests, Economic and Social Council, United Nations, 2000, Paragraph 3(c.i). – Internet resource – https://www.un.org/esa/forests/wp-content/uploads/2013/09/2000_35_E.pdf [Last visit: 30.04.2023]

1744 *Eikermann* [n. 1574], p. 75; *Humphreys*. The Elusive Quest [n. 1635], p. 4.

1745 *Eikermann* [n. 1574], p. 75; *Humphreys*. The Elusive Quest [n. 1635], p. 4.

1746 *Eikermann* [n. 1574], p. 75; *Humphreys*. The Elusive Quest [n. 1635], p. 4.

1747 *Eikermann* [n. 1574], p. 75; *Humphreys*. The Elusive Quest [n. 1635], p. 4.

However, market-based approaches to the environmental regulation have proved to be inefficient, especially in relation to the promotion of good forest governance.[1748] The participation of private actors tends to be low and the most of projects are currently reliant on public funding.[1749]

It is submitted that those private actors, public funds, and international institutions that participate in forest-related activities require legal certainty and transparency that may be delivered only by strict regulations and stronger compliance mechanisms. For instance, the GCF has developed its own regulations, including an Operational Framework, in order to provide clearer regulations in relation to REDD+ finance.[1750] Although the UNFCCC COP failed to adopt stronger provisions related to the implementation of REDD+, the GCF was constrained to develop them for the purpose of a more harmonious implementation of new mechanisms on the ground.[1751]

Therefore, both developing and developed States can be deemed responsible for the failure of all the previous attempts to elaborate a legally binding instrument on forests. Moreover, international forest negotiations should not be regarded anymore as a dispute between forest-rich States opposing a legally binding forest convention and developed States pleading for its adoption.

One can argue that developed States have achieved a certain level of prosperity by destroying their forest ecosystems and, consequently, from the point of view of the intra- and intergenerational justice, the developing world should not undergo any commitments to conserve their forests. However, it should be noted that deforestation in developed States had taken place in the period of time when the knowledge about the environmental functions of forests was either unavailable or very scarce. Moreover, the large-scale deforestation was unavoidable due to the lack of modern technologies. Nowadays, new technologies are available and new scientific knowledge has been acquired. It is also noteworthy that forest-rich States can also benefit from the conservation of forest ecosystems under their jurisdiction. Therefore, developed and developing States should both commit themselves to the promotion of forest protection.

It should be mentioned that all the previous attempts to attract the attention of the international community and of the society in general to the problem of

---

1748 *Humphreys*, The Elusive Quest [n. 1635], p. 4.

1749 Ibid.

1750 *Recio, Mariá Eugenia*, Dancing Like a Toddler? The Green Climate Fund and REDD+ International rule-Making, in: RECIEL, Volume 28, Issue 2, 2019, pp. 128 f.

1751 Ibid.

deforestation failed to convince all relevant stakeholders that a legally binding treaty is needed. However, it can be argued that the development of international environmental law and of law in general strongly depends on current socio-economic and political developments. It is submitted that the perception of the environmental role of forests has grown since the first Earth Summit in 1992. Moreover, more States are interested now in precise and clear regulations of international timber and forest product markets. For example, a multilateral agreement on the conservation and sustainable use of all types of forests, which would tackle those issues that are related to the deforestation caused by the production of vegetable oils, can be more acceptable for palm oil producing States than unilateral bans and restrictions imposed by the European Union.[1752]

Consequently, changing circumstances and the better understanding of multiple functions of forests can potentially facilitate the adoption of a legally binding forest-specific document. However, it cannot be expected that this document will be adopted in the near future, while short-term economic and political interests continue to prevail. Notwithstanding this, the hypothetical positive impact of such a treaty should be analyzed.

II. Possible Benefits of a Legally Binding Forest Convention

The adoption of a legally binding forest convention can bring certain benefits, as this treaty would contribute to the resolution of conflicts between the provisions of different international conventions that affect forest issues.[1753] Inter alia, according to the principle *lex specialis*, a hypothetical forest convention could enjoy the priority of application over the provisions of the UNFCCC or other conventions, at least in certain cases. Admittedly, the conflicts with the international legal regime for combating climate changes would continue to arise, as the climate change mitigation/adaption still could be prioritized in certain cases over the conservation of forests.[1754] However, it would be more difficult to justify these activities from the point of view of international law in case when a legally

---

1752 *Keating, David*, EU Labels Biofuel from Palm Oil as Unsustainable, Bans Subsidies, in: Forbes, March 14, 2019. – Internet resource – https://www.forbes.com/sites/davekeating/2019/03/14/eu-labels-biofuel-from-palm-oil-as-unsustainable-bans-subsidies/#232f4c39c9da [Last visit: 30.04.2023]

1753 *Humphreys*, The Elusive Quest [n. 1635], p. 2.

1754 *Humphreys*, The Elusive Quest [n. 1635], p. 2; *Skala-Kuhmann, Astrid*, Legal Instruments to enhance the conservation and sustainable management of forest resources at the international level, 1996, pp. 10 ff.

binding forest convention would establish a duty to protect forest ecosystems. In case of eventual international disputes on forests, a forest convention could deliver the point of reference.

Moreover, a forest convention would be able to address a wide range of forest-related issues and to integrate multiple forest functions in the mechanisms established under its framework. For example, the role of forests in the formation and distribution of atmospheric precipitation could be addressed by a forest convention.

In this regard, the recognition of the forest's ability to influence atmospheric precipitations could encourage the negotiation and adoption of an international forest convention. As it is already well known, forests maintain and promote fresh water resources. The international community of States should also recognize that forests play a significant role in the formation and distribution of atmospheric precipitations.[1755] The existence of these functions of forests delivers a counterargument to the statement, according to which there is no need for a legally-binding forest convention as all the major globally relevant forest issues can be tackled by the UNFCCC and the CBD.[1756] It is plausible that the preservation of multiple forest functions cannot be effectively promoted without ensuring the conservation of forest ecosystems in their integrity. Following this logic, a comprehensive legally binding forest convention must be adopted in order to promote a holistic approach to the conservation of forests.

A legally-binding forest convention could also promote the conservation of natural forests by establishing a duty to protect them and creating more effective mechanisms endeavored to deliver financial resources for this purpose. In order to achieve this goal, a forest convention could promote the cooperation between developed and developing States and attract further financial resources, while promoting legal certainty and transparency. It can be argued also that such a convention would facilitate the development and implementation of new financial mechanisms for the conservation of forests. Such financial mechanisms should go beyond REDD+ or ABS in order to promote the multiple functions of forests in a holistic manner. Consequently, in spite of the fact that there are legitimate concerns in relation to the potential shortage of financial resources for the implementation of a hypothetical forest convention, the very adoption of this treaty could deliver new incentives and legal certainty that would help to raise additional funds for the conservation of forests. It is noteworthy that already in

1755 See Subsection A.II.3 of Chapter 1.

1756 *Davenport* [n. 1715], p. 108.

1999, the Costa Rica-Canada Initiative that was active under the umbrella of the IFF outlined that the adoption of a legally binding forest convention could significantly enhance the amounts of financial resources provided yearly for the purpose of the conservation of forests.[1757]

In order to be effective, a forest convention, or any other new forest-related document, would need to consider those challenges that can arise in the process of the application of forest-related regulations on the ground. Among other things, it is necessary to consider the available technological and financial resources in forest-rich States.[1758] To be effective, international provisions and national legislation have to consider local circumstances and needs, while considering technical and financial limitations to the implementation of relevant provisions. Therefore, a forest convention would have to tackle effectively a lack of resources available in developing States. Otherwise, negotiators would encounter the risk that forest users and owners, as well as local authorities would ignore the relevant provisions of a forest convention, while undermining the rule of law.[1759] Moreover, the international and national legal provisions dealing with forests must be appropriate for the achievement of goals pursued by lawmakers but should not exceed the limits of necessity and proportionality; otherwise, they can become socially unacceptable. The provisions in question should not impose restrictions on activities that do not represent any threat to forests. Moreover, they should always consider local land tenure systems and established local norms, traditions and practices. In the cases where local governmental and non-governmental institutions would have to be reformed, the international legal regime for forests would have to provide sufficient incentives for this purpose. Importantly, a hypothetical forest convention should also strengthen the role of local civil society groups by involving the representatives of forest-dependent people in the process of policy-making.[1760]

Furthermore, a forest convention should be flexible enough in order to facilitate its effective implementation in different regions across the globe, where local circumstances respectively vary. This fact can render the elaboration and

---

1757 *Costa Rica-Canada Initiative*, The final results of the 10 international meetings in support of the Programme of Work of the Intergovernmental Forum on Forests, 2000, p. 167. – Internet resource – https://cfs.nrcan.gc.ca/pubwarehouse/pdfs/10484.pdf [Last visit: 30.04.2023]

1758 *MacKenzie*, Lessons from Forestry [n. 1605], p. 116.

1759 Ibid, p. 117.

1760 Ibid, p. 118.

adoption of such a treaty more difficult. However, this circumstance does not mean that a forest convention cannot be effective. As it was demonstrated in previous chapters, international provisions that would tackle many crucial forest-relevant issues are either absent or insufficiently robust. A legally binding convention could deliver further clarity on these issues and promote transparency and legal certainty. Therefore, such a document should not be regarded as an inherently flawed one even if its development and implementation would be a complicated process. Instead, international negotiators and national lawmakers should develop an adequate approach to its elaboration and implementation. In the case of a forest convention, an adequate approach could consist in the elaboration of a framework convention with regional protocols.[1761] However, in order to foster its implementation, a basic framework convention would have to contain precise targets to be achieved in the course of its implementation, and not just stipulate general goals and objectives. As the example of the UNFCCC has shown, only weak provisions in the main text of a convention often result in its delayed and insufficient implementation. Clear and precise goals, objectives, and the mechanism for ensuring compliance with the provisions in question should be included in a framework convention. In fact, a number of important issues, including the illegal harvesting of and the illegal trade in forest timber and the expansion of unsustainable monoculture plantations can be combated only with the participation of all key States. Consequently, these issues should be regulated in the main text of a convention and not in its protocols. On their turn, regional protocols should promote the implementation of a forest convention by considering local circumstances.

The adoption of a forest convention could also be appreciated by businesses and other private stakeholders, while stronger international regulations do not necessarily restrict the freedom of private actors that are currently playing a significant role in forest governance and in the management of forest resources. Instead, such regulations can promote transparency and legal certainty. Indeed, as it was demonstrated in previous chapters, it was the lack of clear regulations and stronger safeguards that resulted in the scarcity of private investments in the implementation of REDD+ projects.

A forest convention would be able to improve the position of ILCs by establishing the requirement to implement any forest-related project based on MAT and after PIC is obtained. At the moment, this requirement is absent in those legal documents that guide the implementation of a bulk of forest-related

1761 Ibid.

mechanisms. For example, this is the case with the REDD+-mechanism. Ideally, the representatives of forest-dependent people should be involved directly in the process of negotiations and in the activities of international institutions established under a forest convention.

It can be argued that the main obstacle that hinders the development of a legally binding forest convention is a lack of political will on behalf of key States to overcome short-term political and economic issues in order to elaborate, adopt and implement a comprehensive international treaty. For this reason, other possible approaches to the development of the international legal regime for forests should be examined at this place.

III. Other Approaches to the Development of the International Legal Regime for Forests

1. Adoption of Forest-Specific Protocols to the Existing Conventions

While the adoption of a legally binding forest convention is often regarded as a time-consuming and complicated option, the adoption of a protocol specifically dealing with forests to one of the existing international treaties is sometimes perceived as a plausible solution.[1762] Indeed, the adoption of a protocol to one of the existing conventions can be regarded as a more realistic option as its elaboration can be less complicated due to the fact that the general consensus on the adoption of an initial treaty has already been achieved.[1763] The governing and scientific bodies of the convention concerned can be used as a negotiation platform for the elaboration of a protocol so that less expenditures are to be expected. A protocol would not only clarify and specify the provisions of the basic convention, but also establish new legal obligations.[1764]

However, it should not be forgotten that any protocol is bound by the main goal and the objectives of the treaty, under whose umbrella it was negotiated. The initial treaty text can be amended by a protocol, but such wide-going modification that would alter its core provisions is very unlikely. In fact, it would jeopardize the very existence of the basic treaty, as the consensus that was once achieved can be undermined. Consequently, only such a convention, the main goal of which would fully coincide with the necessities of forest conservation

1762 *MacKenzie*, Future Prospects [n. 1557], p. 253; *Anthony, Aust*, Modern Treaty Law and Practice, 2007, p. 120.

1763 *Anthony* [n. 1762], p. 120

1764 *MacKenzie*, Future Prospects [n. 1557], p. 253.

may deliver an adequate platform for the elaboration of a protocol specific to forests.[1765]

For instance, such a protocol could be adopted under the umbrella of the UNFCCC, as it was once proposed by the Netherlands in anticipation of the Rio Summit 1992.[1766] However, the promotion of forest protection through the adoption of new documents under the umbrella of the UNFCCC is further hindered by the fact that climate change mitigation activities can be harmful for forest ecosystems and their biodiversity. The issue of the compatibility of climate change mitigation policies with the conservation of forest biodiversity is not resolved in international environmental law, as it is demonstrated in the case of REDD+. For example, the cultivation of palm oil seeds for the purpose of producing biofuels can be harmful for the environment. However, the relevant issue is not regulated by the UNFCCC.

Indeed, a bulk of natural rainforests has already been cleared in order to expand the areas of palm oil plantations. At the same time, the great share of palm oil seeds is used for the production of biofuels, which were supposed to be one of the solutions to the problem of climate change.[1767] For instance, in 2017 more than 50% of palm oil consumed in Germany was used for the needs of the energy sector.[1768] The existing certification schemes that are deemed to promote the "sustainable palm oil production" fail to promote the protection of natural rainforests due to a significant number of inconsistences and a lack of good governance.[1769] The current experience from such regions as Peruvian Amazonia and Papua New Guinea has shown that those palm oil plantations that were supposedly managed sustainably can have significant adverse impacts on forest ecosystems and on forest-dependent people.[1770]

1765 Ibid.

1766 See Subsection B.II of Chapter 4.

1767 *Meo Carbon Solution Team*, Der Palmölmarkt in Deutschland im Jahr 2017, 25 Oktober 2018. – Internet resource – https://www.forumpalmoel.org/imglib/Palmoelstudie%202017_Meo_FONAP_ho.pdf [Last visit: 30.04.2023].

1768 Ibid.

1769 *Dauvergne, Peter*, The Global Politics of the Business of "Sustainable" Palm Oil, in: Global Environmental Politics, Volume 18, Issue 2, May 2018, pp. 18 ff.

1770 *Bennet, Aoife/Ravikumar, Ashwir/Paltán Homero*, The Political Ecology of Oil Palm Company-Community Partnerships in the Peruvian Amazon: Deforestation Consequences of the Privatization of Rural Development, in: World Development, Volume 109, September 2018, p. 38; *Chao, Sophie*, The Truth About "Sustainable" Palm Oil, in: Sapiens, 13 June 2019. – Internet resource – https://www.sapiens.org/culture/palm-oil-sustainable/ [Last visit: 30.04.2023]; See also: *Kilvert, Nick*, "Sustainable

It can be evaluated positively that the EU decided to label palm oil as unsustainable, meaning that its Member States are not any more allowed to account the biofuel produced from palm oil in their renewable transport targets.[1771] However, a bulk of palm oil for fuel is consumed also in the USA, China and other major States outside the EU. Therefore, it can be advised to adopt a similar decision on the international level, as well as to prohibit any climate change mitigation activities that are harmful to forest ecosystems and to forest biodiversity. However, the most of States support the maintenance of the independent nature of the UNFCCC and oppose the inclusion of stronger conservationist provisions in this treaty.[1772] Consequently, the UNFCCC in its current form cannot serve as a base for the adoption of a forest-specific protocol. Once adopted, such a protocol would be subordinated to the main goal of the UNFCCC. Consequently, it would not be able to embrace all issues related to the conservation and sustainable utilization of forests.

Alternatively, the CBD can potentially serve as a base for the adoption of a forest-specific protocol, as the conservation of biodiversity is directly linked to the protection of forests. The adoption of such a protocol could also foster the implementation of the CBD by enhancing the general value of this Convention and by attracting additional financial resources.

However, it can be problematic that the consensus that made the adoption of the CBD possible is limited to the conservation of biodiversity and the fair and equitable sharing of benefits arising from the utilization of genetic resources, whereas the promotion of forest conservation requires implementing measures in multiple areas. The wording of the CBD's provisions dealing with the conservation and sustainable utilization of biodiversity is vague and unprecise. A forest-specific protocol to the CBD could deliver further clarity and legal certainty. However, its provisions would not go beyond the initial scope of the CBD.

Palm Oil" May Not Be So Sustainable after all, in: ABC News, 15 June 2018. – Internet resource – https://www.abc.net.au/news/science/2018-06-16/orangutan-video-comes-as-sustainable-palm-oil-questioned/9811642 [Last visit: 30.04.2023]; *Gatti, Roberto, Cazzolla/Liang, Jingjing/Velichevskaya, Alena/Zhou, Mo*, Sustainable Palm Oil May Not Be So Sustainable, in: Science of The Total Environment, Volume 652, 2019 pp. 48 ff.

1771 *Keating, David*, EU Labels Biofuel from Palm Oil as Unsustainable, Bans Subsidies, in: Forbes, March 14, 2019. – Internet resource – https://www.forbes.com/sites/davekeating/2019/03/14/eu-labels-biofuel-from-palm-oil-as-unsustainable-bans-subsidies/#232f4c39c9da [Last visit: 30.04.2023]

1772 See Section B of Chapter 5.

Thus, it is highly questionable if the adoption of such a protocol would permit to raise additional technological and financial resources for the protection of forests.

Consequently, none of the existing treaties provides an adequate base for the adoption of a protocol specific to forests. It can be argued that such a protocol adopted under the umbrella of any of these treaties would include only weak commitments that would not permit to develop further the existing legal regime for forests.

Moreover, a protocol specific to forests would hardly contribute to the resolution of potential conflicts between the provisions of different forest-related conventions. Indeed, such a protocol, which would be subordinated to the main goal of one of the treaties in question, would not represent any kind of *lex specialis* in relation to the conservation of forests.

2. Adoption of an Overarching Coordinative Agreement on Forests

Another approach to the development of the international legal regime for forests could consist in the adoption of an overarching coordinative agreement on forests endeavored to regulate the relationship between different forest-related treaties in order to achieve a more effective and comprehensive implementation thereof.[1773] Such a coordinative convention is often proposed as an alternative to a legally binding forest treaty. However, it should be considered that the promotion of cooperation between treaties is limited due to the fact that States prefer to maintain the independent nature of each convention.[1774] As the practice has shown, States are reluctant to adopt any regulation that would make the fulfillment of their rights and obligations derived from one of the existing treaties dependent on the fulfillment of the provisions of other legal documents.

One argument for the adoption of a coordinative convention is that any forest convention would not provide a remedy to the highly fragmented nature of the international legal regime for forests.[1775] It is further argued that such a forest convention would rather worsen the situation by adding a new layer of governance to the highly fragmented and intricate system of international legal documents dealing with forests.[1776] However, it should be pointed out that a forest convention could still deliver a guidance on how international institutions and

1773 *Eikermann* [n. 1574], p. 153.
1774 See Section B of Chapter 5.
1775 *Eikermann* [n. 1574], p. 153.
1776 *Eikermann* [n. 1574], p. 153.

courts should resolve conflicts that arise between the conservation of forests on one side and the goals of other environmental treaties on the other side. It would be the case also when the application of a forest convention with regard to legal disputes arising from conflicting provisions of different forest-related treaties would not be obligatory. Moreover, a forest convention could establish a duty to apply the provisions and instruments of other treaties in a manner consistent with the conservation of forests.

In what concerns the potential conflicts with other international treaties, a hypothetical forest convention would not be in direct contradiction with the provisions of the CBD and the UNCCD. The implementation of these two conventions is deeply related to the promotion of forest protection. Moreover, the memorandum of cooperation could be concluded between them and a hypothetical forest treaty in order to ensure their consistent and comprehensive implementation. On the other hand, conflicts would likely arise between the provisions of a hypothetical forest convention, on one side, and the regulations adopted under the umbrella of the UNFCCC and the ITTA, on the other side.

Therefore, it is submitted that, if properly designed, a legally binding forest convention could contribute to the resolution of the problem of fragmentation of the international legal regime for forests. The potential benefits of such a hypothetical treaty should not be underestimated.

It is questionable how an overarching coordinative convention on forest issues could contribute to the promotion of the conservation and sustainable utilization of forests in the current situation, when a number of crucial issues are not covered by the international legal regime for forests. Even the duty to act with due diligence is not yet established in relation to forests.[1777] The promotion of enhanced cooperation between the existing treaties cannot remedy the lack of provisions that would effectively tackle driving forces of deforestation on all the levels. It is submitted that the existing set of forest-related initiatives, instruments and policies, which are scattered among numerous international environmental treaties, is not sufficiently robust in order to provide a solid base for the promotion of the conservation of global forests. The adoption of an overarching cooperative convention would not tackle all the gaps and defects that are inherent to the existing international legal regime for forests.

It is noteworthy that the cooperation between different treaties could also be regulated by a hypothetical forest convention, which could simultaneously deliver further clarity in relation to the major forest issues. It is noteworthy that

---

1777 See Subsection C.II of Chapter 2.

the achievement of a consensus on the adoption of a cooperative convention on forest issues would likely be time- and resource-consuming to the same extent as the negotiation of a forest convention itself. Thus, the elaboration of a forest-specific convention may be more effective than the adoption of an overarching convention.

3. Strengthening the Role of the UNFF

Another approach to the development of the international legal regime for forests consists in the enhancement of the role of the UNFF.[1778] This approach can permit to facilitate the coordination between different forest-relevant conventions; however, in order to empower the UNFF in such a way that this institution would acquire competences that would go beyond its initial advisory and coordinative role, the international community would have to conduct such a profound and thorough reform thereof that would be comparable with the adoption of a forest convention. On the other hand, the UNFF could serve as a platform for the negotiation of a legally binding forest convention. Once adopted, a hypothetical forest convention could either assign new powers to the UNFF or establish a new body for monitoring forest issues. Therefore, the promotion of the UNFF's role cannot be seen as an autonomous tool for the development of the international legal regime for forests, but rather as an option that is directly related to the adoption and implementation of legally binding documents specific to forests.

4. Promoting Regional Initiatives

Regional initiatives can also contribute to the promotion of the international legal regime for forests.[1779] In fact, the elaboration and adoption of regional forest treaties can be a less complicated process than the negotiation of a legally binding forest convention. Indeed, the consensus can be easier achieved on the regional than on the global level.

States are closer linked to each other on the regional level and, consequently, are more likely to fulfill their political commitments in order to prove their good will. Following this logic, it is submitted that a legally binding form plays a less relevant role in the case of regional arrangements when compared to universal international treaties.[1780] As the States of one region are characterized by a higher

1778 *MacKenzie*, Future Prospects [n. 1557], p. 254.
1779 *MacKenzie, Future Prospects* [n. 1557], p. 255.
1780 Ibid.

level of political, economic, and cultural similarities, regional treaties can effectively consider the national circumstances of forest-rich States. By tackling more effectively organizational challenges that impede the effective communication between governments, regional arrangements can contribute to the promotion of monitoring, reporting, and verifying forest conditions.[1781] The promotion of monitoring on the regional level can permit to combat more effectively the illegal logging of and the trade in illegally harvested forest products. Additionally, regional arrangements can potentially facilitate the implementation of international forest-relevant mechanisms and initiatives by adopting the international legal framework to regional circumstances.

However, it is submitted that regional initiatives cannot substitute an international forest convention because major forest issues have already acquired a global dimension. Relevant issues cannot be tackled on a regional level. Even those regional arrangements that are endeavored to contribute to the resolution of global deforestation crisis by tackling some of the driving forces of deforestation, as the FLEGT initiative of the EU does,[1782] cannot be implemented effectively without the collaboration with developing States, where these programs are supposed to be implemented. Therefore, it can be affirmed that forest-related initiatives should be adopted and implemented at the global or at least at the interregional level. Moreover, there is a risk that the development of regional arrangements can lead to the leakage of deforestation into the areas where no effective regional initiatives have yet been established.[1783]

A number of regional forest-related initiatives have already been established worldwide. The progress achieved by the Ministerial Conference on the protection of forests in Europe should be particularly mentioned in this regard. In 2011, it was agreed to launch the negotiations of a Legally Binding Agreement on Forests in Europe.[1784] The Intergovernmental Negotiating Committee, which

---

1781 Ibid.

1782 *Myers, Rodd/Rutt, Rebecca/McDermott, Constance/Maryudi, Ahmad/Acheampong, Emmanuel/Camargo, Marisa/Cầm, Hoàng*, Imposing Legality: Hegemony and Resistance under the EU Forest Law Enforcement, Governance, and Trade (FLEGT) Initiative, in: Journal of Political Ecology, Volume 27, 2020, pp. 126 ff.

1783 See *Garret, R./Lambin, Eric F./le Polain de Waroux, Yan*, To Eliminate Deforestation in South America, Reduce Differences in Regulations Across Regions and Actors, Institute for the Study of International Development, Policy Brief PB-2017-05, 2017. – Internet resource – https://lendev.weebly.com/uploads/1/0/7/0/107005033/pb-2017-05__le_polain_de_waroux_.pdf [Last visit: 30.04.2023]

1784 Ministerial Conference on the protection of forests in Europe, Oslo Ministerial Mandate for Negotiating a Legally Binding Agreement on Forests in Europe, Oslo 14–16

is active since the 27 February 2012, has not yet been successful in establishing a clear consensus on forest issues among European States.[1785] In 2015, the Madrid Ministerial Decision recognized the draft Version of the European forest convention presented by the Committee and committed itself to analyze possible ways for the achievement of final consensus on the adoption of the European forest treaty at latest by 2020.[1786] However, no legally-binding agreement has yet been adopted. Another example of the successful regional collaboration on forest issues is the Central American Regional Convention for the Management and Conservation of the Natural Forest Ecosystems and the Development of Forest Plantations, which came into force in 1999.[1787]

These and other regional forest-related initiatives are often criticized for that they add another level of governance to the highly fragmented system of the international legal regime for forests, which is already overloaded with ineffective provisions, without contributing substantially to forest conservation.[1788] However, regional initiatives can have a positive impact as they can promote the implementation of international forest-related treaties, and consider regional circumstances. Moreover, regional agreements can serve as an interim option for the promotion of forest conservation in the current situation, when the relevant initiatives on the global level remain low effective.[1789]

Thus, regional initiatives should be promoted and strengthened in accordance with existing international documents dealing with forests. They could be further promoted by a legally binding forest treaty. They should not contain any

---

June 2011. – Internet resource – https://www.foresteurope.org/docs/MC/MC_oslo_lba_mandate.pdf [Last visit: 30.04.2023]

1785 Ibid, Article 20.

1786 Ministerial Conference on the protection of forests in Europe, Madrid Ministerial Decision, Extraordinary Ministerial Conference, Madrid, 2015, p. 1. – Internet resource – https://www.foresteurope.org/sites/default/files/ELMEMC_Madrid2015_MinisterialDecision_adopted.pdf [Last visit: 30.04.2023]

1787 Convenio regional para el manejo y conservación de los ecosistemas naturales forestales y el desarrollo de plantaciones forestales, 1999. – Internet resource – http://www2.ecolex.org/server2neu.php/libcat/docs/TRE/Full/Sp/TRE-001208.pdf [Last visit: 30.04.2023]

1788 *MacKenzie*, Future Prospects [n. 1557], pp. 254–255.

1789 See: *Balsiger, Jörg/Prys, Miriam/Steinhoff, Niko*, The Nature and Role of Regional Agreements in International Environmental Politics: Mapping Agreements, Outlining Future Research, GIGA Working Paper, Number 208, 2012, pp. 4 ff.; See also *Halle, Mark*, Life after Rio: A commentary by Mark Halle, The International Institute for Sustainable Development, 2012, p. 5.

provisions and regulations that are weaker than those that already exist on the global level. Additionally, the harmonization between forest-related regulations should be promoted on all the levels.

Against this backdrop, it can be argued that any approach to the development of the international legal regime for forests, which does not foresee the adoption of a legally binding forest convention, is likely to serve only as an interim measure for promoting forest conservation. It is highly questionable if such approaches would be considerably more effective than the already existing forest-related mechanisms such as REDD+.

Consequently, it can be recommended to strengthen the environmental safeguards for the existing mechanisms, including REDD+, and to promote their implementation. At the same time, the negotiations of a legally binding forest treaty should be launched.

5. *Promoting those Financial Mechanisms that Foresee a Holistic Approach to Forests*

An ideal financial mechanism endeavored to promote forest conservation should be properly designed in order to enable the valorization of the multiple functions of forests. Such a mechanism should not only enable the valorisation of carbon-related forest functions or their role as biodiversity reservoirs but also promote the valorization of those functions of forests that are related to the formation and distribution of atmospheric precipitations. For example, it should promote the valorization of the role of forests in relation to the protection of soils. At the same time, the existing mechanisms aimed at promoting forest finance, such as REDD+, can provide a base for the elaboration of more comprehensive tools and instruments for the promotion of forest protection. It would be important to ensure that a new financial mechanism would be acceptable both for developing States and for developed ones. Such a mechanism should also consider the needs and equity concerns of forest-rich States and of forest dependent people, as well as the expectations of investors. In order to raise funds, the mechanism in question should effectively combine non-market-based approaches on one side and market-based ones on the other side. It should consider the existing land tenure systems or the need to improve them. Finally, but equally importantly, forest conservation would be promoted only if the provisions and tools for the implementation of a new forest financial mechanism would ensure the integrity of forest ecosystems and the conservation of natural forests and forest biodiversity.

Another question is whether a new forest-related financial mechanism could be compatible with the already existing financial mechanisms, including

REDD+. The fact that REDD+ is now implemented also outside the UNFCCC process points to the possibility of a wider cooperation with a hypothetical new financial mechanism. However, it is also imaginable that REDD+ would become redundant if such a comprehensive financial mechanism came into being.

In order to ensure security and transparency, a hypothetical forest financial mechanism should be established in accordance with strong legally binding provisions that could be delivered by a legally binding forest convention. The question arises on how additional funds can be raised in order to ensure its implementation. For this purpose, a hypothetical mechanism should be flexible enough and foresee the development of projects that deliver benefits for investors and create new income opportunities for forest-dependent people, while permitting to reduce the anthropogenic pressure on forests. For example, the investments in power and energy sector can permit to reduce the dependence of local communities on wood fuel and charcoal in rural areas. The adoption and implementation of new forest-related financial instruments and tools would become possible if a wide range of the multiple functions of forests was recognized by civil society and the international community of States. The increased participation of developed States in the promotion of a new mechanism as well as the enhanced utilization of resources available in the global South should both be promoted. The establishment of a Global Forest Fund for the purpose of the promotion of forest conservation can be welcomed. However, the option of creating a new financial mechanism is likely to encounter significant opposition from the side of donor States. The development of a new financial mechanism should also be accompanied by the adoption of agreements that would establish effective policies that would permit to reduce the global demand for timber, forest products, and agricultural produce that are cultivated and harvested unsustainably in forests and on converted forestlands.

E. Protection of Forest Ecosystems as a "Common Concern of Humankind"

Independently from whether an international forest convention will be adopted or the international community will choose other approaches for the further development of the international legal regime for forests, it is very unlikely that the concept of the "common heritage of mankind" would be applied to forests in any future agreement. It is more likely that the conservation of forest ecosystems may be recognized as a "common concern of humankind". In this regard, it is important to note that the very nature of the concept of "common concern" means that relevant issues cannot be successfully resolved by a single state, but a collective response of the international community has to be developed.[1790] The ranking of any environmental good as a "common concern of humankind" permits to take into account the sovereignty and equity concerns of forest-rich States and to establish the base for the enhanced cooperation aimed at the conservation and sustainable utilization of certain environmental goods.[1791] It can be argued that this concept reaffirms sovereign rights and at the same time restricts them by underlining the primary responsibility of States to protect forests under their jurisdiction.[1792] Those states, which are deprived of large forest areas due to climatic conditions or which have already implemented the conservation of forest ecosystems on their sovereign territories, can fulfill their relevant duties by participating in international mutually beneficial cooperative mechanisms for the promotion of forest conservation.

It is generally recognized that the concept of common concern of humankind, which is sometimes referred to as a principle, foresees that each State must prohibit those activities that can cause a serious damage or threat to a protected environmental good. It should be done even if there is no direct harm to territories and goods under the jurisdiction of other States.[1793] Even if deforestation does not cause serious damage to territories beyond national jurisdiction, a State that promotes unsustainable forest management would be in breach with the commitments derived from the concept of "common concern", according to which the international community commits itself to cooperate in order to resolve certain environmental problems. It is plausible that the concept itself would not foresee any sanctions for such a breach. However, the ranking of forest conservation as a "common concern of humankind" would provide a basis for the development of a mechanism for ensuring compliance.

What concerns the concept of "common heritage of mankind", its applicability to environmental goods under national jurisdiction has been constantly

1790 *Soltau, Friedrich*, Common Concern of Humankind, in: *Gray, Kevin R./Tarasofsky, Richard G./Carlarne, Cinnamon P.* (eds.), The Oxford Handbook of International Climate Change Law, 2016, pp. 205 ff.

1791 *Horn, Laura*, Globalisation, Sustainable Development and the Common Concern of Humankind, in: Macquarie Law Journal, Volume 7, 2007, pp. 55 ff.; *Castillo-Winckels, Nadia Sanchez*, why "Common Concern of Humankind" Should Return to the Work of the International Law Commission on the Atmosphere, in: Georgetown Environmental Law Review, Volume 29, Issue 1, 2017, pp. 135 f.

1792 *Castillo-Winckels* [n. 1791], pp. 136 f..

1793 *Soltau* [n. 1790], p. 208.

denied.[1794] The unique exception is the UNESCO World Heritage Convention. However, the legal content of this concept as applied in this Convention defers substantially from its application to the objects and areas beyond national jurisdiction, such as the Area of "the seabed and ocean floor and subsoil thereof".[1795] In the case of the UNESCO Convention, this concept entrusts each Party to preserve cultural and natural goods on its territory for the benefits and with the support of the entire international community. Such interpretation of the concept that emphasizes the State's responsibility and the duty to cooperate tends to be more appropriate in the case of forest ecosystems. It should be noted, however, that States have always reaffirmed their right to utilize forest resources. In contrast, the UNESCO Convention stipulates that the duty to preserve the cultural and natural heritage of humankind always prevails over the right to exploit it.[1796]

Moreover, the concept of "common heritage of mankind" is sometimes criticized as an arbitrary attempt to violate the sovereignty rights of developing States and to promote the utilization of their natural resources without sharing benefits delivered by forests and other natural ecosystems with the States of origin of these resources.[1797] On the other hand, it is also argued that the concerns of those developing States that don't have technologies needed in order to exploit natural resources have catalyzed the introduction of the concept of common heritage in relation to the seabed resources.[1798] From this point of view, it is submitted that this last concept tends to regulate the access to resources and the sharing of

---

1794 See also *Taylor, Prue*, Common Heritage of Mankind and Common Concern of Humankind, in: *Krämer, Ludwig/Orlando, Emanuela* (eds.), Elgar Encyclopedia of Environmental Law, Volume VI: Principles of Environmental Law, 2018, pp. 303 ff.; *Lee, Seokwoo/Kim, Jeong Woo*, Applying the Principle of the Common Heritage of Mankind: An East Asian Perspective, in: *Zou, Keyuan* (eds.), Global Commons and the Law of the Sea, 2018, pp. 15 ff.; *Taylor, Prue*, Common Heritage of Mankind Principle, in: *Bosselman, Klaus/Fogel, Daniel/Ruhl, J.B.* (eds.), The Encyclopedia of Sustainability, Volume 3: The Law and Politics of Sustainability, 2011, pp. 64 ff.

1795 *Taylor* [n. 1794], pp. 64 ff.

1796 Convention for the Protection of the World Cultural and Natural Heritage, adopted on 23 November 1972 and entered into force on 17 December 1975, 1037 UNTS 151, Article 4.

1797 *Scholtz, Werner*, Common heritage: saving the environment for humankind or exploiting resources in the name of eco-imperialism? in: Comparative and International Law Journal of Southern Africa, Volume 41, Issue 2, 2008, p. 290.

1798 *Soltau* [n. 1790], p. 206.

benefits arising from their utilization, whereas the concept of "common concern of humankind" incorporates a more conservation-oriented approach.[1799]

Consequently, the concept of "common concern of humankind" underlines the commitment to the international cooperation on the conservation and sustainable exploitation of forests. It can also strengthen the conservational issues in international forestry policies.

On the other hand, if the legal nature of the "common heritage" concept was further clarified and conservational aspects were further accentuated, it would be reasonable to apply it to certain types of forest ecosystems, for example, to natural forests. However, it is to be expected that forest-rich States would be reluctant to accept that a significant part of their forests would be recognized as "common heritage of mankind".

Consequently, it would be plausible to apply the concept of "common concern of humankind" to forest ecosystems. It should be noted that the very extension of this last concept on forest ecosystems would represent a step forward, as it has yet never been applied in relation to forests in any international treaty. Such extension could promote the role of international community in the conservation of forests and could contribute to the development of a holistic approach to their protection from the point of view of comprehensiveness that this concept incorporates.

F. Conclusions

Thus, the current level of the development and implementation of the international legal regime for forests is unsatisfactory from the point of view of the promotion of forest conservation. The main issues that constitute the core of the global consensus on forests have found their way into the existing international documents dealing with forests so that it is possible to draw conclusions about the content of an ideal forest-specific agreement. However, existing provisions are vague and superficial and cannot deliver a solution to the problem of deforestation and forest degradation. Many forest-related topics are only briefly mentioned in the forest-related international documents. Such important forest-related issues include the promotion of the balance between the conservation of forests on one side and their utilization on the other side. They also include the preservation of the existing global forest area; the promotion of forest finance; etc.[1800]

1799 *Soltau* [n. 1790], p. 206.

1800 See Section C of this Chapter.

At the same time, the international legal provisions dealing with these issues have not been further developed in last years. While following a soft-law market-based approach to the promotion of forest protection and while attempting to achieve certain progress on forest issues by developing forest-related initiatives of other treaties, the international community has been distracted from the core topics of forest governance. However, the attempts to deliver stronger forest-relevant regulations by building linkages between the international legal regime for forests, on one side, and the one for combating climate change, on the other side, have resulted in that the relevant provisions developed under the UNFCCC regime acquired the generally vague character of international forest law. The conclusion can be made that the only way to achieve progress on forest protection is to adopt documents specific to forests that would tackle the core issues of the international legal regime for forests. Unless forest specific problems are effectively resolved, those alternative approaches to the development of the international legal regime for forests that foresee the achievement of forest protection as a side effect of other environmental policies are likely to fail. At the same time, the failure to promote forest conservation is likely to have negative consequences for the achievement of the main goal of the UNFCCC and of other environmental treaties that tend to regulate forest-related issues as a side effect of their implementation process.

The effectiveness of any forest-related mechanism depends on the clarity, robustness, and legal enforceability of relevant provisions. It is noteworthy that scarce achievements in relation to the implementation of REDD+ can be explained by the vague and superficial nature of the pertinent international regulations. Against this backdrop, it can be argued that the option of the elaboration and adoption of legally binding forest documents, such as a legally binding forest convention, should return on the agenda of international forest talks. It is submitted that it were not only the sovereignty issues but also the equity concerns of developing States and the lacking political will of developed States that determined the failure of international forest negotiations in the past. Subsequently, the attitude of States to forest issues has changed. This means that the positions of different States at international forest-related negotiations cannot be anymore explained by the North-South antagonism. The recognition of environmental functions of forests is growing, and the new scientific evidence on the ecological role of forests have been acquired. The time for the adoption of a legally binding forest convention has not yet come, but current developments can give certain hope for the adoption of such a document in the future. However, such a convention should be properly designed in order to ensure that it would serve for the purpose of forest conservation and not only for the promotion of the short-term economic interests of States. Unless States are able to agree on such a treaty text that would include all important forest-relevant topics and establish the duty to protect forests, it would be better to refuse on the option of the adoption of a legally binding forest treaty specific to forests.

As long as a properly designed forest convention that would consider multiple forest functions and prioritize the conservation and restoration of forests – especially of natural ones –, over their utilization, is absent, there will be strong doubts concerning the general ability of international environmental law to tackle the problem of deforestation. However, the development and implementation of existing international regulations and mechanisms that have already had positive outcomes for forest conservation should continue. The pertinent provisions and initiatives of the CITES, the CBD and the CCD should be promoted and supported. REDD+ and ABS should be implemented, but their environmental safeguards should be strengthened in order to ensure the integrity of forest ecosystems and the protection of the rights of forest-dependent people.

Consequently, the existing forest-relevant policies and instruments should be promoted as an interim measure in order to avoid the worsening of the conditions of global forests. The growing involvement of civil society in the global forest gives hope that the implementation of the existing forest-related instruments will move into a more environmentally-friendly direction. However, considering the high rates of deforestation, it can be argued that the international community should foster its efforts dedicated to the elaboration and implementation of an effective and legally binding forest convention that would establish the duty to protect remaining natural forests. The mechanisms for promoting the implementation of such a convention and for ensuring the compliance with it should be established.

reverse the loss of forests worldwide, new approaches should emerge and robust legally binding regulations ought to be adopted. Relevant provisions and mechanisms should also contribute to the reduction of the dependence of the human society on timber and non-timber forest products, such as wood, charcoal, biofuel, and agricultural produce originating from converted forestlands, whose harvesting and cultivating leads to the increased pressure on forest ecosystems. For this purpose, international and national regulations should promote the development and application of alternative materials and energy sources; the restoration of degraded lands in order to establish new croplands and forest areas; scientific, efficient, and more productive agricultural practices on the existing agricultural lands in order to avoid forestland conversions. This means that the international legal regime for forests ought to embrace issues that are not traditionally covered by its scope.

Moreover, the conflict between forest conservation and climate change mitigation ought to be definitely resolved in a way that would ensure that those mitigation and adaptation activities that are harmful for forest ecosystems are not eligible under the UNFCCC and the Paris Agreement. For this purpose, a relevant legally binding regulation should be incorporated in the international legal regime for climate change. It is submitted that such a regulation can be adopted only after the international consensus on forest issues is established. The consensus in question is also relevant for the adoption of a legally binding forest convention. Therefore, such a regulation will hardly be adopted unless the consensus on the adoption of a legally binding forest convention is elaborated.

At the same time, the evaluation of genetic resources of forests and of other forest goods and ecosystem services, whose utilization is compatible with the conservation of natural forests, should be promoted. An obligation to redirect a part of benefits from the utilization of genetic resources into the protection of forests ought to be established.

Considering the recent acceleration of deforestation rates worldwide, which is further exacerbated by the economic impacts of the Covid-Pandemics, there is a need to pursue robust and effective policies for the conservation of forests.[1802] It is submitted that the excessive reliance on pure soft-law and market-based approaches in the period 2000–2020, as well as the lack of a holistic approach to

---

1802 *Gross, Anna/Schipani, Andres/Palma, Stefania/Findlay, Stephanie*, Global deforestation accelerates during pandemic, in: Financial Times, 9 August 2020. – Internet resource – https://www.ft.com/content/b72e3969-522c-4e83-b431-c0b498754b2d [Last visit: 30.04.2023]

forests since 1960s, resulted in the loss of previous opportunities to deliver robust forest conservation mechanisms, which are urgently needed in the light of recent developments. The growing occurrence of droughts and forest fires due to the changes in global and regional precipitation patterns that ought to be attributed not only to the emissions of greenhouse gases, but also to deforestation itself, should serve as an alarm bell.[1803] At any time, only the future development of international environmental law will show if the international community will be able to tackle effectively the problem of deforestation.

1803 *Zanon, Sibélia/Cataldo, Roberto,* Deforestation in the Amazon is drying up the rest of Brazil: Report, in: Mongabay, 7 August 2020. – Internet resource – https://news.mongabay.com/2020/08/deforestation-in-the-amazon-is-drying-up-the-rest-of-brazil-report/ [Lasst visit: 30.04.2023]

List of Used International Treaties, Court Decisions, and Other International Documents

International Treaties and Non-binding Documents

Convention on the Recognition and Enforcement of Foreign Arbitral Awards, adopted on 10 June 1958 and entered into force on 7 June 1959, in: 330 UNTS 3.

International Convention on the Elimination of all Forms of Racial Discrimination, adopted on 21 December 1965 and entered into force on 4 January 1969, in: 660 UNTS 195.

International Covenant on Civil and Political Rights, adopted on 16 December 1966 and entered into force on 23 March 1976, in: 999 UNTS 171.

First Optional Protocol to the International Covenant on Civil and Political Rights, adopted on 16 December 1966 and entered into force on 23 March 1976, in: 999 UNTS 171.

Vienna Convention on the Law of Treaties (with annex), adopted on 23 May 1969 and entered into force on 27 January 1980, in: 1155 UNTS 331.

Convention on Wetlands of International Importance Especially as Waterfowl Habitat, adopted on 2 February 1971 and entered into force on 21 December 1975, in: 996 UNTS 245.

Convention for the Protection of the World Cultural and Natural Heritage, adopted on 23 November 1972 and entered into force on 17 December 1975, in: 1037 UNTS 151.

Convention on International Trade in Endangered Species of Wild Fauna and Flora, adopted on 3 March 1973 and entered into force on 1 July 1975, in: 993 UNTS 243

Convention on the Conservation of Migratory Species of Wild Animals, adopted on 23 June 1979 and entered into force on 1 November 1983, in: 1651 UNTS 333.

United Nations Convention on the Law of the Sea, adopted on 10 December 1982 and entered into force on 16 November 1994, in: 1833 UNTS 3.

International Tropical Timber Agreement, adopted on 18 November 1983 and entered into force on 1 April 1985, in: 1393 UNTS 67.

Non-Legally Binding Authoritative Statement of Principles for a Global Consensus on the Management, Conservation and Sustainable Development of All Types of Forests, adopted on 14 June 1992, in: ILM, Volume 31, 1992, pp. 881 ff.

United Nations Framework Convention on Climate Change, adopted on 9 May 1992 and entered into force on 21 March 1994 in: 1771 UNTS 107.

Convention on Biological Diversity, adopted on 5 June 1992 and entered into force on 29 December 1993, in: 1760 UNTS 79.

Rio Declaration on Environment and Development, United Nations Conference on Environment and Development, Rio de Janeiro, Brazil, 3–14 June, 1992, in: ILM, Volume 31, 1992, pp. 874 ff.

Marrakesh Agreement establishing the World Trade Organization (with final act, annexes and protocol), adopted on 15 April 1994 and entered into force on 1 January 1995, in: 1867 UNTS 3.

TRIPS Agreement: Agreement on Trade-Related Aspects of Intellectual Property Rights, adopted on 15 April 1994 and entered into force on 1 January 1995, Marrakesh Agreement Establishing the World Trade Organization, Annex 1C, in: 1869 UNTS 299.

United Nations. Agenda 21, United Nations Conference on Environment and Development, Rio de Janeiro, Brazil, 3–14 June, 1992. – Internet resource: – https://sustainabledevelopment.un.org/content/documents/Agenda21.pdf [Last visit: 30.04.2023].

Convention to Combat Desertification in those Countries Experiencing Serious Drought and/or desertification, Particularly in Africa, adopted on 17 June 1994 and entered into force on 26 December 1996, in: 1954 UNTS 3.

International Tropical Timber Agreement, adopted on 26 January 1994 and entered into force on 1 January 1997, in: 1955 UNTS 81.

Kyoto Protocol to the United Nations Framework Convention on Climate Change, adopted on 11 December 1997 in Kyoto, Japan, in: 2303 UNTS 162.

International Tropical Timber Agreement, adopted on 27 January 2006 and entered into force on 7 December 2011, in: 2797 UNTS 75.

Non-Legally Binding Instrument on All Types of Forests, GA/RES/62/98, 17 December 2007. – Internet resource – https://documents-dds-ny.un.org/

doc/UNDOC/GEN/N07/469/65/PDF/N0746965.pdf?OpenElement [Last visit: 30.04.2023]

Optional Protocol to the International Covenant on Economic, Social, and Cultural Rights, adopted on 10 December 2008 and entered into force on 5 May 2013, in: 2922 UNTS No. 14531.

The Paris Agreement, adopted on 12 December 2015 and entered into force on 4 November 2016, in: UNTS No. 54113.

Documents Adopted by the UN General Assembly

United Nations A/RES/55/2, United Nations Millennium Declaration, General Assembly Resolution, 18 September 2000. – Internet resource – https://undocs.org/A/RES/55/2 [Last visit: 30.04.2023]

United Nations A/RES/56/83, Resolution adopted by the General Assembly on 12 December 2001, Responsibility of States for internationally wrongful acts, in: Report of the International Law Commission on the work of its fifty-third session, in: Yearbook of the international Law Commission, Volume 2, Part 2, United Nations, New York and Geneva, 2001.

United Nations A/RES/66/288, Resolution adopted by the General Assembly on 27 July 2012: The future we want, 2012. – Internet resource – https://undocs.org/A/RES/66/288 [Last visit: 30.04.2023]

United Nations A/RES/70/1, Resolution adopted by the General Assembly on 25 September 2015: Transforming our world: the 2030 Agenda for Sustainable Development, 2015. – Internet resource – https://undocs.org/en/A/RES/70/1 [Last visit: 30.04.2023]

International Law Commission

United Nations, UN Doc. A/56/10, Introductory commentary by Special Rapporteur J. Crawford on Part III/Chapter II of the Draft Articles on State Responsibility adopted by the International Law Commission ('ILC Draft Articles'), Report of the International Law Commission on its 53rd Session, 29, 2001.

United Nations A/CN.4/L.682, Fragmentation of International Law: Difficulties arising from the Diversification and Expansion of International Law, Report of the Study Group of the International Law Commission, International Law Commission, Fifty-eighth Session, Geneva, 1 May-9 June and 3 July-11 August

2006, 2006. – Internet resource – https://undocs.org/en/A/CN.4/L.682 [Last visit: 30.04.2023]

United Nations A/CN.4/671, Second report on subsequent agreements and subsequent practice in relation to the interpretation of treaties by Georg Nolte, Special Rapporteur, International Law Commission, Sixty-sixth session, Geneva, 5 May-6 June and 7 July-8 August 2014, 2014. – Internet resource – https://undocs.org/en/A/CN.4/671 [Last visit: 30.04.2023]

UNFCCC Preparatory Process

United Nations A/AC.237/Misc.1/Add.1, Preparation of a Framework Convention on Climate Change: Set of Informal Papers Provided by Delegations, Related to the Preparation of a Framework Convention on Climate Change, Geneva, 19–28 June 1991. – Internet resource – https://undocs.org/en/a/ac.237/misc.1/add.1 [Last visit: 30.04.2023]

United Nations A/AC.237/Misc.1/Add.3, Preparation of a Framework Convention on Climate Change: Set of Informal Papers Provided by Delegations, Related to the Preparation of a Framework Convention on Climate Change, Note by the Secretariat, Addendum 1. Geneva, 19–28 June 1991. – Internet resource – https://undocs.org/en/a/ac.237/misc.1/add.1 [Last visit: 30.04.2023]

United Nations, Statement by his H. E. Dato' Seri Dr. Mahathir bin Mohamad, Prime Minister of Malaysia, in: Report of the United Nations Conference on Environment and Development, Volume 3, Statements Made by Heads of State or Government at the Summit Segment of the Conference, Rio De Janeiro, 3–14 June 1992.

Documents Adopted by the ECOSOC

United Nations E/1995/95, Resolutions and decisions of the Economic and social Council, Economic and social Council, Official records, New York, 1996. – Internet resource – https://undocs.org/E/1995/95 [Last visit: 30.04.2023]

United Nations E/CN.17/1995/36, Commission on Sustainable Development, Report on the Third Session (11–28 April 1995), Economic and Social Council, Official Records, Supplement No.12, United Nations, New York, 1995. – Internet resource – https://undocs.org/E/CN.17/1995/36 [Last visit: 30.04.2023]

United Nations E/CN.17/1997/12, Report of the Ad Hoc Intergovernmental Panel on Forests, Economic and Social Council, New York, 20 March 1997. – Internet resource – https://undocs.org/E/CN.17/1997/12 [Last visit: 30.04.2023]

United Nations E/CN.17/2000/14, Report of the Intergovernmental Forum on Forests on its Fourth Session, E/CN.17/2000/14, Economic and Social Council, 20 March 2000. – Internet resource – https://undocs.org/E/CN.17/2000/14 [Last visit: 30.04.2023]

United Nations E/CN.17/2000/20, Report on the eighth session of the Commission on Sustainable Development, Economic and Social Council, Official Records, 2000. – Internet resource – https://undocs.org/E/CN.17/2000/20 [Last visit: 30.04.2023]

ECOSOC Resolution 2000/35, Report on the forth Session of the Intergovernmental Forum on Forests, Economic and Social Council, United Nations, 2000. – Internet resource – https://www.un.org/esa/forests/wp-content/uploads/2013/09/2000_35_E.pdf [Last visit: 30.04.2023]

United Nations E/CN.18/2006/18, United Nations Forum on Forests, Report of the sixth Session (27 May 2005 and 13 to 24 February 2006), Economic and Social Council, Official Records, New York, 2006. – Internet resource – https://undocs.org/E/CN.18/2006/18 [Last visit: 30.04.2023]

United Nations E/CN.18/2007/8, United Nations Forum on Forests, Report on the seventh Session (24 February 2006 and 16–27 April 2007), Economic and Social Council, Official Records, 2007. – Internet resource – https://www.undocs.org/en/E/CN.18/2007/8 [Last visit: 30.04.2023]

United Nations E/CN.18/2013/18, United Nations Forum on Forests, Report on the tenth Session (4 February 2011 and 8 to 19 April 2013), Economic and Social Council, Official Records, New York, 2013. – Internet resource – https://undocs.org/E/CN.18/2013/18 [Last visit: 30.04.2023]

United Nations E/CN.18/AEG/2014/INF/1, Compilation of Views and Proposals by Member States and Other Relevant Stakeholders on the International Arrangement on Forests, Economic and Social Council, New York, 12 February 2014. – Internet resource – https://undocs.org/E/CN.18/AEG/2014/INF/1 [Last visit: 30.04.2023]

United Nations E/CN.18/2015/14, United Nations Forum on Forests, Report on the eleventh Session (19 April 2013 and 4 to 15 May 2015), Economic and Social Council, Official Records, New York, 2015. – Internet resource – http://undocs.org/E/CN.18/2015/14 [Last visit: 30.04.2023]

United Nations E/CN.18/2015/9, Letter dated 6 January 2015 from the permanent representative of China to the United States addressed to the Secretary-General,

Economic and Social Council, 21 January 2015. – Internet resource – http://undocs.org/E/CN.18/2015/9 [Last visit: 30.04.2023]

United Nations E/CN.18/2015/11, United Nations Forum on Forests, eleventh Session, (4–15 May 2015), Report on the first meeting of the Open-ended Intergovernmental Ad Hoc Expert Group on the International Arrangement on Forests, Item 3 of the provisional agenda, Economic and Social Council, New York, 2015. – Internet resource – https://undocs.org/E/CN.18/2015/11 [Last visit: 30.04.2023]

United Nations E/CN.18/2015/12, Note verbale dated 23 March 2015 from the Permanent Mission of Switzerland to the United Nations addressed to the Secretary-General, Economic and Social Council, New York, 4–15 May 2015. Internet resource – https://undocs.org/en/E/CN.18/2015/12 [Last visit: 30.04.2023]

United Nations E/CN.18/2005/18, United Nations Forum on Forests, Report of the fifth session (14 May 2004 and 16 to 27 May 2005, Economic and Social Council, Official Records, New York, 2005. – Internet resource – https://undocs.org/E/CN.18/2005/18 [Last visit: 30.04.2023]

United Nations E/RES/2017/4, Resolution adopted by the Economic and Social Council on 20 April 2017: United Nations Strategic Plan for Forests 2017–2030 and Quadrennial Programme of Work of the United Nations Forum on Forests for the Period 2017–2020, Economic and Social Council, New York, 7 July 2017. – Internet resource – https://undocs.org/E/RES/2017/4 [Last visit: 30.04.2023]

Documents of the Human Rights Committee

United Nations CCPR/C/95/D/1457/2006, Ángela Poma Poma v. Peru, Communication No. 1457/2006, Ninety-fifth Session of the Human Rights Committee, 16 March to 9 April 2009, 2009. – Internet resource – https://undocs.org/en/CCPR/C/95/D/1457/2006 [Last visit: 30.04.2023]

UN General Assembly

United Nations General Assembly Resolution A/RES/37/7, Resolution adopted by the General Assembly on 28 October 1982, World Charter for Nature, United Nations, New York, 1982. – Internet resource – https://undocs.org/en/A/RES/37/7 [Last visit: 30.04.2023]

United Nations A/RES/55/2, United Nations Millennium Declaration, General Assembly Resolution, 18 September 2000. – Internet resource – https://undocs.org/A/RES/55/2 [Last visit: 30.04.2023]

United Nations A/RES/56/83, Resolution adopted by the General Assembly on 12 December 2001, Responsibility of States for internationally wrongful acts, United Nations, 2001. – Internet resource – https://undocs.org/en/A/RES/56/83 [Last visit: 30.04.2023]

United Nations A/RES/66/288, Resolution adopted by the General Assembly on 27 July 2012: The Future We Want UNCSD Rio+20, The future we want, Outcome document of the United Nations Conference on Sustainable Development, United Nations, New York, 2012. – Internet resource – https://undocs.org/A/RES/66/28 [Last visit: 30.04.2023]

United Nations A/RES/70/1, Resolution Adopted by the General Assembly on 25 September 2015: Transforming our World: the 2030 Agenda for Sustainable Development, United Nations, New York, 21 October 2015. – Internet resource – http://www.un.org/ga/search/view_doc.asp?symbol=A/RES/70/1&Lang=E [Last visit: 30.04.2023]

United Nations A/RES/71/285, Resolution adopted by the General Assembly on 27 April 2017: United Nations Strategic Plan for Forests 2017–2030, United Nations, New York, 1 May 2017. – Internet resource – https://undocs.org/E/RES/2017/4 [Last visit: 30.04.2023]

United Nations A/RES/72/277, Resolution adopted by the General Assembly on 10 May 2018: Towards a Global Pact for the Environment, United Nations, New York, 2018. – Internet resource – https://undocs.org/A/RES/72/277 [Last visit: 30.04.2023]

KP CMP Decisions

KP CMP Decision 3/CMP.1, "Modalities and procedures for a clean development mechanism as defined in Article 12 of the Kyoto Protocol", in: United Nations FCCC/KP/CMP/2005/8/Add.2, Report of the Conference of the Parties serving as the meeting of the Parties to the Kyoto Protocol on its first session, held at Montreal from 28 November to 10 December 2005, 2006. – Internet resource – https://unfccc.int/resource/docs/2005/cmp1/eng/08a03.pdf [Last visit: 30.04.2023]

KP CMP Decision 5/CMP.1, "Modalities and procedures for afforestation and reforestation project activities under the clean development mechanism in the first commitment period of the Kyoto Protocol", in: United Nations FCCC/KP/CMP/2005/8/Add.2, Report of the Conference of the Parties serving as the meeting of the Parties to the Kyoto Protocol on its first session, held at Montreal from 28

November to 10 December 2005, 2006. – Internet resource – https://unfccc.int/resource/docs/2005/cmp1/eng/08a03.pdf [Last visit: 30.04.2023]

KP CMP Decision 16/CMP.1 "Land use, land-use change and forestry", para. 1(d), in: United Nations: UNFCCC, Decision 16/CMP.1 "Land use, land-use change and forestry", in: United Nations FCCC/KP/CMP/2005/8/Add.2, Report of the Conference of the Parties serving as the meeting of the Parties to the Kyoto Protocol on its first session, held at Montreal from 28 November to 10 December 2005, United Nations, 2006. – Internet resource – https://unfccc.int/resource/docs/2005/cmp1/eng/08a03.pdf [Last visit: 30.04.2023]

Documents and Decisions Adopted by the UNFCCC COP

United Nations FCCC/AGBM/1997/INF.2, Compilation of Responses from Parties on Issues Related to Sinks, Ad Hoc Group on the Berlin Mandate, Eighth Session, Agenda item 3, Kyoto, 30 November 1997. – Internet resource – https://unfccc.int/resource/docs/1997/agbm/inf02.pdf [Last visit: 30.04.2023]

UNFCCC COP Decision 11/CP.7, "Land use, land-use change and forestry", in: United Nations FCCC/CP/2001/13/Add.1, Report of the Conference of the Parties on its seventh Session, held at Marrakesh from 29 October to 10 November 2001, United Nations, 2002. – Internet resource – https://unfccc.int/sites/default/files/resource/docs/cop7/13a01.pdf?download [Last visit: 30.04.2023]

UNFCCC COP Decision 17/CP.7, "Modalities and procedures for a clean development mechanism as defined in Article 12 of the Kyoto Protocol", in: United Nations FCCC/CP/2001/13/Add.2, Report of the Conference of the Parties on its seventh Session, held at Marrakesh from 29 October to 10 November 2001, United Nations, 2002. – Internet resource – https://unfccc.int/resource/docs/cop7/13a01.pdf [Last visit: 30.04.2023]

UNFCCC COP Decision 28/CP.7, "Guidelines for the preparation of national adaptation programmes of action", in: United Nations FCCC/CP/2001/13/Add.1, Report of the Conference of the Parties on its seventh Session, held at Marrakesh from 29 October to 10 November 2001, United Nations, 2002. – Internet resource – https://undocs.org/FCCC/CP/2001/13/Add.4 [Last visit: 30.04.2023]

UNFCCC COP Decision 19/CP.9 "Modalities and procedures for afforestation and reforestation project activities under the clean development mechanism in the first commitment period of the Kyoto Protocol" in: United Nations FCCC/CP/2003/6/Add.2, Report of the Conference of the Parties on its ninth Session, held at Milan from 1 to 12 December 2003, United Nations, 2004. – Internet

1995. – Internet resource – https://www.cbd.int/decisions/cop/?m=cop-02 [Last visit: 30.04.2023]

CBD COP Decision III/8 “Memorandum of understanding between the Conference of the Parties to the Convention on Biological Diversity and the Council of the Global Environment Facility”, Decisions adopted by the Conference of the Parties to the Convention on Biological Diversity at its third Session, held in Buenos Aires from 4 to 15 November 1996, United Nations, 1996. – Internet resource – https://www.cbd.int/decision/cop/?id=7104 [Last visit: 30.04.2023]

CBD COP 4 Decision IV/7 “Forest Biological Diversity”, Decision adopted by the Conference of the Parties to the Convention on Biological Diversity at its fourth Session, held at Bratislava from 4 to 15 May 1998, United Nations, 1998. – Internet resource – https://www.cbd.int/doc/decisions/cop-11/cop-11-dec-06-en.pdf [Last visit: 30.04.2023]

CBD COP Decision VII/1 “Forest biological diversity”, Decision adopted by the Conference of the Parties to the Convention on Biological Diversity at its seventh Session, held at Kuala Lumpur from 9 to 20 and on 27 February 2004, United Nations, 2004. – Internet resource – https://www.cbd.int/doc/decisions/cop-07/cop-07-dec-01-en.pdf [Last visit: 30.04.2023]

CBD COP Decision VII/26 “Cooperation with other conventions and international organizations and initiatives”, Decision adopted by the Conference of the Parties to the Convention on Biological Diversity at its seventh Session, held at Kuala Lumpur from 9 to 20 and on 27 February 2004, United Nations, 2004. – Internet resource – https://www.cbd.int/decision/cop/default.shtml?id=7763 [Last visit: 30.04.2023]

CBD COP Decision VIII/16, Cooperation with other conventions and international organizations and initiatives, Decision adopted by the Conference of the Parties to the Convention on Biological Diversity at its eighth Session, held at Curitiba, Brazil, from 20 till 31 March 2006, United Nations, 2006. – Internet resource – https://www.cbd.int/doc/decisions/cop-08/cop-08-dec-16-en.pdf [Last visit: 30.04.2023]

CBD COP Decision VIII/28, “Impact assessment: Voluntary guidelines on biodiversity-inclusive impact assessment”, Decisions adopted by the Conference of the Parties to the Convention on Biological Diversity at its eighth Session, held at the Curitiba, from 20 to 31 March 2006. – Internet resource – https://www.cbd.int/doc/decisions/cop-08/cop-08-dec-28-en.pdf [Last visit: 30.04.2023]

CBD COP Decision IX/5, "Forest Biodiversity", Decisions adopted by the Conference of the Parties to the Convention on Biological Diversity at its ninth Session, held in Bonn from 19 to 30 May 2008, United Nations, 2008. – Internet resource – https://www.cbd.int/doc/decisions/cop-09/cop-09-dec-05-en.pdf [Last visit: 30.04.2023]

CBD COP Decision IX/16 "Biodiversity and climate change", Decision adopted by the ninth Session of the COP, held in Bonn from 19 to 30 May 2008, United Nations, 2008. – Internet resource – https://www.cbd.int/doc/decisions/cop-09/cop-09-dec-16-en.pdf [Last visit: 30.04.2023]

CBD COP Decision X/2, "The Strategic Plan for Biodiversity 2011–2020 and the Aichi Biodiversity Targets", Decisions adopted by the Conference of the Parties to the Convention on Biological Diversity at its tenth Session, held at Nagoya, Japan, from 18 to 29 October 2010, United Nations, 2010. – Internet resource – https://www.cbd.int/doc/decisions/cop-10/cop-10-dec-02-en.pdf [Last visit: 30.04.2023]

CBD COP Decision X/36, "Forest biodiversity", Decision adopted by the tenth Session of the COP, held in Nagoya, Japan, from 18 to 29 October 2010, United Nations, 2010. – Internet resource – https://www.cbd.int/doc/decisions/cop-10/cop-10-dec-36-en.pdf [Last visit: 30.04.2023]

CBD COP Decision XI/6 "Cooperation with other conventions, international organizations, and initiatives", Decision adopted by the Conference of the Parties to the Convention on Biological Diversity at its eleventh Session, held at Kuala Lumpur from 9 to 20 and on 27 February 2004, United Nations, 2012. – Internet resource – https://www.cbd.int/doc/decisions/cop-11/cop-11-dec-06-en.pdf [Last visit: 30.04.2023]

CBD COP Decision XI/19, "Biodiversity and climate change related issues: advice on the application of relevant safeguards for biodiversity with regard to policy approaches and positive incentives on issues relating to REDD+", Decisions adopted by the Conference of the Parties to the Convention on Biological Diversity at its eleventh Session, held at Hyderabad, India, from 8 to 19 October 2012, United Nations, 2012. – Internet resource –https://www.cbd.int/doc/decisions/cop-11/cop-11-dec-19-en.pdf [Last visit: 30.04.2023]

CBD COP Decision XI/21, "Biodiversity and climate change: integrating biodiversity considerations into climate-change related activities", Decisions adopted by the Conference of the Parties to the Convention on Biological Diversity at its eleventh Session, held at Hyderabad, India, from 8 to 19 October 2012, United

Nations, 2012. – Internet resource – https://www.cbd.int/doc/decisions/cop-11/cop-11-dec-21-en.pdf [Last visit: 30.04.2023]

CBD COP Decision XII/20, "Biodiversity and climate change and disaster risk reduction", Decisions adopted by the Conference of the Parties to the Convention on Biological Diversity at its twelfth Session, held at PyeongChang, Korea, from 6 to 17 October 2014, United Nations, 2014. – Internet resource – https://www.cbd.int/doc/decisions/cop-12/cop-12-dec-20-en.pdf [Last visit: 30.04.2023]

CBD COP Decision XIII/4, "Biodiversity and climate change", Decisions adopted by the Conference of the Parties to the Convention on Biological Diversity at its thirteenth Session, held at Cancun, Mexico, from 4 to 17 December 2016, United Nations, 2016. – Internet resource – https://www.cbd.int/doc/decisions/cop-13/cop-13-dec-04-en.pdf [Last visit: 30.04.2023]

CBC COP Decision XIV/5, "Biodiversity and Climate Change", Decisions adopted by the Conference of the Parties to the Convention on Biological Diversity at its fourteenth Session, held at Sharm El-Sheikh, Egypt, from 17 to 29 November 2018, United Nations, 2018. – Internet resource – https://www.cbd.int/doc/c/9860/44b3/042fbf32838cf31a771bb145/cop-14-l-23-en.pdf [Last visit: 30.04.2023]

Documents and Decisions Adopted by the COP-MOP of the Nagoya Protocol

NP COP-MOP Decision 1/10, The need for and modalities of a global multilateral benefit-sharing mechanism (Article 10), First meeting of the COP-MOP of the Nagoya Protocol, PyeongChang, Republic of Korea, 13–17 October 2014. – Internet resource – https://www.cbd.int/doc/decisions/np-mop-01/np-mop-01-dec-10-en.pdf [Last visit: 30.04.2023]

NP COP-MOP Decision 2/10, The need for and modalities of a global multilateral benefit-sharing mechanism (Article 10), Second meeting of the COP-MOP of Nagoya Protocol in Cancun, Mexico, 4–17 December 2016. – Internet resource – https://www.cbd.int/doc/decisions/np-mop-02/np-mop-02-dec-10-en.pdf [Last visit: 30.04.2023]

NP COP-MOP Decision 3/10, Global multilateral benefit-sharing mechanism (Article10), Third meeting of the COP-MOP of Nagoya Protocol in Sharm El-Sheikh, Egypt, 17–29 November 2018. – Internet resource – https://www.cbd.int/doc/decisions/np-mop-02/np-mop-02-dec-10-en.pdf [Last visit: 30.04.2023]

Documents Adopted by the CBD SBSTTA

CBD SBSTTA Information Note 8/10 "Joint Work Programme between the Secretariat of the Convention on biological Diversity and the Secretariat of the Convention on the biological diversity of dry and sub-humid lands", eighth Meeting, held in Montreal from 10 to 14 March 2003, United Nations, 2003. Internet resource –https://www.cbd.int/doc/meetings/sbstta/sbstta-08/information/sbstta-08-inf-10-en.pdf [Last visit: 30.04.2023]

CBD SBSTTA Information Note 9/10, Options for enhanced cooperation among the three Rio Conventions, SBSTTA, ninth Meeting, held at Bangkok from 7 to 11 February 2004, United Nations, 2004. – Internet resource – https://www.cbd.int/doc/meetings/sbstta/sbstta-10/information/sbstta-10-inf-09-en.pdf [Last visit: 30.04.2023]

CBD/UNCCD, Memorandum of Understanding between the Convention on Biological Diversity and the United Nations Convention to Combat Desertification Regarding the Joint Work Plan 2011–2012, 03 September 2011. Internet resource – https://www.cbd.int/doc/agreements/agmt-unccd-2011-09-03-mou-en.pdf [Last visit: 30.04.2023]

CBD SBSTTA Information Note 7/3, Review of the status and trends of, and major threats to the forest biological diversity, prepared by the Ad Hoc Technical Expert Group on Forest Biological Diversity, seventh meeting, held in Montreal from 12 to 16 November 2001, United Nations, 2001. – Internet resource – https://www.cbd.int/doc/meetings/sbstta/sbstta-07/information/sbstta-07-inf-03-en.pdf [Last visit: 30.04.2023]

Documents Adopted by the UNCCD COP

UNCCD, Enhanced scientific cooperation and knowledge exchange between the Committee on Science and Technology and the scientific subsidiary bodies of the United Nations Framework Convention on Climate Change and the Convention on Biological Diversity, the Scientific and Technical Advisory Panel of the Global Environment Facility and relevant specialized agencies of the United Nations, UNCCD Cop 10th session, Changwon, 11–13 October 2011. – Internet resource – https://www.unccd.int/sites/default/files/sessions/documents/ICCD_COP10_5/cstINF5eng.pdf [Last visit: 30.04.2023]

Documents Adopted Jointly by the CBD and the UNCCD

Secretariat of the UNCCD/Secretariat of the CBD, Memorandum of Understanding between the Convention on Biological Diversity and the United Nations Convention to Combat Desertification, 31 July 1998. – Internet resource – https://www.cbd.int/doc/agreements/agmt-unccd-1998-07-31-moc-web-en.pdf [Last visit: 30.04.2023]

Secretariat of the UNCCD/Secretariat of the CBD, Memorandum of Understanding between the Convention on Biological Diversity and the United Nations Convention to Combat Desertification Regarding the Joint Work Plan 2011–2012, 03 September 2011. – Internet resource – https://www.cbd.int/doc/agreements/agmt-unccd-2011-09-03-mou-en.pdf [Last visit: 30.04.2023]

Decisions Adopted by the CITES COP

CITES DOC 9.52, Interpretation and Implementation of the Convention, Trade in Plant Specimens, Implementation of the Convention for timber species, 9th meeting of the CITES COP, 1994, Paragraph 9. – Internet resource – https://www.cites.org/sites/default/files/eng/cop/09/doc/E9-Doc-52-53.pdf [Last visit: 30.04.2023]

CITES COP 6, Amendments to the Appendices I and II of the Convention adopted at the sixth meeting of the Conference of the Parties in Ottawa, Canada, 12 to 24 July 1987, 1987, p. 99. Internet resource – https://cites.org/sites/default/files/eng/cop/06/E06-Amendments-to-Appendices.pdf [Last visit: 30.04.2023]

CITES COP 8, Proposals 8.94 and 8.95, in: Amendments to the Appendices I and II of the Convention, Other Proposals, 1992, pp. 111 ff. Internet resource – https://cites.org/sites/default/files/eng/cop/08/prop/E08-Prop-94_Swietenia.pdf [Last visit: 30.04.2023]

CITES COP 9, Proposal 9.101, in: Amendments to the Appendices I and II of the Convention, Other Proposals, 1994, pp. 101 ff. – Internet resource – https://cites.org/sites/default/files/eng/cop/09/prop/E09-Prop-101_Swietenia.PDF [Last visit: 30.04.2023]

See CITES Resolution Conf. 9.5 (Rev. CoP13), Trade with States not Party to the Convention, 1994. – Internet resource – https://cites.org/eng/cop/09/doc/index.php [Last visit: 30.04.2023]

CITES COP 10, Proposal 10.69., in: Consideration of Proposals for the Amendment of Appendices I and II of the Convention, Other Proposals, 1997, pp. 1

ff. – Internet resource – https://cites.org/sites/default/files/eng/cop/10/prop/E-CoP10-P-06.pdf [Last visit: 30.04.2023]

CITES COP Conf. 10.13, Implementation of the Convention for timber species, in: CITES COP 10, Resolutions of the Conference of the Parties, 1997. – Internet resource – http://www.ciesin.columbia.edu/repository/entri/docs/cop/CITES_COP010_res013.pdf [Last visit: 30.04.2023]

CITES COP 12, Amendments to Appendices I and II of the Convention, adopted by the Conference of the Parties at its 12th meeting, Santiago, Chile, 3 to 15 November 2002, 2002. – Internet resource – https://cites.org/sites/default/files/eng/cop/12/Adopted_Amendments.pdf [Last visit: 30.04.2023]

CITES COP 14 Document 23, Interpretation and implementation of the Convention, Compliance and enforcement issues: Guidelines for Compliance with the Convention, 2007. – Internet resource – https://cites.org/sites/default/files/eng/cop/14/doc/E14-23.pdf [Last visit: 30.04.2023]

CITES Resolution Conf. 14.3, CITES Compliance Procedures, Annex: Guide to CITES Compliance Procedures, 2007 – Internet resource – https://www.cites.org/eng/res/14/14-03C15.php [Last visit: 30.04.2023]

CITES COP 17, Amendments to Appendices I and II of the Convention, adopted by the Conference of the Parties at its 17th Meeting, Johannesburg (South Africa), 24 September – 4 October 2016, 2016. – Internet resource – https://cites.org/sites/default/files/notif/E-Notif-2016-063.pdf [Last visit: 30.04.2023]

Decisions of International Conferences

Atmospheric pollution and climate change Ministerial Conference held at Noordwyk, the Netherlands, on 6 and 7 November 1989, The Noordwyk Declaration on Climate Change, November 1989. – Internet resource – https://ntrl.ntis.gov/NTRL/dashboard/searchResults/titleDetail/PB90210196.xhtml [Last visit: 30.04.2023]

Ministerial Conference on the Protection of Forests in Europe

Ministerial Conference on the protection of forests in Europe, Oslo Ministerial Mandate for Negotiating a Legally Binding Agreement on Forests in Europe, Oslo 14–16 June 2011. – Internet resource – https://www.foresteurope.org/docs/MC/MC_oslo_lba_mandate.pdf [Last visit: 30.04.2023]

Ministerial Conference on the protection of forests in Europe, Madrid Ministerial Decision, Extraordinary Ministerial Conference, Madrid, 2015. – Internet resource – https://foresteurope.org/wp-content/uploads/2016/11/V.-EMC_MadridMinisterialDecision.pdf [Last visit: 30.04.2023]

ICJ Judgments

Legality of the Threat or Use of Nuclear Weapons, Advisory Opinion of 8.07.1996, ICJ Reports 1996, pp. 226–267.

Pulp Mills on the River Uruguay (Argentina v. Uruguay), Judgment of 20.04.2010, ICJ Reports 2010, pp. 14–107.

Questions relating to the Obligation to Prosecute or Extradite (Belgium v. Senegal), Judgment of 20 July 2012, ICJ Reports 2012, pp. 422–463.

Whaling in the Antarctic (Australia v. Japan: New Zealand intervening), Judgement of 31.03.2014, ICJ Report 2014, pp. 226–300.

International Arbitration Courts

Trail Smelter Arbitration (United States v. Canada), Arbitral Judgements of 16.04.1938 and 11.03.1941, Rewards of International Arbitral Awards, Volume III, pp. 1905–1982

Texaco Calasiatic v. Libyan Arab Republic (Merits), in International Legal Materials, Volume 17, 1978, pp. 1–37.

National Legislation

India, Intellectual Property, The Patents Act with Amendments, first adopted in 1970. – Internet resource – https://ipindia.gov.in/writereaddata/Portal/IPOAct/1_113_1_The_Patents_Act_1970_-_Updated_till_23_June_2017.pdf [Last visit: 30.04.2023].

Republic of the Philippines, Executive Order N. 247 of the Philippines "Prescribing Guidelines and Establishing a Regulatory Framework for the Prospecting of Biological and Genetic Resources, their By-Products and Derivatives, for Scientific and Commercial Purposes; and for other Purposes", in: Official Gazette, May 18, 1995.

Republic of Costa Rica, Biodiversity Law, No. 7788, 1998.

Federal Democratic Republic of Ethiopia, Proclamation NO. 482/2006, A Proclamation to Provide for Access to Genetic Resources and Community Knowledge and Community Right, in: Federal Negarit Gazeta 13th Year, No. 13, Addis Ababa, 27 February, 2006.

Bibliography

Abidin, Handa, The Protection of Indigenous Peoples and Reduction of Forest Carbon Emissions: The REDD-Plus Regime and International Law, Brill-Nijhoff, Leiden, 2015.

Adhikari, Bhim, Market-Based Approaches to Environmental Management: A Review of Lessons from Payment for Environmental Services in Asia, Working Paper 134, Asian Development Bank Institute, Tokyo, 2009.

Agung, Putra/Galudra, Gamma/Van Noordwijk, Meine/Maryani, Retno, Reform or Reversal: The Impact of REDD+ Readiness on Forest Governance in Indonesia, in: Climate Policy, Volume 14, Issue 6, 2014, pp. 748–768.

Albizu, Oreina Orrantia/Hidalgo, M. Mercedes Ortega/Madrigal, Olman Quirós/Arregui Javier Loidi, Servicios Ambientales del Bosque: Ensayo en una Cuenca Atlántica Europea con Base en la Experiencia de Centroamérica, in: Revista de Biología Tropical, Volume 56, Issue 4, December 2008, pp. 287–296.

Altea, Maria Pia M., Principle 21 of the Stockholm Declaration: A Customary Norm of International Environmental Law, in: Ateneo Law Journal, Volume 41, Issue 2, 1997, pp. 429–443.

Ambrose Jillian, Converting Coal Plants to Biomass Could Fuel Climate Crisis, Scientists Warn, in: The Guardian, 2019. Internet resource https://www.theguardian.com/environment/2019/dec/16/converting-coal-plants-to-biomass-could-fuel-climate-crisis-scientists-warn?CMP=share_btn_fb&fbclid=IwAR2o7PAn_db_uObZkicQqKleJ-XEpdYmrxysRAmDMlm1UK-WBrF_jkpN51s [Last visit: 30.04.2023]

Anandi, Cut Augusta Mindry/Resosudarmo, Ida Aju Pradnja/Ekaputri, Andini Desita/Komalasari, Mella/Astri Pangestuti/Aryani, Riza, Ulu Masen REDD+ Initiative, Aceh, Indonesia, in: *Sills, O Erin/Atmadja, Stibniati/de Sassi Claudio/Duchelle, Amy E/Kweka, Demetrius L./ Resosudarmo, Ida Aju Pradnja/Sunderlin, William D.* (eds.), REDD+ on the Ground: A Case Book of Subnational Initiatives Across the Globe, CIFOR, 2014, pp. 380–401.

Anderson, Patrick, Free, Prior, and Informed Consent: Principles and Approaches for Policy and Project Development, GIZ, 2011.

Anderson, Nicholas, REDDy or Not? The Effects on Indigenous Peoples in Brazil of a Global Mechanism for Reducing Emissions from Deforestation and Degradation, in: Journal of Sustainable Development, Volume 2, Issue 3, 2009, pp. 18–28.

Andersson, Kriter, Understanding Decentralized Forest Governance: An Application of the Institutional Analysis and Development Framework, in: Sustainability: Science, Practice and Policy, Volume 2, Issue 1, 2006, pp. 25–35.

Anthony, Aust, Modern Treaty Law and Practice, Cambridge University Press, Cambridge, 2007.

Asia Indigenous Peoples Pact, REDD+ Implementation in Asia and the Concerns of Indigenous Peoples; Asia Indigenous Peoples Pact: Chiang Mai, Thailand, 2010. Internet resource – http://www.iwgia.org/iwgia_files_publications_files/0654_REDD_Plus_Implementation_in_Asia_and_the_Concerns_of_Indigenous_Peoples.pdf [Last visit: 30.04.2023]

Azoulay, P./Ding, W./Stuart T., The Impact of Academic Patenting on the Rate, Quality and Direction of (Public) Research Output, in: Journal of Industrial Economics, Volume 57, 2009, pp. 637–676.

Baier, Tina, Wie Klimaschutz der Natur schadet, in: Süddeutsche Zeitung, 20. August 2019. Internet resource – https://www.sueddeutsche.de/wissen/klimawandel-tierschutz-energie-1.4558398?reduced=true [Last visit: 30.04.2023]

Balch, Oliver, Indonesia's Forest Fires: Everything You Need to Know, in: The Guardian, 11 November 2015. Internet resource – https://www.theguardian.com/sustainable-business/2015/nov/11/indonesia-forest-fires-explained-haze-palm-oil-timber-burning [Last visit: 30.04.2021]

Balsiger, Jörg/Prys, Miriam/Steinhoff, Niko, The Nature and Role of Regional Agreements in International Environmental Politics: Mapping Agreements, Outlining Future Research, GIGA Working Paper, Number 208, 2012.

Barber, Victor Charles, 25 Years of So-Called "Sustainable Forest Management" in Sarawak, World Resources Institute, 19 November 2015. Internet resource – http://www.wri.org/blog/2015/11/25-years-so-called-sustainable-forest-management-sarawak [Last visit: 30.04.2023]

Barbosa, Luiz C., Guardians of the Brazilian Amazon Rainforest: Environmental Organizations and Development, Routledge, 2015.

Bastin, Jean-Francois/Finegold, Yelena/Garcia, Claude/Mollicone, Daniloe/Rezende, Marcelo/Routh, Devin/Zohner, Constantin M./Crowther, Thomas W., The Global Tree Restoration Potential, in: Science, Volume 365, Issue 6448, 2019, p. 78.

Bayrak, Mucahid Mustafa/Marafa, Lawal Mohammed, Ten Years of REDD+: A Critical Review of the Impact of REDD+ on Forest-Dependent Communities, in: Sustainability, Volume 8, Issue 7, 2016, pp. 620–641.

Bennet, Aoife/Ravikumar, Ashwir/Paltán Homero, The Political Ecology of Oil Palm Company-Community Partnerships in the Peruvian Amazon: Deforestation Consequences of the Privatization of Rural Development, in: World Development, Volume 109, 2018, pp. 29–41.

Bettwy, Mike, Tropical Deforestation Affects Rainfall in the U.S. and Around the Globe, NASA, 13 September 2005. Internet resource – https://www.nasa.gov/centers/goddard/news/topstory/2005/deforest_rainfall.html [Last visit: 30.04.2023]

Beyerlin, Ulrich, Internationaler Waldschutz am Scheideweg? Von den "Rio-Waldprinzipien" zu "REDD plus", in: *Hecker, Bernd/Henlder, Reinhard/Proelß, Alexander/Reiff, Peter* (eds.), Jahrbuch des Umwelt- und Technikrecht, Schriftenreihe des Instituts für Umwelt- und Technikrecht der Universität Trier, 2012, pp. 171–201.

Beyerlin, Ulrich, Umweltvölkerrecht, Verlag C. H. Beck, München, 2000.

Beyerlin, Ulrich/Marauhn, Thilo, International Environmental Law, Hart Publishing, Oxford, 2011.

Beyerlin, Ulrich/Stoll Peter-Tobias/Wolfrum, Rüdiger, Ensuring Compliance with Multilateral Environmental Agreements: A Dialogue between Practitioners and Academia, Martinus Nijhoff Publishers, the Hague, 2006.

Birdsall, Nancy/Kuczynski, Pedro Pablo, Climate Talks: Rich Countries Should Pay to Keep Tropical Forests Standing, in: The Guardian, 2 December 2015. Internet resource – https://www.theguardian.com/global-development/2015/dec/02/climate-talks-rich-countries-should-pay-to-keep-tropical-forests-standing [Last visit 30.04.2023]

Birnie, Patricia/Boyle, Alan/Redgewell, Catherine, International Law and the Environment, 3rd ed., Oxford University Press, Oxford, 2009.

Blaser, Juergen/Chipeta, Mafa Evaristus/Illueca, Jorge/Lobovikov, Maxim/Umali, Ricardo Martinez, Independent Assessment of the International Arrangement on Forests (IAF), Report of the Team of Independent Consultants, United Nations, New York, 2014.

Blomquist, Robert F., Ratification Resisted: Understanding America's Response to the Convention on Biological Diversity, 1989–2002, in: Golden Gate University Law Review, Volume 32, Issue 4 "Rio's Decade: Reassessing the 1992 Earth Summit", 2002, pp. 493–586

Blundell, Art, A Review of the CITES Listing of Big-leaf Mahogany, in: ORYX, Volume 38, Issue 1, 2004.

Bodansky, Daniel, The History and Legal Structure of the Global Climate Change Regime, in: *Sprinz, Detlef/Luterbacher, Urs* (eds.), International Relations and Global Climate Change, the Potsdam Institute for Climate Impact Research, The MIT Press, Cambridge, Massachusetts, 2001, pp. 11–31.

Bodansky, Daniel, The Legal Character of the Paris Agreement, in: RECIEL, Volume 25, Issue 2, 2016, pp. 142–150.

Bodansky, Daniel/Brunnee, Jutta/ Rajamani, Lavanya, International Climate Change Law, Oxford University Press, Oxford, 2017.

Böhringer, Ayşe-Martina, Die Kooperationsvereinbarungen der Sekretariate multilateraler Umweltschutzübereinkommen, Mohr Sieber, Tübingen, 2014.

Bolle, Ulrike, Das Intergovernmental Panel on Climate Change (IPCC): eine völkerrechtliche Untersuchung, Mohr Siebeck, Tübingen, 2011.

Börner, Jan/Baylis, Kathy/Corbera, Esteve/Ezzine-de-Blas, Driss/Honey-Rosés, Jordi/Persson, U. MartinPersson/Wunder, Sven, The Effectiveness of Payments for Environmental Services, in: World Development, Volume 96, 2017, pp. 1 ff.

Boucher, Doug, How Brazil Has Dramatically Reduced Tropical Deforestation, in: Solutions, Volume 5, Issue 2, 2014, pp. 66–75.

Boucher, Doug/Elias, Pipa/Faires, Jordan/Smith, Sharon, Deforestation Success Stories: Tropical Nations Where Forest Protection and Reforestation Policies Have Worked, Union of Concerned Scientists, 2014.

Boucher, Doug/Roquemore, Sarah/Fitzhugh, Estrellita, Brazil's Success in Reducing Deforestation, in: Tropical Conservation Science, Special Issue, Volume 6, Issue 3, 2013, pp. 426–445.

Boulay, Axelle/Tacconi, Luca, The Drivers of Contract Eucalypt Farming in Thailand, in: International Forestry Review, Volume 14, Issue 1, 2012, pp. 1–12.

Boyd, William, Ways of Seeing in Environmental Law: How Deforestation Became an Object of Climate Governance, in: Ecology Law Quarterly, Volume 37, Issue 3, 2010, pp. 843–916.

Bradshaw, Corey J. A./Sodhi, Navjot S./Peh, Kevin S./Brook, Barry W., Global Evidence That Deforestation Amplifies Flood Risk and Severity in the Developing World, in: Global Change Biology, Volume 13, 2007, pp. 2379–2395.

Brownlie, Ian, Principles of Public International Law, 7. Edition, Oxford University Press, Oxford, 2008.

Brunnée, Jutta, A Conceptual Framework for an International Forests Convention: Customary Law and Emerging Principles, in: Canadian Council on International Law (ed.), Global Forests and International Law, Kluwer Law International, London, 1996, pp. 41–77.

Brunnengraeber, Achim/Dietz, Kristina/Hirschl, Bernd/Walk, Heike/Weber, Melanie, Das Klima neu denken: Eine sozial-ökologische Perspektive auf die lokale, nationale und internationale Klimapolitik, Dampfboot-Verlag, Münster, 2008.

Buck, Matthias/Hamilton, Clare, The Nagoya Protocol on Access to Genetic Resources and the Fair and Equitable Sharing of Benefits Arising from their

Utilization to the Convention on Biological Diversity, in: RECIEL, Volume 20, Issue 1, 2011, pp. 47–61.

Bull, Garry Q./Boedhihartono, Agni K./Bueno, G./Cashore, B./Elliot, C./Langston, James D./Riggs, R. A./Sayer, J., Global Forest Discourses Must Connect with Local Forest Realities, in: International Forestry Review, Volume 20, Issue 2, 2018, pp. 160–166.

Burns, William C., The International Convention to Combat Desertification: Drawing a Line in the Sand?, in: Michigan Journal of International Law, Volume 16, Issue 3, 1995, pp. 831–882.

Bush, Jonah/Engelmann, Jens, Tropical Forests Offer up to 24–30 Percent of Potential Climate Mitigation, Center for Global Development, 4 November 2014. Internet resource – https://www.cgdev.org/blog/tropical-forests-offer-24%E2%80%9330-percent-potential-climate-mitigation [Last visit: 30.04.2023]

Butler, Rhett A., Consequences of deforestation, in: Mongabay, 22 July 2012. Internet resource – http://rainforests.mongabay.com/0901.htm [Last visit: 30.04.2023]

Cadman, Timothy/Maraseni, Tek/Ma, Hwan Ok/Lopez-Casero, Federico, Five Years of REDD+ Governance: The Use of Market Mechanisms as a Response to Anthropogenic Climate Change, in: Forest Policy and Economics, Volume 79, 2017, pp. 8–16.

Cameron, James/Robinson, Jonathan, The Uses of Trade Provisions in International Environmental Agreements and Their Compatibility with the GATT, in: Yearbook of International Environmental Law, Volume 2, Issue 1, 1991, pp. 3–30.

Campbell, Bruce. M., Beyond Copenhagen: REDD+, Agriculture, Adaptation Strategies and Poverty, in: Global Environmental Change, Volume 19, Issue 4, 2009, pp. 397–399.

Campbell, John L., Why Would Corporations Behave in Socially Responsible Ways? An Institutional Theory of Corporate Social Responsibility, in: The Academy of Management Review, Volume 32, Issue 3, 2007, pp. 946–967.

Cançado Trindade, Antônio Augusto, International Law for Humankind: Towards a New Jus Gentium, Brill-Nijhoff, the Hague, 2010.

Cannon, John C., Stop Importing Illegal Timber, PNG Activists Tell China at APEC Summit, in: Mongabay, 26 November 2018. Internet resource – https://www.eco-business.com/news/china-should-stop-importing-illegal-timberactivists/ [Last visit: 30.04.2023]

Caplow, Susan/Jagger, Pamela/Lawlor, Kathleen/Sills, Erin, Evaluating Land Use and Livelihood Impacts of Early Forest Carbon Projects: Lessons for Learning

About REDD+, in: Environmental Science & Policy, Volume 14, Issue 2, 2011, pp. 152–167.

Capstick, Stuart Bryce/Pidgeon, Nicholas Frank, What Is Climate Change Scepticism? Examination of the Concept Using a Mixed Methods Study of the UK Public, in: Global Environmental Change, Volume 24, 2014, pp. 389–401.

Carbon Market Watch, The Clean Development Mechanism: Local Impacts of a Global System, Carbon Market Watch, 2018.

Carraro, Carlo, International Environmental Agreements on Climate Change, Kluwer Academic Publishers, Dordrecht, The Netherlands, Boston, 1999.

Carrington, Damian, "Silver bullet" to Suck CO2 from Air and Halt Climate Change Ruled Out, in: The Guardian, 1 February 2018. Internet resource https://www.theguardian.com/environment/2018/feb/01/silver-bullet-to-suck-co2-from-air-and-halt-climate-change-ruled-out [Last visit: 30.04.2023]

Casey, Frank/Vickerman, Sara/Hummon, Cheryl/Taylor, Bruce, Incentives for Biodiversity Conservation: An Ecological and Economic Assessment, Defenders of Wildlife, Washington, 2006.

Castillo-Winckels, Nadia Sanchez, Why "Common Concern of Humankind" Should Return to the Work of the International Law Commission on the Atmosphere, in: Georgetown Environmental Law Review, Volume 29, Issue 1, 2017, pp. 131–151

Chao, Sophie, The Truth about "Sustainable" Palm Oil, in: Sapiens, 13 June 2019. Internet resource – https://www.sapiens.org/culture/palm-oil-sustainable/ [Last visit: 30.04.2023]

Chazdon, Robin L., Beyond Deforestation: Restoring Forests and Ecosystem Services on Degraded Lands, in: Science, Volume 320, Issue 5882, 2008, pp. 1458–1460.

Chazdon, Robin L.,Guariguata, Manuel R., Natural Regeneration as a Tool for Large-Scale Forest Restoration in the Tropics: Prospects and Challenges, in: Biotropica, Volume 48, Issue 6, 2016, pp. 716–730.

Chazdon, Robin L., Bodin, Blaise, Guariguata, Manuel, Lamb, David, Walder, Bethanie, Chokkalingam, Unna, Shono, Kenichi, Partnering with Nature: The Case for Natural Regeneration in Forest and Landscape Restoration, FERI Policy Brief, Montreal, Canada, 2017.

Chazdon, Robin L., Brancallion, Pedro H. S., Laestadius Lars, Bennett-Curry Aoife, Buckingham, Kathleen, Kumar Chetan, Moll-Rocek, Julian, Vieira, Célia Guimarães Ima, Wilson, Sarah Jane, When Is a Forest a Forest? Forest Concepts and Definitions in the Era of Forest and Landscape Restoration, in: Ambio, 2016, Volume 45, Issue 5, 2016, pp. 538–550.

Chen, Ling, Realizing the Precautionary Principle in Due Diligence, in: Dalhousie Journal of Legal Studies, Volume 25, 2016, pp. 1–23.

Chetail, Vincent, the Legal Personality of Multinational Corporations, State Responsibility and Due Diligence: The Way Forward, in: *Chetail/de Frouville/Vinuales* (eds.) Unity and Diversity of International Law. Essays in Honour of Prof. Pierre-Marie Dupuy, 2014, pp. 105–130.

Chevallier, Romy, The State of Community-Based Natural Resource Management in Southern Africa, South African Institute of International Affairs, 13 Sep 2016.

Chhatre, Ashwini/Lakhanpal, Shikha/Larson, Anne M., Nelson, Fred/Ojha, Hemant, Social Safeguards and Co-Benefits in REDD+: A Review of the Adjacent Possible, in: Current Opinion in Environmental Sustainability, Volume 4, Issue 6, December 2012, pp. 654–660.

Chiarolla, Claudio, The Role of Private International Law under the Nagoya Protocol, in: *Morgera/Buck/Tsioumani* (eds.), The 2010 Nagoya Protocol on Access and Benefit-sharing in Perspective: Implications for International Law and Implementation Challenges, Martinus Nijhof Publishers, the Hague, 2010, pp. 423–449.

Ching Liu, Corsa Lok/Kuchma, Oleksandra/Krutovsky, Konstantin V., Mixed-species versus Monocultures in Plantation Forestry: Development, Benefits, Ecosystem Services and Perspectives for the Future, in: Global Ecology and Conservation, Volume 15, 2018, e00419.

Chomba, Susan/Kariuki, Juliet/Lunda, Jens Friis/Sinclair, Fergus, Roots of Inequity: How the Implementation of REDD+ Reinforces Past Injustices, in: Land Use Policy, Volume 50, 2016, pp. 202–213.

Chong, Joanne, Ecosystem-based Approaches to Climate Change Adaptation: Progress and Challenges, in: International Environmental Agreements, Volume 14, Issue 4, February 2014, pp. 391–405.

Churches, Christopher E./Wampler, Peter J./Sun, Wanxiao/Smith, Andrew J. Smith, Evaluation of Forest Cover Estimates for Haiti Using Supervised Classification of Landsat Data, in: International Journal of Applied Earth Observation and Geoinformation, Volume 30, 2014, pp. 203–216.

Churchill, Robin R./Ulfstein, Geir, Autonomous Institutional Arrangements in Multilateral Environmental Agreements: A Little-Noticed Phenomenon in International Law, in: The American Journal of International Law, Volume 94, Issue 4, 2000, pp. 623–659.

CIFOR, Integrated Conservation and Development: An Overview, CIFOR, 2007. Internet resource – http://www.cifor.org/conservation/publications/pdf_fi

les/cambodia/Integrated%20conservation%20and%20development.pdf [Last visit: 30.04.2023]

Claire, Asher, Brazil Soy Trade Linked to Widespread Deforestation, Carbon Emissions, in: Mongabay, 3 April 2019. Internet resource – https://news.mongabay.com/2019/04/brazil-soy-trade-linked-to-widespread-deforestation-carbon-emissions/ [Last visit: 30.04.2023]

Colchester, Marcus/Boscolo, Marco/Contreras-Hermosilla, Arnoldo/Del Gatto, Filippo/Dempsey, Jessica/Lescuyer, Guillaume/Obidzinski, Krystof/Pommier, Denis/Richards, Michael/Sembiring, Sulaiman N./Tacconi, Luca/Rios, Maria Teresa Vargas/Wells, Adrian, Justice in the Forest: Rural Livelihoods and Forest Law Enforcement, Center for International Forestry Research, Bogor, Indonesia, 2006.

Corbera, Esteve, Problematizing REDD+ as an Experiment in Payments for Ecosystem Services, in: Current Opinion in Environment Sustainability, Volume 4, Issue 6, 2012, pp. 612–619.

Costa Rica-Canada Initiative, The Final Results of the 10 International Meetings in Support of the Programme of Work of the Intergovernmental Forum on Forests, CRCI, 2000, p. 167. Internet resource – https://cfs.nrcan.gc.ca/pubwarehouse/pdfs/10484.pdf [Last visit: 30.04.2023]

Cramer, Wolfgang/Bondeau, Alberte/Schaphoff, Sibyll/Lucht, Wolfgang/Smith, Benjamin/Sitch, Stephen, Tropical Forests and the Global Carbon Cycle: Impacts of Atmospheric Carbon Dioxide, Climate Change and Rate of Deforestation, in: Philosophical Transactions of the Royal Society Biological Sciences, Volume 359, Issue 1443, 2004, DOI 10.1098.

Cuesta, Francisco/Becerra, María Teresa, Guidelines for the Sustainable Management of BioTrade Products: Resource Assessment, UNCTAD, United Nations, New York and Geneva, 2013.

Cullet, Philippe/Kameri-Mbote, Patricia, Joint Implementation and Forestry Projects, Conceptual and Operational Fallacies, in: International Affairs, Volume 74, Issue 2, 1998, pp. 393–408.

Curci, Jonathan, The Protection of Biodiversity and Traditional Knowledge in International Law of Intellectual Property, Cambridge University Press, Cambridge, 2010.

D'Amato, Anthony, The Coerciveness of International Law, in: German Yearbook of International Law, Volume 52, 2009, pp. 437–460.

Das, Madhumita/Chatterjee, Bani, Ecotourism and Empowerment: A Case Analysis of Bhitarkanika Wildlife Sanctuary, Odisha, India, in: IIM Kozhikode Society & Management Review, Volume 4, Issue 2, 2015, pp. 136–145.

Das, Madhumita/Chatterjee, Bani, Ecotourism: A Panacea or a Predicament?, in: Tourism Management Perspectives, Volume 14, 2015, pp. 3–16.

Dauvergne, Peter, The Global Politics of the Business of "Sustainable" Palm Oil, in: Global Environmental Politics, Volume 18, Issue 2, May 2018, pp. 34–52.

Davenport, Deborah S., An Alternative Explanation for the Failure of the UNCED Forest Negotiations, in: Global Environmental Politics Volume 5, Issue 1, 2005, pp. 105–130.

Davenport, Deborah S./Wood, Peter, Finding the Way Forward for the International Arrangement on Forests: UNFF-5, -6 and -7, in: RECIEL, Volume 15, Issue 3, 2006, pp. 316–326.

Dederer, Hans Georg, Staatenverantwortlichkeit („State responsibility") und Haftung („liability") im Bereich der „ultrahazardous activities", in: *Hecker, Bernd/Henlder, Reinhard/Proelß, Alexander/Reiff, Peter* (eds.), Verantwortlichkeit und Haftung für Umweltschäden, 28. Trierer Kolloquium zum Umwelt- und Technikrecht vom 6. bis 7. September 2012, 2013, pp. 13–51.

Delgado, Gian Carlo, Biopiracy and Intellectual Property as the Basis for Biotechnological Development: The Case of Mexico, in: International Journal of Politics, Culture and Society, Volume 16, Issue 2, 2002, pp. 297–318.

Diamond, Jared, Collapse: How Societies Choose to Fail or Survive, Penguin Books, New York, 2005.

Dickie, Glory, The US Is the Only Country That Hasn't Signed On to a Key International Agreement to Save the Planet, in: Quartz, December 25, 2016. Internet resource – https://qz.com/872036/the-us-is-the-only-country-that-hasnt-signed-on-to-a-key-international-agreement-to-save-the-planet/ [Last visit: 30.04.2023]

Dobrovidova, Olga, Russia Tops 2000–2012 Deforestation Charts, in: Climate Home News, 15.11.2013. Internet resource – https://www.climatechangenews.com/2013/11/15/russia-lost-more-forest-cover-in-2012-than-any-other-country/ [Last visit: 30.04.2023]

Doelle, Meinhard, Linking the Kyoto Protocol and Other Multilateral Environmental Agreements: From Fragmentation to Integration?, in: Journal of Environmental Law and Practice, Volume 14, 2004, pp. 75–107.

Doelle, Meinhard, Integration among Global Environmental Regimes: Lessons Learned from Climate Change Mitigation, in: *Chircop, Aldo/McDorman, Theodore/Rolston, Susan* (eds.), The Future of Regime-Building in the Law of the Sea: Essays in Tribute to Douglas M. Johnston, Martinus Nijhof Publishers, the Hague, 2009.

Dooley, Kate/Ozinga, Saskia, Building on Forest Governance Reforms Through FLEGT: The Best Way of Controlling Forests' Contribution to Climate Change?, in: RECIEL, Volume 20, Issue 2, 2011, pp. 163–170.

Dross, Miriam/Wolff, Franziska, New Elements of the International Regime on Access and Benefit-Sharing of Genetic Resources-the Role of Certificates of Origin, Bundesamt für Naturschutz, 2005.

Duchelle, Amy E./de Sassi, Claudio/Jagger, Pamela/Cromberg, Marina/Larson, Anne M./Sunderlin, William M./Atmadja, Stimiati S./Resosudarmo, Ida Aju Pradnja/Pratama, Christy Desta, Balancing Carrots and Sticks in REDD+: Implications for Social Safeguards, in: Ecology and Society, Volume 22, Issue 3, 2017

Duchelle, Amy E./Simonet, Gabriela/Sunderlin, William D./Wunder, Sven, What Is REDD+ Achieving on the Ground?, in: Current Opinion in Environmental Sustainability, Volume 32, 2018, pp. 134–140.

Dudley, Nigel, Forests in Trouble: A Review of the Status of Temperate Forests Worldwide, Gland, Switzerland: WWF- World Wildlife Fund for Nature, 1992.

Dudley, Nigel/Jeanrenaud, Jeau-Paul/Sullivan, Francis, Bad Harvest? The Timber Trade and the Degradation of the World's Forests, Earthscan, London, 1995.

Dutschke, Michael/Michaelow, Axel, Joint Implementation as Development Policy – The Case of Costa Rica, Hamburg Institute of International Economics, Discussion Paper 49, 1997.

Earth Negotiations Bulletin, Summary of the Eleventh Session of the United Nations Forum on Forests, Volume 13, Issue 199, 2015. Internet resource – https://enb.iisd.org/download/pdf/enb13189e.pdf [Last visit: 30.04.2023]

EASAC, Negative Emission Technologies: What Role in Meeting Paris Agreement Targets?, EASAC Policy Report 35, February 2018.

Eikermann, Anja, Forests in International Law: Is There Really a Need for an International Forest Convention? Springer, 2015.

Ellis, Allexandra, The Anthropocene: Deforestation and Its Consequences in Haiti, in: Foreign Affairs Review, 10 April 2013. Internet resource – http://foreignaffairsreview.co.uk/2013/04/haiti/ [Last visit:30.04.2023]

Ellison, David/Morris, Cindy E./Locatelli Bruno/Sheil, Douglas/Cohen, Jane/Murdiyarso, Daniel, Trees, Forests and Water: Cool Insights for a Hot World, in: Global Environmental Change, Volume 43, 2017, pp. 51–61.

Enquete-Kommission „Vorsorge zum Schutz der Erdatmosphäre, Schutz der Erde – Eine Bestandsaufnahme mit Vorschlägen zu einer neuen Energiepolitik. Dritter Bericht der Enquete-Kommission des 11. Deutschen Bundestages, Deutscher Bundestag, Bonn, 1990.

Epiney, Astrid/Scheylie, Martin, Strukturprinzipien des Umweltvölkerrechts, Nomos Verlag-Gesellschaft, Baden-Baden, 1998.

EUROFOR, Europe and the Forest, External Study carried out for European Parliament, European Parliament, 1997

European Commission Press Release IP/04/21, Commission encourages international solidarity when utilizing exotic plants, Brussels, 7 January 2004. Internet resource – https://ec.europa.eu/commission/presscorner/detail/en/IP_04_21 [Last visit: 30.04.2023]

Evans Meredydd/Legro, Susan/Popov, Ilya, The Climate for Joint Implementation: Case Studies from Russia, Ukraine, and Poland, in: Mitigation and Adaptation Strategies for Global Change, Volume 5, Issue 4, 2000, pp. 319–336.

Falkner, Robert, The Paris Agreement and the New Logic of International Climate Politics, in: International Affairs, Volume 92, Issue 5, 2016, pp. 1107–1125.

FAO, State of the World's Forests 2011, FAO, United Nations, Rome, 2011.

FAO, Emerging approaches to Forest Reference Emission Levels and/or Forest Reference Levels for REDD+, FAO, United Nations, Rome, 2015.

FAO, Global Forest Assessment 2015, FAO, United Nations, Rome, 2016.

FAO, Restoring Forest Landscapes through Assisted Natural Regeneration (ANR) – A practical Manual, FAO, United Nations, Rome, 2019.

Farley, Joshua/Costanza, Robert, Payments for Ecosystem Services: From Local to Global, in: Ecological Economics, Volume 69, Issue 11, 2010, pp. 2060–2068

Favre, David S., International Trade in Endangered Species: A Guide to CITES, Martinus Nijoff Publishers, the Hague, 1989

Fearnside, Philip M., Saving Tropical Forests as a Global Warming Countermeasure: An Issue that Divides the Environmental Movement, in: Ecological Economics, Volume 39, Issue 3, 2001, pp. 167–184

Federal Ministry for Economic Cooperation and Development (BMZ), REDD+ in the State of Acre, Brazil: Rewarding a Pioneer in Forest Protection and Sustainable Livelihood Development, KfW Development Bank, 2017, p. 1. Internet resource – https://www.kfw-entwicklungsbank.de/PDF/Entwicklungsfinanzierung/Themen-NEU/REDD-Early-Movers-Acre-Fact-Sheet.pdf [Last visit: 30.04.2023]

Federative Republic of Brazil, Intended Nationally Determined Contribution, Towards Achieving the Objective of the United Nations Framework Convention on Climate Change, 2015, p. 4. Internet resource – https://www4.unfccc.int/sites/ndcstaging/PublishedDocuments/Brazil%20First/BRAZIL%20iNDC%20english%20FINAL.pdf [Last visit: 30.04.2023]

Federle, Christine, Biopiraterie und Patentrecht, Nomos, Baden-Baden, 2005.

Fehr, Ernst/Falk, Armin, Psychological Foundations of Incentives, in: European Economic Review, Volume 46, Issue 4–5, 2002, pp. 687–724.

Ferraro, Paul/Kiss, Agnes, Direct Payments to Conserve Biodiversity, in: Science, Volume 298, Issue 5599, 2002, pp. 1718–1719.

Fischer, Wolfgang, Die Klimapolitik der Staaten zwischen globaler Verantwortung und nationalen Interessen, in: *Borsch, Peter/Wiedemann, Peter M.* (Hrsg.), Was wird aus unserem Klima? Fakten, Analysen und Perspektiven, Verlag Bonn, München, 1992, pp. 200–224.

Flejzor, Lauren, Reforming the International Tropical Timber Agreement, in: RECIEL, Volume 14, Issue 1, 2005.

Fodella, Allessandro, Recent Developments on Access and Benefit Sharing Relating to Genetic Resources (ABS) in International Law, in: *Casonato/Busatta/Penasa/Piciocchi/Tomasi* (a cura di), Il biodiritto e i suoi confini: Definizioni, dialoghi, interazioni, Università degli Studi di Trento, Trento, 2014, pp. 91–125.

Fon, Vincy/Parisi, Francesco, Stability and Change in International Customary Law, in: Supreme Court Economic Review, Volume 17, 2009, pp. 279–309.

Freestone, David/Streck, Charlotte (eds.), Legal Aspects of Implementing the Kyoto Protocol Mechanisms: Making Kyoto Work, Oxford University Press, Oxford, 2009.

Fröhlich, M./Quednau, H. D., Statistical analysis of the distribution pattern of natural regeneration in forests, in: Forest Ecology and Management, Volume 73, 1995, pp. 45–57.

Fry, Ian, Twists and Turns in the Jungle: Exploring the Evolution of Land Use, Land–Use Change and Forestry Decisions within the Kyoto Protocol, in: RECIEL, Volume 11, Issue 2, 2002, pp. 159–168.

FSC Watch, The 10 Worst Things about the Forest Stewardship Council, FSC Watch. Internet resource – https://fsc-watch.com/2014/06/01/the-10-worst-things-about-the-forest-stewardship-council/ [Last visit 30.04.2023]

Fujisaki, Taiji/Hyakumura, Kimihiko/Scheyvens, Henry/Cadman, Tim, Does REDD+ Ensure Sectoral Coordination and Stakeholder Participation? A Comparative Analysis of REDD+ National Governance Structures in Countries of Asia-Pacific Region, in: Forests, Volume 7, 2018, Article Number 195.

Fuss, Sabine/Lamb, William F./Callaghan, Max W./Hilaire, Jerome/Creutzig, Felix/Amann, Thorben/Beringer, Tim/Carcia, Wagner de Oliveira/Hartmann, Jens/Khanna, Tarun/Luderer, Gunnar/Nemet, Gregory F./Rogelj, Joeri/Smith, Pete/Vicente, Jose Luis Vicente/Wilcox, Jennifer/Dominguez, Maria del Mar Zamora/Minx, Jan C., Negative Emissions – Part 2: Costs, Potentials and Side Effects, in: Environmental Research Letters, Volume 13, Issue 6, 2018, pp. 1 ff.

Gabatiss, Josh, Alarming photos reveal devastating scale of rainforest destruction in Papua New Guinea, in: Independent, 21 March 2018. Internet resource – https://www.independent.co.uk/environment/papua-new-guinea-rainforest-destruction-photos-deforestation-global-witness-illegal-logging-a8265451.html [Last visit: 30.04.2023]

Gabble, Ravinder/Kohler, Jillian Clare, To Patent or Not to Patent? The Case of Novartis' Cancer Drug Glivec in India, in: Global Health, Volume 10, Issue 3, 2014, e10.1186.

Gagnon, Jennifer, A Global Conservation Assessment of Temperate Forests: Status and Protection, University of New Orleans Theses and Dissertations 45, 2003.

Gale, Fred P., The Tropical Timber Trade Regime, Palgrave Macmillan, London, 1998.

Gamfeldt, Lars/Snäll, Tord/Bagchi, Robert/Jonsson, Micael/Gustafsson, Lena/Kjellander, Petter/Ruiz-Jaen, María C./Fröberg, Mats/Stendahl, Johan/Philipson, Christopher D./Mikusiński, Grzegorz/Andersson, Erik/Westerlund, Bertil/Andrén, Henrik/Moberg, Fredrik/Moen, Jon/Bengtsson, Jan, Higher Levels of Multiple Ecosystem Services Are Found in Forests with More Tree Species, in: Nature communications, Volume 4, 2013, pp. 1340–1348.

Gardner, Toby Alan/Burgess, Neil David/Aguilar-Amuchastegui, Naikoa/Barlow, Jos/Berenguer, Erika/Clements, Tom/Danielsen, Finn/Ferreira, Joice/Foden, Wendy/Kapos, Valerie/Khan, Saiful M./Lees, Alexander C./Parry, Luke/Roman-Cuesta, Rosa Maria/Schmitt, Christine B./Strange, Niels/Theilade, Ida/Vieira, Ima C. G., A framework for Integrating Biodiversity Concerns into National REDD+ Programmes, in: Biological Conservation, Volume 154, October 2012, pp. 61–71.

Garret, R./Lambin, Eric F./le Polain de Waroux, Yan, To Eliminate Deforestation in South America, Reduce Differences in Regulations Across Regions and Actors, Institute for the Study of International Development, Policy Brief PB-2017-05, 2017. Internet resource – https://lendev.weebly.com/uploads/1/0/7/0/107005033/pb-2017-05__le_polain_de_waroux_.pdf [Last visit: 30.04.2023]

Gatti, Roberto, Cazzolla/Liang, Jingjing/Velichevskaya, Alena/Zhou, Mo, Sustainable Palm Oil May Not Be So Sustainable, in: Science of The Total Environment, Volume 652, 2019 pp. 48–51.

Gaworecki, Mike, New Report Examines Drivers of Rising Amazon Deforestation on Country-by-Country Basis, in: Mongabay, 23 May 2019. Internet resource – https://news.mongabay.com/2019/05/new-report-examines-drivers-of-rising-amazon-deforestation-on-country-by-country-basis/ [Last visit: 30.04.2023]

Gaynutdinova, Tamilla/Juncurt, Stefan, Workshop on "Forests and Forest Ecosystems: Promoting Synergy in the Implementation of the Three Rio Conventions", in: Earth Negotiations Bulletin, International Institute for Sustainable Development, Volume 4, Issue 174, 2004. Internet resource – http://enb.iisd.org/download/pdf/enb04174e.pdf [Last visit: 30.04.2023]

Gebara, Maria Fernanda, Can REDD+ help Brazil roll back rising deforestation rates?, in: Forest News, 23 Jun 2017. – Internet resource – https://forestsnews.cifor.org/50288/can-redd-help-brazil-roll-back-rising-deforestation-rates?fnl=en [Last visit: 30.04.2023]

Gebara, Maria Fernanda, Agrawan, Arul, Beyond Rewards and Punishments in the Brazilian Amazon: Practical Implications of the REDD+ Discourse, in: Forests, Volume 8, Issue 3, 2017, pp. 66–94.

Gehl Sampath, Regulating Bioprospecting: Institutions for Drug Research, Access and Benefit-Sharing, United Nations University Press, Tokyo, 2005.

Gießen, Lucas, Reviewing the Main Characteristics of the International Forest Regime Complex and Partial Explanations for Its Fragmentation, in: International Forestry Review, Volume 15, Issue 1, 2013, pp. 60–70.

Gilbert, Jérémie, Indigenous Peoples, Human Rights, and Cultural Heritage: Towards a Right to Cultural Integrity, in: *Xanthaki, Alexandra/Valkonen, Sanna/Heinämäki, Leena/Nuorgam, Piia Kristiina* (eds), Indigenous Peoples and Cultural Heritage: Rights, Debates, Challenges, Brill-Nijhoff, the Hague, 2017, pp. 20–38.

Gilmour, Don, Forty Years of Community-Based Forestry, a Review of Its Extent and Effectiveness, FAO Forestry Paper 176, FAO, United Nations, Rome, 2016.

Glowka, Lyle/Burhenne-Guilmin Françoise/Synge, Hugh, A Guide to the Convention on Biological Diversity, IUCN, Gland, Switzerland, 1994.

Gneezy, Uri/Rustichini, Aldo, Pay Enough or Don't Pay at All, in: The Quarterly Journal of Economic, Volume 115, Issue 3, 2000, pp. 791–810.

Göcke, Katja, The Case of Ángela Poma Poma v. Peru before the Human Rights Committee, The Concept of Free Prior and Informed Consent and the Application of the International Covenant on Civil and Political Rights to the Protection and Promotion of Indigenous Peoples' Rights, in: *Bogdandy, Armin von/Wolfrum, Rüdiger* (eds.), Max Planck Yearbook of United Nations Law, Volume 14, 2010, pp. 337–370.

Goodman, Rosa C./Herold, Martin, Why Maintaining Tropical Forests Is Essential and Urgent for a Stable Climate, CGD Working Paper 385, 2014, pp. 1 ff. Internet resource – https://www.cgdev.org/sites/default/files/CGD-Climate-Forest-Paper-Series-11-Goodman-Herold-Maintaining-Tropical-Forests.pdf [Last visit: 30.04.2023]

Goodwin, Edward J., The World Heritage Convention, the Environment, and Compliance, Colorado Journal of International Environmental Law and Policy, Volume 20, Issue 2, 2009.

de Gouvenain, Roland C/Silander, John, Temperate Forests, in: *Roitberg, Bernard D./Cotter, Paul D.* (eds.), Reference Module in Life Sciences, Elsevier, 2017, pp. 1–19.

Graf Vitzthum, Wolfgang/Proelß, Alexander, Völkerrecht, 7. überarbeitete und aktualisierte Auflage, Walter de Gruyter GmbH, Berlin, 2016.

Greiber, Thomas/Moreno, Sonia Peña/Mattias Åhrén/Carrasco, Jimena Nieto/Kamau, Evanson Chege/Medaglia, Jorge Cabrera/Oliva, Maria Julia/Perron-Welch, Frederic/Ali, Natasha/Williams, China, An Explanatory Guide to the Nagoya Protocol on Access and Benefit-Sharing, IUCN, Gland, Switzerland, 2012.

Griffiths, Tom/Martone, Francesco, Seeing "REDD"? Forests, Climate Change Mitigation and the Rights of Indigenous Peoples and Local Communities; Update for Poznan (UNFCCC COP 14), Forest Peoples Programme, Moreton-in-Marsh, UK, 2009.

Grubb, Michael/Koch, Matthias/Thomson, Koy/Munson, Abby/Sullivan, Francis, The "Earth Summit" Agreements: A Guide and Assessment, an Analysis of the Rio' 92 UN Conference on Environment and Development, Earthscan, London, 1993.

Grubb, Michael/Vrolijk, Christiaan/Brack, Duncan, The Kyoto Protocol: A Guide and Assessment, Earthscan, London, 1999.

Gulbrandsen, Lars H./Humphreys, David, International Initiatives to Address Tropical Timber Logging and Trade, A Report for the Norwegian Ministry of the Environment, Fridtjof Nansens Institutt, Oslo, 2006.

Gupta, Joyeeta, The Climate Change Convention and Developing Countries – from Conflict to Consensus?, Springer, Berlin, 1997.

Gupta, Joyeeta/Sanchez, Nadia, Global Green Governance: Embedding the Green Economy in a Global Green and Equitable Rule of Law Polity, in: RECIEL, Volume 21, Issue 1, 2012, pp. 12–22

Gupta, Joyeeta/van der Grijp, Nicolien/Kuik, Onno, Climate Change, Forests and REDD+: Lessons for Institutional Design, Routledge, London/New York, 2013.

Haberl, Helmut/Sprinz, Detlef/Bonazountas, Marc/Cocco, Pierluigi/Desaubies, Yves/Henze, Mogens et al., Correcting a Fundamental Error in Greenhouse Gas Accounting Related to Bioenergy, in: Energy Policy, Volume 45, 2012, pp. 18 ff.

Hajost, Scott A., The Role of the United States, in: *Campidoglio, Luigi/Pineschi, Laura/Siniscalco, Domenico/Treves, Tullio* (eds.), The Environment after Rio, International Law and Economics, International Environmental Law and Policy Series, 1994, pp. 15–23.

Halkka, Antti/Lappalainen, Iiris, Insight into Europe's Forest Protection, WWF Report, WWF, 2001.

Hall, Bronwyll H., Patents and Patent Policy, in: Oxford Review of Economic Policy, Volume 23, Issue 4, 2007, pp. 570 ff.

Halle, Mark, Life after Rio: A Commentary by Mark Halle, The International Institute for Sustainable Development, 2012.

Hance, Jeremy, Is the World Moving Backwards on Protected Areas?, in: Mongabay, 6 November 2014. Internet resource – https://news.mongabay.com/2014/11/is-the-world-moving-backwards-on-protected-areas/ [Last visit: 30.04.2023]

Hance, Jeremy, Monoculture Tree Plantations Are "green deserts" Not Forests, Say Activists, in: Mongabay 19 September, 2008. Internet resource – https://news.mongabay.com/2008/09/monoculture-tree-plantations-are-green-deserts-not-forests-say-activists/ [Last visit: 30.04.2023]

Harmer, Ralph/Morgan, Geoff, Development of Quercus Robur Advance Regeneration Following Canopy Reduction in an Oak Woodland, in: Forestry: An International Journal of Forest Research, Volume 80, Issue 2, 2007, pp. 137–149.

Harvey, Celia A./Dickson, Barney/Kormos, Cyril, Opportunities for achieving biodiversity conservation through REDD, in: Conservation Letters, Volume 3, Issue 1, 2010, pp. 53–61.

Hein, Jonas/Guarin, Alejandro/Frommé, Ezra/Pauw, Pieter, Deforestation and the Paris Climate Agreement: An Assessment of REDD+ in the National Climate Action Plans, in: Forest Policy and Economics, Volume 90, 2018, pp. 7–11.

Heller, Michael A./Eisenberg, Rebecca S., Can Patents Deter Innovation? The Anticommons in Biomedical Research, in: Science, Volume 280, 1998, pp. 698–701.

Henne, Gudrun, Genetische Vielfalt als Ressource: Die Regelung Ihrer Nutzung, 1. Auld., Nomos, Baden-Baden, 1998.

Hennen, Gudrun, "Mutually Agreed Terms" in the CBD: Requirements under Public International Law, in: *Mugabe, John/Barber, Charles Victor/La Vinay, Antonio* (eds.), Access to Genetic Resources – Strategies for Sharing Benefits, IUCN, 1997.

Hodjas, David R., Biodiversity and Climate Change Laws: A Failure to Communicate?, 3rd Colloquium of the IUCN Academy of Environmental Law, Macquarie University, Sydney, Australia, 10–15 July 2005, p. 17.

Hohmann, Harald, Ergebnisse des Erdgipfels von Rio-Weiterentwicklung des Umweltvölkerrechts durch die UN-Umweltkonferenz von 1992, in: Neue Zeitschrift für Verwaltungsrecht, 12. Jahrgang, Heft 4, 1993, pp. 311–319.

Hönerbach, Frank, Verhandlungen einer Waldkonvention: Ihr Ansatz und ihr Scheitern, WZB Discussion Paper, No. FS II 96-404, 1996.

Honey, Wahl, Vereinte Nationen und Umwelt, UN Basis-Informationen, United Nations, 1993.

Horn, Laura, Globalisation, Sustainable Development and the Common Concern of Humankind, in: Macquarie Law Journal, Volume 7, 2007, pp. 53–81.

Huettner, Michael, Risks and Opportunities of REDD+ Implementation for Environmental Integrity and Socio-Economic Compatibility, in: Environmental Science Policy, Volume 15, Issue 1, 2012, pp. 4–12.

Hughes, Donald/Thirgood, J. V., Deforestation, Erosion, and Forest Management in Ancient Greece and Rome, in: Journal of Forest History, Volume 26, Issue 2, 1982, pp. 60–75.

Hulme, Max, H., Preambles in the Treaty Interpretation, in: University of Pennsylvania Law Review, Volume 164, 2016, pp. 1281–1343.

Humphreys, David, Forest Politics: The Evolution of International Cooperation, Routledge, London/New York, 1996.

Humphreys, David, The Elusive Quest for a Global Forests Convention, in: RECIEL, Volume 14, Issue 1, 2005, pp. 1–10.

Humphreys, David, Logjam: Deforestation and the Crisis of Global Governance, Earthscan, London, 2006.

Humphreys, David, Discourse as Ideology: Neoliberalism and the Limits of International Forest Policy, in: Forest Policy and Economics, Volume 11, Issues 5–6, October 2009, pp. 319–325.

Humphreys, David, From Corporate Social Responsibility to the Democratic Regulation of Transnational Corporations, in: The International Journal of Environmental, Cultural, Economic and Social Sustainability, Volume 5, Issue 4, 2009, pp. 207–218.

Humphreys, David, Negotiating the Future under the Shadow of the Past: The Eleventh Session of the United Nations Forum on Forests and the 2015 Renewal of the International Arrangement on Forests, in: International Forestry Review, Volume 17, Issue 4, 2015, pp. 385–399.

Hutton, Jon/Dickson, Barbara, Endangered Species, Threatened Convention: The Past, Present and Future of CITES, the Convention on International Trade in Endangered Species of Wild Flora and Fauna, Routledge, London, 2000.

Ianni, Elena/Mattenet, Mauricio/Geneletti, Davide/Malizia, Lucio R., Community-Based Forest Management in the Yunga's Biosphere Reserve, Northern Argentina, in: Environment, Development and Sustainability, Volume 12, 2010, pp. 631–644

Institute for the Study of International Development, To Eliminate Deforestation in South America, Reduce Differences in Regulations across Regions and Actors, ISID, 2017. Internet resource – https://lendev.weebly.com/uploads/1/0/7/0/107005033/pb-2017-05__le_polain_de_waroux_.pdf [Last visit: 30.04.2023]

Ireland, Perin, Canada's Boreal Forest Is a Carbon Bomb Unless We Keep It Intact, Natural Resource Defense Council, October 12, 2018. Internet resource – https://www.nrdc.org/stories/canadas-boreal-forest-carbon-bomb-unless-we-keep-it-intact [Last visit: 30.04.2023]

Isozaki, Hiroji, Enforcement of ABS Agreements in User States, in: *Kamau, Evanson, C./Winter, Gerd* (eds.), Genetic Resources, Traditional Knowledge, and the Law: Solutions for Access and Benefit Sharing, Earthscan, London, 2009.

ITTO, Criteria and Indicators for the Sustainable Management of Tropical Forests. ITTO Policy Development Series No. 21. International Tropical Timber Organization, Yokohama, Japan, 2016.

Iyer, Vijayan Gurumurthy, Strategic Environmental Assessment (SEA) Process for Sustainable Mining and Mineral Management Development, in: Open Access Library Journal, Volume 4, Issue 04, 2017. Internet resource – https://m.scirp.org/papers/75449 [Last visit: 30.04.2023]

Jack, B. Kelsey/Kousky, Carolyn/Sims, Katharine R. E., Designing Payments for Ecosystem Services: Lessons from Previous Experience with Incentive-based Mechanisms, in: Proceedings of the National Academy of Sciences of the United States of America, Volume 105, Issue 28, 2008, pp. 9465–9472.

Jacquemont, Frédéric/Caparrós, Alejandro, The Convention on Biological Diversity and the Climate Change Convention 10 Years After Rio: Towards a Synergy of the Two Regimes?, in: RECIEL, Volume 11, Issue 2, 2002, pp. 169–180.

Jahrmarkt, Lena, Internationales Klimaschutzrecht, Der Weg zu einem Weltklimavertrag im Sinne gemeinsamer, aber differenzierter Verantwortlichkeit, Nomos, Baden-Baden, 2016.

Jaramillo, Lorena/Kane, Malick/Rimmer, Lawrence/Garcia, Beatriz/Frondisi, Isaura/Siqueira, Fernanda, Training Manual on Developing Joint BioTrade and REDD+ Projects, UNCTAD, United Nations, New York and Geneva, 2016

Johanne, Pelletier/Gélinas, Nancy/Skutsch, Margaret, The Place of Community Forest Management in the REDD+ Landscape, in: Forests, Volume 7, Issue 8, 2016, pp. 170–193.

Johnson, Alaine, Why Is Palm Oil Such a Problem Now?, in: Sentient Media, 30 April 2020. Internet resource – https://sentientmedia.org/why-is-palm-oil-such-a-problem-now/ [Last visit: 30.04.2023]

Jong, Hans Nicholas/Hanafiah, Junaidi, Deforestation Wanes in Indonesia's Aceh and Leuser Ecosystem, But Threats Remain, NGO Says, in: Mongabay, 2018. Internet resource – https://news.mongabay.com/2018/02/deforestation-wanes-in-indonesias-aceh-and-leuser-ecosystem-but-threats-remain-ngo-says/ [Last visit: 30.04.2023]

Jospeh, Reji K., International Regime on Access and Benefit Sharing: Where Are We Now?, in: Asian Biotechnology and Development Review, Volume 12, Issue 3, 2010, pp. 77–94.

Kaiser, Karl/Weizsäcker, Ernst Ulrich von/Raimund, Bleischwitz/Comes, Stefan, Internationale Klimapolitik. Eine Zwischenbilanz und ein Vorschlag um Abschluß einer Klimakonvention, (Arbeitspapiere zur internationalen Politik 65), Bonn, 1991.

Kamal, Baher, Arable Land Lost at Unprecedented Rate: 33,000 Hectares… a Day!, in: Inter Press Service, 2016. Internet resource – http://www.globalissues.org/news/2016/08/16/22410 [Last visit: 30.04.2023]

Kaplan, Jed O./Krumhardt, Kristen M./Zimmermann, Niklaus, The Prehistoric and Preindustrial Deforestation of Europe, in: Quarterny Science Reviews, Volume 28, 2009, pp. 3016–3034.

Keating, David, EU Labels Biofuel from Palm Oil as Unsustainable, Bans Subsidies, Forbes, March 14, 2019. Internet resource – https://www.forbes.com/sites/davekeating/2019/03/14/eu-labels-biofuel-from-palm-oil-as-unsustainable-bans-subsidies/#232f4c39c9da [Last visit: 30.04.2023]

Kill, Jutta, REDD+: A Lost Decade for International Forest Conservation, Heinrich Böll Sttiftung, Berlin, 11. January 2019. Internet resource – https://www.boell.de/en/2019/01/11/redd-lost-decade-international-forest-conservation-0 [Last visit: 30.04.2023]

Kilvert, Nick, "Sustainable Palm Oil" May Not Be So Sustainable after All, in: ABC News, 15 June 2018. Internet resource – https://www.abc.net.au/news/science/2018-06-16/orangutan-video-comes-as-sustainable-palm-oil-questioned/9811642 [Last visit: 30.04.2023]

King, Ed, Desertification Crisis Affecting 168 Countries Worldwide, Study Shows, in: The Guardian, 17 April 2013. Internet resource – https://www.theguardian.com/environment/2013/apr/17/desertification [Last visit: 30.04.2023]

Kiss, Alexander/Shelton, Dinah, International Environmental Law, Third Edition, Transnational Publishers, New York, 2000.

Klein, Eckart/Schmahl, Stefanie, Die Internationalen und Supranationalen Organisationen, in: *Proelß, Alexander/Graf Vitzthum, Wolfgang* (hrsg.), Völkerrecht, 2016, pp. 247–360.

Klein, Daniel/Carazo, María Pía/Doelle, Meinhard/Bulmer, Jane/Higham, Andrew, The Paris Agreement on Climate Change: Analysis and Commentary, Oxford University Presss, Oxford, 2017.

Koester, Veit, Compliance Committees Within MEAs and the Desirability and Feasibility of Establishing Special Compliance Bodies under CITES, SC54 Doc.Inf.3, 2004.

Kortetmääki, Teea, Anthropocentrism versus Ecocentrism Revisited: Theoretical Confusions and Practical Conclusions, in: SATS: Northern European Journal of Philosophy, Volume 14, Issue 1, 2013, pp. 21–37.

Koskenniemi, Martti, Solidarity Measures: State Responsibility as a New International Order?, in: British Yearbook of International Law, Volume 72, Issue 1, 2002, pp. 337–356.

Koutouki, Konstantia/ von Bieberstein, Katharina Rogalla, The Nagoya Protocol: Sustainable Access and Benefits-Sharing for Indigenous and Local Communities, in: Vermont Journal of Environmental Law, Volume 13, 2012, pp. 513–535.

Krohn, Susan Nicole, Die Bewahrung tropischer Regenwälder durch völkerrechtliche Kooperationsmechanismen, Duncker & Humboldt, Berlin, 2002.

Kunzmann, Katharina, The Non-legally Binding Instrument on Sustainable Management of All Types of Forests Towards a Legal Regime for Sustainable Forest Management?, in: German Law Journal, Volume 09, Issue 08, 2008, pp. 981–1005.

Kürzinger, Edith/Schipule, Peter Hans, Desertifikationskonvention: Ein Lehrstück für den Rio-Folgeprozeß?. in: Entwicklung und Zusammenarbeit, Band 37, Heft 1, 1996.

Laitos, Jan/Okulski, Julian, Why Environmental Policies Fail?, Cambridge University Press, Cambrigde, 2017.

Lang, Chris, After Seven Years, Norway's US$1 Billion REDD Deal in Indonesia Is Still Not Stopping Deforestation, REDD-Monitor, 2017. Internet resource – http://www.redd-monitor.org/2017/12/28/after-seven-years-norways-us1-billion-redd-deal-in-indonesia-is-still-not-stopping-deforestation/ [Last visit: 30.04.2023]

Lang, Chris, REDDheads: The People behind REDD and the Climate Scam in Southeast Asia, Stiftung Asienhaus, 2016. Internet resource – https://www.

asienhaus.de/archiv/asienhaus/kattermann-stiftung/user_upload/REDDheads.pdf [Last visit: 30.04.2023]

Lang, Chris, The Warsaw Framework for REDD Plus: The Decision on Summary of Information on Safeguards, in: REDD Monitor, 2013. Internet resource – http://www.redd-monitor.org/2013/12/17/thewarsaw-framework-for-redd-plus-the-decision-on-summary-of-information-on-safeguards/ [Last visit: 30.04.2023]

de Lassus St-Geniès, Géraud, The Outcome of the Negotiations on the Global Pact for the Environment: A Commentary, in: Sustainability, Volume 12, Issue 3, 2020, pp. 877–889.

Laurance, William F., Reflections on the Tropical Deforestation Crisis, in: Biological Conservation, Volume 91, Issues 2–3, 1999, pp. 109–117.

Lausche, Barbara J., Weaving a Web of Environmental Law: Contributions of the IUCN Environmental Law Programme, IUCN/ICEL, Bonn, 2008.

Lawson, Charles, The Role of Patents in Biodiversity Conservation, in: Nature Biotechnology, Volume 27, 2009, pp. 994–995.*Lee, Seokwoo/Kim, Jeong Woo*, Applying the Principle of the Common Heritage of Mankind: An East Asian Perspective, in: *Zou, Keyuan* (eds.), Global Commons and the Law of the Sea, Brill-Nijhoff, the Hague 2018, pp. 15–49.

Lee, Xuhui, Observed Increase in Local Cooling Effect of Deforestation at Higher Latitudes, in: Nature, Volume 479, Issue 7373, 2011, pp. 384–387.

Leggett, Matthew/Lovell, Heather, Community Perceptions of REDD+: A Case Study from Papua New Guinea, in: Climate Policy, Volume 12, Issue 1, 2012, pp. 115–134.

Lehmann, Susanne, Schutz der Wälder – Nationale Verantwortung tragen und global handeln, BfN-Skripten, Bonn, 2007.

Ling, Chee Yoke, The Rio Declaration on Environment and Development: Assessment, Third World Network, 2012.

Llanos, Roberto Espinoza/Feather, Conrad, The Reality of REDD+ in Peru: Between Theory and Practice; Indigenous Amazonian Peoples' Analyses and Alternatives, Forest Peoples Forest Programme, 2011, p. 1. Internet resource – http://www.forestpeoples.org/sites/fpp/files/publication/2011/11/reality-redd-peru-betweentheory-and-practice-website-english-low-res.pdf [Last visit: 30.04.2023]

Lochen, Tobias, Die völkerrechtlichen Regelungen über den Zugang zu genetischen Ressourcen, Mohr Sieben, Tübingen, 2007.

Loft, Lasse, Erhalt und Finanzierung biologischer Vielfalt – Synergien zwischen internationalem Biodiversitäts- und Klimaschutzrecht, 2009.

Long, Andrew, The Convention on Biological Diversity, in: *Voigt, Christina* (eds.), Research Handbook on REDD+ and International Law, Cheltenham, United Kingdom, 2016, pp. 186–213.

Lund, Jens Friis/Sungusia, Eliezeri/Mabele, Mathew, Bukhi/Scheba, Andreas, Promising Change, Delivering Continuity: REDD+ as Conservation Fad, in: World Development, Volume 89, January 2017, pp. 124–139.

Luyssaert, Sebastiaan/Schulze, E.-Detlef/Börner, Annett/Knohl, Alexander/Hessenmöller, Dominik/Law, Beverly E./Ciais, Philippe/Grace, John, Old-Growth Forests as Global Carbon Sinks, in: Nature, Volume 455, Issue 7210, 2008, pp. 213–215.

Lyster, Rosemary, REDD+, Transparency, Participation and Resource Rights: The Role of Law, in: Environmental Science & Policy, Volume 14, Issue 2, 2011, pp. 118–126.

Lyster, Rosemary, International Legal Frameworks for REDD+: Ensuring Legitimacy, in: *Lyster, Rosemary/MacKenzie, Catherine/McDermott, Constance* (eds.), Law, Tropical Forests, and Carbon: The Case for REDD+, Cambridge University Press, New York, 2013, pp. 3–25.

Lyster, Simon, The Convention on the Conservation of Migratory Species of Wild Animals (The "Bonn Convention"), in: Natural Resources Journal, Volume 29, Issue 4, 1989, pp. 979–1000.

MacKenzie, Catherine P., Future Prospects for International Forest Law, in: The International Forestry Review, Volume 14, Issue 2, 2012, pp. 249–259.

MacKenzie, Catherine P., Lessons from Forestry for International Environmental Law, in: RECIEL, Volume 21, Issue 2, 2012, pp. 114–126.

Mackey, Brendan/DellaSala, Dominick A./Kormos, Cyril/Lindenmayer, David/Kumpel, Noelle/Zimmerman, Barbara/Hugh, Sonia/Young, Virginia/Foley, Sean/Arsenis, Kriton/Watson, James E.M., Policy Options for the World's Primary Forests in Multilateral Environmental Agreements, in: Conservation Letters, Volume 8, Issue 2, 2015, pp. 139–147.

Maclean, Ruth, Chocolate Industry Drives Rainforest Disaster in Ivory Coast, in: Guardian, 13 September 2017. Internet resource – https://www.theguardian.com/environment/2017/sep/13/chocolate-industry-drives-rainforest-disaster-in-ivory-coast [Last visit: 30.04.2023]

Maguire, Rowena, The International Regulation of Sustainable Forest Management: Doctrinal Concepts, Governing Institutions and Implementation, PhD thesis, Queensland University of Technology, 2010.

Maguire, Rowena, Global Forest Governance, Legal Concepts and Policy Trends, Edward Elgar, London, 2013.

Makarieva, Anastasia/Gorshkov, Viktor, The Biotic Pump: Condensation, Atmospheric Dynamics and Climate, in: International Journal of Water, Volume 5, Issue 4, 2010, pp. 365–385.

Makarieva, Anastassia M./Gorshkov, Victor G./ Sheil, Douglas/Nobre, Antonio Donato/Bunyard, Peter/Li, B. Larry, Why Does Air Passage over Forest Yield More Rain? Examining the Coupling between Rainfall, Pressure, and Atmospheric Moisture Content, in: Journal of Hydrometeorology, Volume 15, Issue 1, 2014, pp. 411–426.

Maljean-Dubois, Sandrine, The Paris Agreement: A New Step in the Gradual Evolution of Differential Treatment in the Climate Regime?, in: RECIEL, Volume 25, Issue 2, 2016, pp. 151–160.

Martin, Pamela L., Global Governance from the Amazon: Leaving Oil Underground in Yasuní National Park, Ecuador, in: Global Environmental Politics, Volume 11, Issue 4, 2011, pp. 22–42.

Massarella, Kate/Sallu, Susannah M./Ensor, Jonathan E./Marchant, Rob, REDD+, Hype, Hope and Disappointment: The Dynamics of Expectations in Conservation and Development Pilot projects, in: World Development, Volume 109, September 2018, pp. 375–385.

Matsumoto, Ken'ichi/Hasegawa, Tomoko/Morita, Kanako/Fujimori, Shinichiro, Synergy Potential Between Climate Change Mitigation and Forest Conservation Policies in the Indonesian Forest Sector: Implications for Achieving Multiple Sustainable Development Objectives, in: Sustainability Science, Volume 14, 2018, pp. 1657–1672.

McAfee, Kathleen/Shapiro-Garza, Elizabeth, Payments for Ecosystem Services in Mexico: Nature, Neoliberalism, Social Movements and the State, in: Annals of the Association of American Geographers, Volume 100, 2010, pp. 579–599.

McCluskey, Kevin/Katharine B. Barker/Barton, Hazel/Boundy-Mills, Kyria, The U.S. Culture Collection Network Responding to the Requirements of the Nagoya Protocol on Access and Benefit Sharing, in: mBio, Volume 8, Issue 4, 2017. Internet resource – https://mbio.asm.org/content/mbio/8/4/e00982-17.full.pdf [Last visit: 30.04.2023]

McConnell, Fiona, The Biodiversity Convention. A Negotiating History, A Personal Account of Negotiating the United Nations Convention on Biological Diversity – and After, Kluwer Law International, London/The Hague/Boston, 1996.

McDermot, Constance L./Levin, Kelly/Cashore, Benjamin, Building the Forest-Climate Bandwagon: REDD+ and the Logic of Problem Amelioration, in: Global Environmental Politics, Volume 11, Issue 3, August 2011, pp. 85–103.

McGrath, Matt, Climate Change: US Formally Withdraw from Paris Agreement, BBC, 4 November 2020. Internet resource – https://www.bbc.com/news/science-environment-54797743 [Last visit: 30.04.2023]

McIntyre, Owen/Mosedale, Thomas, The Precautionary Principle as a Norm of Customary International Law, in: Journal of Environmental Law, Volume 9, Issue 2, 1997, pp. 221–241.

Meo Carbon Solution Team, Der Palmölmarkt in Deutschland im Jahr 2017, 25 Oktober 2018. Internet resource – https://www.forumpalmoel.org/imglib/Palmoelstudie%202017_Meo_FONAP_ho.pdf [Last visit: 30.04.2023]

Meyer, Chris, Clarifying the Role of Non-Carbon Benefits in REDD+, Environmental Defense Fund Talks Global Climate, 2013. Internet resource – http://blogs.edf.org/climatetalks/2013/08/19/clarifying-the-role-of-non-carbon-benefits-in-redd/ [Last visit: 30.04.2023]

Michaelowa, Alex/Dutschke, Michael, Climate Policy and Development, Edward Elgar Publishing, Cheltenham, UK, 2000.

Michelot, Agnes/Aseeva, Anna, From Ecosystem Services to Ecological Solidarity, in: *Westra, Laura/Gray, Janice/Gottwald, Franz-Theo* (eds.), The Role of Integrity in the Governance of the Commons, Governance, Ecology, Law, Ethics, Springer, Heidelberg, 2017, pp. 37–50.

Miles, Lera/Trumper, Kate/Osti, Mattea/Munroe, Robert/Santamaria, Catalina, REDD+ and the 2020 Aichi Biodiversity Targets. Promoting synergies in international forest conservation efforts, UN-REDD Programme Policy Brief, Issue 5, 2012.

Miller, Daniel C./Agrawal, Arun/Roberts J. Timmons, Biodiversity, Governance, and the Allocation of International Aid for Conservation, in: Conservation Letters, Volume 6, Issue 1, 2013, pp. 12–20.

Milne, Sarah/Mahanty, Sangoty/To, Phuc/Dressler, Wolfram/Kanowski, Peter/Thavat, Maylee, Learning From "Actually Existing" REDD+: A Synthesis of Ethnographic Findings, in: Conservation & Society, Volume 17, Issue 1, 2019, pp. 84–95.

Moeliono, Moira/Wollenberg, Eva/Limberg, Godwin, The Decentralization of Forest Governance: Politics, Economics and the Fight for Control of Forests in Indonesian Borneo, Earthscan, London, 2009.

Mongabay.com, Facing Legacy of Deforestation and Corruption, Sarawak May Cease Granting New Logging Concessions, in: Mongabay, 18 January 2015. Internet resource – https://news.mongabay.com/2015/01/facing-legacy-of-deforestation-and-corruption-sarawak-may-cease-granting-new-logging-concessions/ [Last visit: 30.04.2023]

Montzka, Stephen A./Kuijpers, Lambert/Battle, Mark O./Aydin, Murat/Verhulst, Kristal/Salzman, Eric S./Fahey, David W., Recent Increases in Global HFC-23 Emissions, in: Geophysical Research Letters, Volume 37, 2010, pp. 1 ff.

Moreales-Hidalgo, David/Oswalt, Sonja N./Somanathan, E., Status and Trends in Global Primary Forest, Protected Areas, and Areas Designated for Conservation of Biodiversity from the Global Forest Resources Assessment 2015, in: Forest Ecology and Management, Volume 352, 2015, pp. 68–77.

Morgera, Elisa, Faraway, So Close: A Legal Analysis of the Increasing Interactions between the Convention on Biological Diversity and Climate Change Law, Edinburgh School of Law Working Paper Series, University of Edinburgh, Number 05, 2011.

Morgera, Elisa/Razzaque, Jona, Biodiversity and Nature Protection Law, Elgar Encyclopaedia of International Law, Volume 3, Edward Elgar Publishing, 2017.

Morgera, Elisa/Tsioumani, Elsa/Buck, Matthias, Unraveling the Nagoya Protocol, A Commentary on the Nagoya Protocol on Access and Benefit-sharing to the Convention on Biological Diversity, Brill, 2014.

Morita, Kanako/Matsumoto, Ken´ichi, Synergies among Climate Change and Biodiversity Conservation Measures and Policies in the Forest Sector: A Case Study of Southeast Asian Countries, Forest Policy and Economics, Volume 87, February 2018, pp. 59–69.

Müller, Robert/Pacheco Pablo/Montero, Juan Carlos, The Context of Deforestation and Forest Degradation in Bolivia: Drivers, Agents and Institutions, Center for International Forestry Research (CIFOR), Bogor, Indonesia, Occasional Paper 108, 2014.

Muñoz-Piña, Carlos/Guevara, Alejandro/Torres, Juan Manuel/Braña, Josephina, Paying for the Hydrological Services of Mexico's Forests: Analysis, Negotiations and Results, in: Ecological Economics, Volume 65, Issue 4, 2008, pp. 725–736.

Muradian, Roldan/Pellegrini, Lorenzo/Aguilar, Bernardo/Agarwal, Bina, Payments for Ecosystem Services and the Fatal Attraction of Win-Win Solutions, in: Conservation Letters, Volume 6, Issue 4, 2013, pp. 274–279.

Muradian, Ronald/Rival, Laura, Between Markets and Hierarchies: The Challenge of Governing Ecosystem Services, in: Ecosystem Services, Volume 1, Issue 1, 2012, pp. 93–100.

Myers, Rodd/Larson, Anne M./Ravikumar, Ashwin/Kowler, Laura F./Yang, Anastasia, Messiness of Forest Governance: How Technical Approaches Suppress Politics in REDD+ and Conservation Projects, in: Global Environmental Change, Volume 50, May 2018, pp. 314–324.

Myers, Rodd/Rutt, Rebecca/McDermott, Constance/Maryudi, Ahmad/Acheampong, Emmanuel/Camargo, Marisa/Cầm, Hoàng, Imposing Legality: Hegemony and Resistance under the EU Forest Law Enforcement, Governance, and Trade (FLEGT) Initiative, in: Journal of Political Ecology, Volume 27, 2020, pp. 125–149.

Myhre, Gunnar/Shindell, Drew/Bréon, François-Marie/Collins, William/Fuglestvedt, Jan/Huang, Jianping/Koch, Dorothy/Lamarque, Jean-François/Lee, David/Mendoza, Blanca/Nakajima, Teruyuki/Robock, Alan/Stephens, Graeme, Takemura, Toshihiko/Zhang, Hua, Anthropogenic and Natural Radiative Forcing, in: *Stocker, Thomas F./Qin, Dahe/Plattner, Gian-Kasper/Tignor, Melinda M. B./Allen, Simon K./Boschung, Judith/Nauels, Alexander/Xia, Yu/Bex, Vincent/Midgley, Pauline M.* (eds.) Climate Change 2013: The Physical Science Basis, Contribution of Working Group I to the Fifth Assessment Report of the Intergovernmental Panel on Climate Change, Cambridge University Press, Cambridge, 2013, pp. 659–740.

Narloch, Ulf/Pascual, Unai/Drucker, Adam G., Collective Action Dynamics Under External Rewards: Experimental Insights from Andean Farming Communities, in: World Development, Volume 40, Issue 10, 2012, pp. 2096–2107.

Nasi, Robert/Frost, Peter, G. H., Sustainable Forest Management in the Tropics: Is Everything in Order but the Patient Still Dying?, in: Ecology and Society, Volume 14, Issue 2, 2009. Internet resource – https://www.ecologyandsociety.org/vol14/iss2/art40/ [Last visit: 30.04.2023]

Navid, Daniel, The International Law of Migratory Species: The Ramsar Convention, in: Natural Resources Journal, Volume 29, Issue 4, 1989, pp. 1001–1016.

Neeff, Till/Göhler, Daniela/Ascui, Francisco, Finding a path for REDD+ between ODA and the CDM, in: Climate Policy, Volume 15, Issue 2, 2014, pp. 149–166.

Neil, Roberts/Woodbridge, Jessie/Davis, Basil A. S./Kaplan, Jed O./Marquer, Laurent/Mazier, Florence/Nielese, Birgitte Anne/Trondman, Anna-Kari/Michelle, Leydet/Sugita, S./Fyfe, R. M./Gaillard, M.-J., Europe's Lost Forests: A Pollen-based Synthesis for the Last 11,000 years, in: Scientific Reports, Volume 8, Issue 1, 2018. Internet resource – https://www.nature.com/articles/s41598-017-18646-7 [Last visit: 30.04.2023]

Nelson, Eric/Polasky, Stephen/Lewis, David J./Plantinga, Andrew J./Lonsdorf, Eric/White. Denis/Bael, David/ Lawler, Joshua J., Efficiency of Incentives to Jointly Increase Carbon Sequestration and Species Conservation on a Landscape, in: Proceedings of the National Academy of Sciences, Volume 105, Issue 28, 2008, pp. 9471–9476.

Nguyen, Tan Quang/Luong, Thi, Truong/Nguyen, Thi Hai Van/Enters, Thomas/Yasim, Yurdi/Vickers, Ben, Evaluation and Verification of the Free, Prior and Informed Consent Process under the UN-REDD Programme in Lam

Dong Province, Vietnam, RECOFTC – The Center for People and Forests, Hanoi, 2010.

Nijar, Guardial Singh, The Nagoya Protocol on Access and Benefit Sharing of Genetic Resources: Analysis and Implementation Options for Developing Countries, Centre of Excellence for Biodiversity Law, University of Malay, 2011.

Nobre, Antonio Donato, The Future Climate of Amazonia, Articulación Regional Amazônica, Brazil, 2014.

Norman, Marigold/Nakhooda, Smita, The State of REDD+ Finance, CGD Climate and Forest Paper Series Number 5, Working Paper 378, 2015.

Novelli, Marina/Gebhardt, Bwana Tucke-Tucke, Kathrin, Community Based Tourism in Namibia: "Reality Show" or "Window Dressing"?, In: Current Issues in Tourism, Volume 10, Issue 5, 2009, pp. 443–479.

Oberthür, Sebastian/Rosendal, G. Kristin, Global Governance of Genetic Resources – Access and Benefit Sharing after the Nagoya Protocol, Routledge, Abingdon-on-Thames, 2015.

OECD, Environmental Taxation – A Guide for Policy Makers, OECD, 2011. Internet resource – https://www.oecd.org/env/tools-evaluation/48164926.pdf [Last visit: 30.04.2023]

Oldham, Paul/Hall, Stephen/Forero, Oscar, Biological Diversity in the Patent System, in: PLoS ONE, Volume 8, Issue 11, 2013, e78737.

Olsson, Roger, To Manage or Protect? – Boreal Forests from a Climate Perspective, in: Air Pollution and Climate Series, Volume 26, 2011.

Orgad, Liav, The Preamble in Constitutional Interpretation, International Journal of Constitutional Law, Volume 8, Issue 4, October 2010, pp. 1–68.

Pagiola, Stefano/Bishop, Joshua/Landell-Mills, Natasha, Selling Forest Environmental Services: Market-based Mechanisms for Conservation and Development, Earthscan, London, 2003.

Paladino, Stephanie/Fisket, Shirley J., The Carbon Fix: Forest Carbon, Social Justice, and Environmental Governance: Routledge, New York, 2017.

Parker, Charlie/Mitchell, Andrew/Trivedi, Mandar/Mardas, Niki, The Little REDD+ Book, Global Canopy Program, Oxford, 2009.

Pavoni, Riccardo, Mutual Supportiveness as a Principle of Interpretation and Law-Making: A Watershed for the WTO-and-Competing-Regimes Debate? in: European Journal of International Law, Volume 21, Issue 3, 2010, pp. 649–679.

Payne, Cymie, Introductory Note to Convention on International Trade in Endangered Species of Wild Flora and Fauna (CITES) Compliance Procedures, in: ILM 46, 2007, pp. 1174–1177.

Pearce, David/Putz, Francis E./Vanclay, Jerome K., Sustainable Forestry in the Tropics: Panacea or Folly?, in: Forest Ecology and Management, Volume 172, Issues 2–3, 2003, pp. 229–247.

Penman, Jim/Gytarsky, Michael/Hiraishi, Taka/Krug, Thelma/Kruger, Dina/Pipatti, Riitta/Buendia, Leandro/Miwa, Kyoko/Ngara, Todd/Tanabe, Kiyoto/Wagner, Fabian, Good Practice Guidance for Land Use,Land-Use Change and Forestry, IPCC National Greenhouse Gas Inventories Programme, Intergovernmental Panel on Climate Change, Good Practice Guidance for Land Use,Land-Use Change and Forestry, IPCC National Greenhouse Gas Inventories Programme, 2003.

Pistorius, Till/Schmitt, Christine/Entenmann, Stephen K. B., Greening REDD+: Challenges and Opportunities for Forest Biodiversity Conservation, Policy Paper, Second Revised Edition, University of Freiburg, Germany, 2011.

Pöhlker, Christopher, Wiedemann, Kenia T.; Sinha, Bärbel/Manabu, Shiraiwa/Gunthe, Sachin S./Smith, Mackenzie/Su, Hang/Artaxo, Paulo/Chen, Qi/Cheng, Yafang/Elbert, Wolfgang/Gilles, Mary K./Kilcoyne, Arthur L. D./Moffet, Ryan/Weigand, Markus/Martin, Scot/Pöschl, Markus/Andreae, Meinrat O., Biogenic Potassium Salt Particles as Seeds for Secondary Organic Aerosol in the Amazon, in: Science, Volume 337, Issue 6098, 2012, pp. 1075–1078.

Polgreen, Lydia, In Niger, Trees and Crops Turn Back the Desert, in: New York Times, 11 February 2007. Internet resource – http://www.nytimes.com/2007/02/11/world/africa/11niger.html [Last visit: 30.04.2023]

Porras, Inna, Fair and Green? Social Impacts of Payments for Environmental Services in Costa Rica, International Institute for Environment and Development, London, 2010.

Proelß, Alexander, Prinzipien des Internationales Umweltrechts in: *Proelß, Alexander* (ed.), Internationales Umweltrecht, Walter de Gruyter GmbH, Berlin, 2017, pp. 69–104.

Radkau, Joachim, Wood and Forestry in German History: In Quest of an Environmental Approach, in: Environment and History, Volume 2, Issue 1, Lammi Symposium special issue, 1996, pp. 63–76.

Rajamani, Lavanja, Ambition and Differentiation in the 2015 Paris Agreement: Interpretative Possibilities and Underlying Politics, in: International and Comparative Law Quarterly, Volume 65, 2016, pp. 493–514.

Ramakrishna, Kilaparti/Young, Oran R., International organizations in a Warming World: Building a Global Climate Regime, in: *Mintzer, Irving M.* (ed.), Confronting Climate Change: Risks, Implications, and Responses, Cambridge University Press, UK, 1992, pp. 253 ff.

Ramsar Handbooks, River Basin Management, Ramsar Handbooks for the Wise Use of Wetlands, 4th Edition, Volume 9, Gland, Switzerland, 2011.

Raustlala, Kal/Victor, David G., Biodiversity Since Rio: The Future of the Convention on Biological Diversity, in: Environment: Science and Policy for Sustainable Development, Volume 38, Issue 4, 1996, pp. 16–27.

Rayner, Jeremie/Buck, Alexander/Katila, Pia, Embracing Complexity in International Forest Governance: A Way Forward, Policy Brief, IUFRO, 2010.

Rayner, Steve, How to Eat an Elephant: A Bottom-up Approach to Climate Policy, in: Climate Policy, Volume 10, Issue 6, 2010, pp. 615 ff.

Read, Ruppert/O'Riordan, Tim, The Precautionary Principle Under Fire, in: Environment: Science and Policy for Sustainable Development, Volume 59, Issue 5, 2007, pp. 4–15.

Recio, Mariá Eugenia, Dancing Like a Toddler? The Green Climate Fund and REDD+ International Rule-Making, in: RECIEL, Volume 28, Issue 2, 2019, pp. 122–136.

Reeve, Rosalind, Policing International Trade in Endangered Species: The CITES Treaty and Compliance, Earthscan, London, 2002.

Rey, Daniela/Roberts, J./Korwin, S./Rivera, L./Ribet, U., A Guide to Understanding and Implementing the UNFCCC REDD+ Safeguards, ClientEarth, 2013.

Rice, Richard E./Gullison, Raymond E./Reid, John W., Can Sustainable Management Save Tropical Forests?, in: Scientific American, Volume 276, Issue 4, 1997, p. 44.

Ripley, Steven/Patry, Mark, World Heritage Forests: Leveraging Conservation at the Landscape Level, in: Proceedings of the 2nd World Heritage Forest Meeting, March 9–11, 2005, World Heritage Reports, Number 21, 2007.

Robinson, Nicholas A., AGENDA 21: Earth's Action Plan, IUCN Environmental Policy and Law Paper No. 27, 1993.

Rodan, Bruce D./Blundell, Arthur G., Can Sustainable Mahogany Stem from CITES Science?, in: Bioscience, Volume 53, 2003, p. 619.

Rodriguez-Ward, Dawn/Larson, Anne M./Ruesta, Gordilla Harold, Top-down, Bottom-up and Sideways: The Multilayered Complexities of Multi-level Actors Shaping Forest Governance and REDD+ Arrangements in Madre de Dios, Peru, in: Environmental Management, Volume 62, Issue 1, 2018, pp. 98–116.

Roque Patrícia Gonçalves/Severo, Eliana Andréa/Dorion, Eric Charles Henri/Roque, Eliel de Souza/de Guimarães, Julio Cesar Ferro, The Dilemma of Environmental Sustainability in a Developing Country: Environmental Crimes in Southern Brazil, in: Business Strategy and Development, Volume 1, Issue 1, 2018, pp. 43–52.

Rosen, Richard A./Guentherb, Edeltraud, The Economics of Mitigating Climate Change: What Can We Know?, in: Technological Forecasting and Social Change, Volume 91, February 2015, pp. 93–106.

Rosenne, Shabtai, Breach of Treaty, Cambridge: Grotius Publications Limited, Cambridge, 1985.

Royal Society, Keeping Science Open: The Effects of Intellectual Property on the Conduct of Science, Royal Society, London, 2003.

Rudenstine, Sasha/Galea, Sandro, The Causes and Behavioural Consequences of Disasters: Models Informed by the Global Experience 1950–2005, Springer, New York, 2012.

Rutt, Rebecca L./Myers, Rodd/Ramcilovic-Suominen, Sabaheta/McDermott, Constance, FLEGT: Another "Forestry Fad"?, in: Environmental Science and Policy, Volume 89, 2018, pp. 266–272.

Sagemueller, Imke, Forest Sinks under the United Nations Framework Convention on Climate Change and the Kyoto Protocol: Opportunity or Risk for Biodiversity?, in: Columbia Journal of Environmental Law, Volume 31, Issue 2, 2006, pp. 189–242.

von Saint André, Susanne, Genetische Ressourcen und traditionelles Wissen: Zugang, Teilhabe und Rechtsdurchsetzung, Universitätsverlag Göttingen, 2013.

Sanchez-Azofeifa, Arturo G./Pfaff, Alexander/Robalino, Juan Andreas/Boomhower, Judson P., Costa Rica's Payment for Environmental Services Program: Intention, Implementation, and Impact., in: Conservation Biology, Volume 21, Issue 5, 2007, pp. 1165–1173.

Sand, Peter H., Das Washingtoner Artenschutzabkommen (CITES) von 1973, in: Gehring, Thomas/Oberthür, Sebastian (eds.), Internationale Umweltregime, Umweltschutz durch Verhandlungen und Verträge, 1997, pp. 166 ff.

Sand, Peter H., Sanctions in Case of Non-compliance and State Responsibility: Pacta Sunt Servanda – Or Else?, in: *Zaelke, Durwood/Kaniaru, Donald/Kružíková, Eva/Cameron, May* (eds.), Making Law Work: Environmental Compliance and Sustainable Development, Volume 1, Cameron May Ltd, London, 2005, pp. 259–271.

Sand, Peter H. Enforcing CITES: The Rise and Fall of Trade Sanctions, in: RECIEL, Volume 22, Issue 3, 2013, pp. 1–13.

Sandbag, Playing with Fire: An Assessment of Company Plans to Burn Biomass in EU Coal Power Stations, 2019. Internet resource – https://ember-climate.org/wp-content/uploads/2019/12/2019-SB-Biomass-report-1.7b_DIGI.pdf [Last visit: 30.04.2023]

Sandker, Marleke, Emerging Approaches to Forest Reference Emission Levels and/or Forest Reference Levels for REDD, FAO, 2014.

Sands, Philippe, Greening International Law, New Press, 1994.

Sands, Philippe, Principles of International Environmental Law, Cambridge University Press, Cambridge, 2012.

Savaresi, Anna, The Role of REDD+ in the Harmonization of Overlapping International Obligations, In: *Hollo, Erkki/Kulovesi, Kati/Mehling, Michael* (eds.), Climate Change and the Law. A Global Perspective. Ius Gentium: Comparative Perspectives on Law and Justice, Volume 21, Springer, Berlin, 2013, pp. 391–418.

Savilaasko, Sini/Cerutti, Paolo Omar/Zumaeta, Javier G. Montoya/Ruslandi/ Mendoula, Edouard E./Tsanga, Raphael, Timber Certification as a Catalyst for Change in Forest Governance in Cameroon, Indonesia, and Peru, in: International Journal of Biodiversity Science, Ecosystem Services & Management, Volume 13, Issue 1, 2017, pp. 116–133.

Sax, Sarah, Brazil to Receive First-ever Results-based REDD+ Payments, But Concerns Remain, in; Mongabay, 1 March 2019. Internet resource – https:// news.mongabay.com/2019/03/brazil-to-receive-first-ever-results-based-redd-payment-but-concerns-remain/ [Last visit: 30.04.2023]

Saunders, Jade, CITES Conference Marks Major Breakthrough in Battle Against Endangered Rosewood Trade, Chatham House, the Royal Institute of International Affairs, 20 October 2016. Internet resource – https://www.chathamho use.org/expert/comment/cites-conference-marks-major-breakthrough-bat tle-against-endangered-rosewood-trade [Last visit: 30.04.2023]

Savaresi, Annalisa, The Paris Agreement: A New Beginning? in: Journal of Energy & Natural Resources Law, Volume 34, Issue 1, 2016, pp. 16–26.

Schneider, Tapio/O'Gorman, Paul A./Levine, Xavier J., Water Vapor and the Dynamics of Climate Changes, in: Review of Geophysics, Volume 48, Issue 3, 2010, pp. 1 f. Internet resource – https://agupubs.onlinelibrary.wiley.com/doi/ epdf/10.1029/2009RG000302 [Last visit: 30.04.2023]

Schneider, Thomas W., A Non-legally Binding Instrument as an Alternative to a Forest Convention, Work Report of the Institute for World Forestry 2006/4, Hamburg, 2006.

Scholz, Imme, A Forest Convention – Yes or No?, Position Paper for the BMZ on the State and the Perspectives for Creating a Legally Binding Instrument for International Forest Policy, Deutsche Institut für Entwicklungspolitik, Bonn, 2004.

Scholtz, Werner, Common Heritage: Saving the Environment for Humankind or Exploiting Resources in the Name of Eco-imperialism?, in: Comparative and

International Law Journal of Southern Africa, Volume 41, Issue 2, 2008, pp. 273–293.

Schröter, Jessica, Strukturprinzipien des Umweltvölkerrechts und ihr Beitrag zur Eindämmung des Klimawandels, Schriftenreihe des Instituts für Umwelt- und Technikrecht der Universität Trier, Band 130, Erich Schmidt Verlag GmbH, Berlin, 2015.

Schütz, Sebastian, The Kyoto Protocol with an emphasis on its flexible instrument: The Clean Development Mechanism, FH Bielefeld, University of Applied Sciences, 2011.

Scott, Schirley V., Does the UNFCCC Fulfil the Functions Required of a Framework Convention? Why Abandoning the United Nations Framework Convention on Climate Change Might Constitute a Long Overdue Step Forward, in: Journal of Environmental Law, Volume 27, Issue 1, 2015, pp. 69–89.

Sdunzig, Tobias, Die UN-Konvention über Biodiversität und ihre Zusatzprotokolle: Verhandlungshistorie, Inhalt, Kritik sowie Analyse der rechtlichen Steuerungsfähigkeit aus völkerrechtlicher und europarechtlicher, 1. Auflage, Nomos Verlag-Gesellschaft, Baden-Baden, 2017.

Secretariat of the CBD, Interlinkages between Biological Diversity and Climate Change. Ad hoc Technical Group on Biological Diversity and Climate Change, CBD Technical Series No. 10, October 2003, pp. 1 ff.

Secretariat of the CBD, Connecting Biodiversity and Climate Change Mitigation and Adaptation: Report of the Second Ad Hoc Technical Expert Group on Biodiversity and Climate Change. CBD Technical Series No. 41, Montreal, 2009.

Secretariat of the CBD, Global Biodiversity Outlook 3, CBD, United Nations, 2010.

Seling, Irene/Spathelf, Peter, Tropical and Subtropical Forests, in: Owes, John N./ Lund, Gyde H. (eds.), Encyclopaedia of Life Support Systems. Forests and Forest Plants, Volume 1, Eolss Publishers Co. LTD, Oxford, 2009, pp. 176–199.

Setear, John K., Can Legalization Last? Whaling and the Durability of National (Executive) Discretion, in: Virginia Journal of International Law, Volume 44, Issue 3, 2004.

Seymour, Frances, What If We Had a Safe and Natural Way to Capture and Store Carbon?, Center for Global Development, 4 November 2014. Internet resource – https://www.cgdev.org/blog/what-if-we-had-safe-and-natural-way-capture-and-store-carbon [Last visit: 30.04.2023]

Seymour, Frances/Busch, Jonah, Why Forests? Why Now? The Science, Economics, and Politics of Tropical Forests and Climate Change, Center for Global Development, Brookings Institution Press, Washington, DC, 2016.

Seymour, Frances, Indonesia Reduces Deforestation, Norway to Pay Up, World Resources Institute, February 21, 2019. Internet resource – https://www.

wri.org/blog/2019/02/indonesia-reduces-deforestation-norway-pay [Last visit: 30.04.2023]

Sheil, Douglas/Bargués-Tobella, Aida/Ilstedt, Ulrik/Ibisch, Pierre L./Makarieva, Anastassia/McAlpine, Clive/Morris, Cindy E./Murdiyarso, Daniel/Nobre, Antonio D./Poveda, Germán/Spracklen, Dominick V./ Sullivan, Caroline A./Tuinenburg, Obbe A./van der Ent, Ruud J., Forest Restoration: Transformative Trees, in: Science, Volume 366, Issue 6463, 2019, pp. 316–317.

Shelton, Dinah L./Kiss, Alex, Strict Liability in International Environmental Law, in: *Ndiaye, Tafsir Malick/Wolfrum, Rüdiger* (eds.), Law of the Sea, Environmental Law and Settlement of Disputes, Brill Academic Publishers, the Hague, 2007, pp. 1131–1151.

Sierra, Rodrigo/Russman, Eric, On the Efficiency of Environmental Service Payments: A Forest Conservation Assessment in the Osa Peninsula, in: Ecological Economics, Volume 59, Issue 1, 2006, pp. 131–141.

Sills, Erin O./Atmadja, Stibniati S./de Sassi, Claudio/Duchelle, Amy E./Kweka, Demetrius L./Resosudarmo, Ida Aju Pradnja/Sunderlin, William D., REDD+ on the Ground, A Case Book of Subnational Initiatives Across the Globe, CIFOR, Bogor, Indonesia, 2014.

Singer, Benjamin/Giessen, Lucas, Towards a Donut Regime? Domestic Actors, Climatization, and the Hollowing-out of the International Forests Regime in the Anthropocene, in: Forest Policy and Economics, Volume 79, 2017, pp. 69–79.

Skala-Kuhmann, Astrid, Legal Instruments to Enhance the Conservation and Sustainable Management of Forest Resources at the International Level, German Federal Ministry for Economic Cooperation and Development and GTZ, 1996.

zu Sodingen, Beate Schulte, Der völkerrechtliche Schutz der Wälder – Nationale Souveränität, multilaterale Schutzkonzepte und unilaterale Regelungsansätze, Springer, Berlin, 2002.

Soltau, Friedrich, Common Concern Of Humankind, in: *Gray, Kevin R./Tarasofsky, Richard G./Carlarne, Cinnamon P.* (eds.), The Oxford Handbook of International Climate Change Law, Oxford University Press, 2016, pp. 202–213.

Somorin, Olufunso O./Visseren-Hamakers, Ingrid J./Arts, Bas/Tiani, Anne-Marie/ Sonwa, Denis, J., Integration through interaction? Synergy between Adaptation and Mitigation (REDD+) in Cameroon, in: Environment and Planning, in: Government and Policy, Volume 34, Issue 3, 2016, pp. 415–432.

Soto, Valverde Max, General Principles of International Environmental Law, in: ILSA Journal of International & Comparative Law, Volume 3, Issue 1, 1996, pp. 193–209.

South, David B./Lee, Xuhu/Messina, Michael G. Will afforestation in temperate zones warm the Earth?, in: Journal of Horticulture and Forestry, Volume 3, Issue 7, 2011.

Spracklen, Dominik V./Arnold, S. R./Taylor, C. M., Observations of Increased Tropical Rainfall Preceded by Air Passage over Forests, in: Nature, Volume 489, Issue 7415, 2012, pp. 282–285.

Steffan, Martin, die Bemühungen um eine Internationale Klimakonvention, Verhandlungen, Interessen, Akteure, (Studien zur internationalen Umweltpolitik), Lit. Verlag, Münster, 1994.

Steiner, Melanie, The Journey from Rio to Johannesburg: Ten Years of Forest Negotiations, Ten Years of Successes and Failures, in: Golden Gate University Law Review, Volume 32, Issue 4: Rio's decade: Reassessing the 1992 Earth Summit, 2002, pp. 629–664.

Steven N. Panfil/Celia A. Harvey, REDD+ and Biodiversity Conservation: A Review of the Biodiversity Goals, Monitoring Methods, and Impacts of 80 REDD+ Projects, in: Conservation Letters, Volume 9, Issue 2, 2016, pp. 143–150.

Stoll, Peter Tobias, Gestaltung der Bioprospektion unter dem Übereinkommen für biologische Vielfalt durch international verbindliche Verhaltensstandards: Hintergründe, Möglichkeiten und Inhalte, Forschungsbericht 297 81 904, UBA-FB 000045, Heidelberg, 2000.

Stoll, Peter-Tobias/Krüger, Hagen, Klimawandel, in: *Proelß, Alexander* (ed.), Internationales Umweltrecht, Walter de Gruyter GmbH, Berlin, 2017, pp. 283–320.

Streck, Charlotte/Conway, Darragh, Forestry and Agriculture under the UNFCCC: A Jigsaw Waiting to be Assembled, in: *Gray, Kevin R./Tarasofsky, Richard/Carlarne, Cinnamon P.* (eds.), The Oxford Handbook of International Climate Change Law, Oxford University Press, Oxford, 2016, pp. 563–588.

Streck, Charlotte/O'Sullivan, Robert/Toby Janson-Smith, Toby/Tarasofsky Richard G., Climate Change and Forests: Emerging Policy and Market Opportunities, Brookings Institution Press, Washington, DC, 2008.

Streck, Charlotte/Scholz, Sebastian M., The Role of Forests in Global Climate Change: Whence We Come and Where We Go, in: International Affairs, Volume 82, Issue 5, 2006, pp. 861–879

Subrata Roy, Chowdhury, Permanent Sovereignty Over Natural Resources in International Law: Principle and Practice, New York: St. Martin's Press, 1984.

Sukhotanang, Anawat/Iadpala, Phanumard/Kamuttachart, Panuwat, Thailand Experience with REDD+, CBD Asia-Pacific Workshop on REDD+, Singapore 15–18 March 2011. Internet resource – https://www.cbd.int/doc/meeti

ngs/for/wscbredd-apac-01/other/wscbredd-apac-01-thailand-en.pdf [Last visit: 30.04.2023]

Suleman, Kassahun Kelifa, Enhancing Biotrade in an Era of Biodiversity Mainstreaming, South African Institute of International Affairs, 18 May 2016. Internet resource – https://saiia.org.za/research/international-day-for-biological-diversity-enhancing-biotrade-in-an-era-of-biodiversity-mainstreaming/ [Last visit: 30.04.2023]

Suliman, Adela, World's Forests "in Emergency Room" After Years of Losses, in: Reuters, 25 April 2019. Internet resource – https://www.reuters.com/article/us-global-environment-climatechange/worlds-forests-in-emergency-room-after-years-of-losses-idUSKCN1S10DZ [Last visit: 30.04.2023]

Sunderlin, William D./De Sassi, Claudio/Sills, Erin O./Duchelle Amy A./Larson, Anne M./Resosudarmo, Ida Aju Pradnja/Awono, Abdon/Kweka, Demetrius Leo/Huynh, Thu B, Creating an Appropriate Tenure Foundation for REDD+: The Record to Date and Prospects for the Future, in: World Development, Volume 106, June 2018, pp. 376–392.

Taib, Fauziah Mohd, Malaysia and UNCED: An Analysis of a Diplomatic Process:1989–1992, Kluwer International Press, London, 1997.

Taki, Hisatomo/Yamaura, Yuichi/Okabe, Kimiko/Maeto, Kaoru, Plantation vs. Natural Forest: Matrix Quality Determines Pollinator Abundance in Crop Fields, in: Scientific Reports, Volume 1, 2011, pp. 132–135.

Taylor, Prue, Common Heritage of Mankind Principle, in: *Bosselman, Klaus/Fogel, Daniel S./Ruhl, J.B.* (eds.), Berkshire Encyclopedia of Sustainability, Volume 3: The Law and Politics of Sustainability, MA. Berkshire Publishing, Great Barrington, 2011, pp. 64–69.

Taylor, Prue, Common Heritage of Mankind and Common Concern of Humankind, in: *Krämer, Ludwig/Orlando, Emanuela* (eds.), Elgar Encyclopedia of Environmental Law, Volume VI: Principles of Environmental Law, Edward Elgar Publishing, 2018, pp. 303–323.

Thomas, Sean C./MacLellan, James, Boreal and Temperate Forests, in: *Owes, John N./Lund, Gyde H.* (eds.), Encyclopedia of Life Support Systems. Forests and Forest Plants, Volume 1, Eolss Publishers Co. LTD, Oxford, 2009, pp. 152–176.

Thomas, Zakir, Common Heritage to Common Concern, Preserving a Heritage and Sharing Knowledge, in: Journal of World Intellectual Property, Volume 8, Issue 3, 2005, pp. 241–270.

Thorsell, Jim/Sigaty, Todd, A Global Overview of Forest Protected Areas on the World Heritage List, a Contribution to the Global Theme Study of the World Heritage Natural Sites, Natural Heritage Programme, IUCN, 1997.

Tladi, Dire, Sustainable Development in International Law: An Analysis of Key Enviro-Economic Instruments, Pretoria University Law Press, Pretoria, South Africa, 2007.

Topping, John C., U. S. Law and Public Policy Concerning Global Warming, in: *Iwama, Turu* (ed.), Politics and Laws in Global Warming: International and Comparative Analysis, Environmental Research Center, Tokyo, 1991, pp 136–137.

Treves, Tullio/Tanzi, Attila/Pineschi, Laura/Pitea, Cesare/Ragni, Chiara/Jacur, Francesca Romanin, Non-Compliance Procedures and Mechanisms and the Effectiveness of International Environmental Agreements, T.M.C. Asser Press, 2009.

Tvedt, Morten W./Young, Tomme, Beyond Access: Exploring Implementation of the Fair and Equitable Sharing Commitment in the CBD, IUCN Environmental Policy and Law Paper No. 67/2, IUCN, Gland, Switzerland, 2007.

UN Office on Drugs and Crime, Transnational Organized Crime in East Asia and the Pacific: A Threat Assessment, UNODC, 2013.

UNEP, Global Environmental Outlook 3, Past, Present and Future Perspectives, Earthscan, London, 2002.

United Nations, Report of the International Law Commission on the work of its fifty-third session, in: Yearbook of the international Law Commission, Volume 2, Part 2, United Nations, New York and Geneva, 2001, pp. 31 ff.

United Nations, Compilation of Views and Proposals by Member States and other relevant stakeholders on the International Arrangement on Forests, United Nations, New York, 12 February 2014.

United Nations, Report of the World Commission on Environment and Development: Our Common Future, United Nations, 1987. Internet resource – https://sustainabledevelopment.un.org/content/documents/5987our-common-future.pdf [Last visit: 30.04.2023]

United Nations Conference on Trade and Development, The Social Responsibility of Transnational Corporations, UNCTAD, United Nations, New York and Geneva, 1999.

United Nations Conference on Trade and Development, Biodiversity and Climate Change: Integrating REDD+ into BioTrade Strategies, UNCTAD, United Nations, 2014. Internet resource – http://unctad.org/en/PublicationsLibrary/ditcted2014d6_en.pdf [Last visit: 30.04.2023]

United Nations Conference on Trade and Development, Training Manual on developing Joint Biotrade and REDD+ projects, UNCTAD, United Nations, 2014. Internet resource – http://unctad.org/en/PublicationsLibrary/ditcted2015d1_en.pdf [Last visit: 30.04.2023]

United States Security and Exchange Commission, Climate Consequences of Current Carbon Accounting Practices for Bioenergy, Case Study on Wood-Pellet Manufacturer Enviva, 2019. Internet resource – https://www.sec.gov/rules/petitions/2019/ptn4-741-exa.pdf [Last visit: 30.04.2023]

Van Asselt, Harro, Integrating Biodiversity in the Climate Regime's Forest Rules: Options and Trade-offs in Greening REDD Design, in: RECIEL, Volume 20, Issue 2, 2011, pp. 139–149.

Van Asselt, Harro, Managing the Fragmentation of International Environmental Law: Forests at the Intersection of the Climate and Biodiversity Regimes, in: New York University Journal of International Law and Politics, Volume 44, 2012, pp. 1205–1276.

Van Asselt, Harro, The Fragmentation of Global Climate Governance, Consequences and Management of Regime Interactions, Edward Elgar Publishing, Cheltenham, UK, 2014.

Van Asslet, Harro/Gupta, Joyeeta/Biermann, Frank, Advancing the Climate Agenda: Exploiting Material and Institutional Linkages to Develop a Menu of Policy Options, in: RECIEL, Volume 14, Issue 3, 2005, pp. 255–264.

Van Asselt, Harro/Sindico, Francesco/Mehling, Michael A., Global Climate Change and the Fragmentation of International Law, in: Law & Policy, Volume 30, Issue 4, 2008, pp. 423–449.

Vanderhaegen, Koen/Verbist, Bruno/Hundera, Kitessa/Muys, Bart, REALU vs. REDD+: Carbon and Biodiversity in the Afromontane Landscapes of SW Ethiopia, in: Forest Ecology and Management, Volume 343, 2015, pp. 22–33.

Verheyen, Roda Leiden, Climate Change Damage and International Law, Brill-Nijhoff, the Hague, 2005.

Vhugen, Darryl/Aguilar, Soledad/Peskett, Leo/Miner, Jonathan, REDD+ and Carbon Rights: Lessons from the Field, Property Rights and Resource Governance Project (PRRGP), 2012. Internet resource – https://www.land-links.org/wp-content/uploads/2016/09/USAID_Land_Tenure_REDD_and_Carbon_Rights_Lessons_from_the_Field.pdf [Last visit: 30.04.2023]

Victor, David G., The Collapse of the Kyoto Protocol and the Struggle to Slow Global Warming, Princeton University Press, Princeton, 2001.

Vierdag, E.W., The Time of the "Conclusion" of a Multilateral Treaty: Article 30 of the Vienna Convention on the Law of Treaties and Related Provisions, in: British Yearbook of International Law, Volume 59, Issue 1, 1988, pp. 75–111.

Vijge, Marjanneke J./Brockhaus, Maria/Di Gregorio, Monica/Muharrom, Efrian, Framing National REDD+ Benefits, Monitoring, Governance and Finance: A

Comparative Analysis of Seven Countries, in: Global Environmental Change, Volume 39, 2016, pp. 57–68.

La Viña, Antonio G. M./de Leon, Alaya/Barrer, Reginald Rex, History and Future of REDD+ in the UNFCCC: Issues and Challenges, in: *Voigt, Christina*, Research Handbook on REDD+ and International Law, Cheltenham, United Kingdom, 2016, pp. 11–29.

Young, Robert N., Importance of Biodiversity to the Modern Pharmaceutical Industry, in: Pure and Applied Chemistry, Volume 71, Issue 9, 1999, pp. 1655–1661.

Young, Tomme Rosanne, Legal Issues Regarding the International Regime: Objectives, Options, and Outlook, in: *Carrizosa, Santiago/Brush, Stephen B./Wright, Brian D./McGuire, Partrick E.* (eds.), Accessing Biodiversity and Sharing the Benefits: Lessons from Implementing the Convention on Biological Diversity, IUCN Environmental Policy and Law Paper No. 5, 2004, pp. 271–294.

Wallbott, Linda, Indigenous Peoples in UN REDD+ Negotiations: "Importing Power" and Lobbying for Rights through Discursive Interplay Management, in: Ecology and Society, Volume, 19, Issue 1, 2014, pp. 21–35

Wang, Pu/Poe, Gregory, L./Wolf, Steven A., Payments for Ecosystem Services and Wealth Distribution, in: Ecological Economics, Volume 132, 2017, pp. 63–68.

Wang, Sen, Towards an International Convention on Forests: Building Blocks versus Stumbling Blocks, in: The International Forestry Review, Volume 3, Issue 4, 2001, pp. 251–264.

Wassermann, Ursula, International Tropical Timber Agreement, in: Journal of World Trade, Volume 18, Issue 1, 1984, pp. 89–91.

Watts, Jonathan, Costa Rica Recognized for Biodiversity Protection, in: The Guardian, 25 October 2010. Internet source – https://www.theguardian.com/environment/2010/oct/25/costa-rica-biodiversity [Last visit: 30.04.2023]

Weiner, James E./Glaskin, Katie, Customary Land Tenure & Registration in Australia and Papua New Guinea: Anthropological Perspectives, in: Pacific Affairs, Volume 81, Issue 1, 2007, pp. 156–158 ff.

Wells, Michael/Scott, Guggenheim/Asmeen, Khan/Wahjudi, Wardojo/Paul, Jepson, Investing in Biodiversity: A Review of Indonesia's Integrated Conservation and Development Projects, The World Bank, East Asia Region, World Bank, Washington, DC, 1998.

Wiersema, Annecoos, The New International Law-Makers? Conferences of the Parties to Multilateral Environmental Agreements, in: Michigan Journal of International Law, Volume 31, Issue 1, 2009, pp. 231–287.

Wiersema, Annecoos, Climate Change, Forests, and International Law: REDD's Descent into Irrelevance, in: Vanderbilt Journal of Transnational Law, Volume 47, Issue 1, 2014, pp. 1–67.

Wijnstekers, Willelm, The Evolution of CITES: A Reference to the Convention on International Trade in Endangered Species of Wild Fauna and Flora, 11th edition, International Council for Game and Wildlife Conservation, 2011.

Wilkie, Mette Løyche/Holmgren, Peter/Castañeda, Froylan, Sustainable Forest Management and the Ecosystem Approach: Two Concepts, One Goal, FAO, United Nations, 2003.

Willy, Liz Alden, Customary Land Tenure in the Modern World, Rights to Resources in Crisis: Five Briefs on Reviewing the Fate of Customary Tenure in Africa, Rights and Resources, Washington, DC, November 2011.

Wilson, George R./Hayward, Matt W./Wilson, Charles, Market-Based Incentives and Private Ownership of Wildlife to Remedy Shortfalls in Government Funding for Conservation, in: Conservation Letters, Volume 10, Issue 4, 2017, pp. 485–492.

Wilson, Mark, The Green Economy: The Dangerous Path of Nature Commoditization, Consilience, in: The Journal of Sustainable Development, Volume 10, Issue 1, 2013; pp. 85–98.

Wolff, Nina/Köck, Wolfgang, 10 Jahre Übereinkommen über die biologische Vielfalt: eine Zwischenbilanz, 1. Auflage, Nomos Verlag-Gesellschaft, Baden-Baden, 2004.

Wolfrum, Rüdiger/Klepper, Gernot /Stoll, Peter-Tobias /Franck, Stephanie L., Genetische Ressourcen, traditionelles Wissen und geistiges Eigentum im Rahmen des Übereinkommens über die biologische Vielfalt, Schlussbericht des F+E Vorhabens „rechtliche Analyse des Übereinkommens über die biologische Vielfalt unter besonderer Berücksichtigung der Fragen des geistigen Eigentums", Bundesamt für Naturschutz, Bonn-Bad Godesberg, 2001.

Wong, Anny, The Roots of Japan's International Environmental Policies, Routledge, London, 2001.

Wood, Harold W., The United Nations World Charter for Nature: The Developing Nations' Initiative to Establish Protections for the Environment, in: Ecology Law Quarterly, Volume 12, Issue 4, 1985, pp. 977–996.

World Wild Fund, Causes of Deforestation, WWF, 2016. Internet resource – http://www.worldwildlife.org/threats/deforestation [Last visit: 30.04.2023]

The World Bank, Enhance Livelihoods of Forest Communities, The World Bank, 16 March 2016. Internet resource – http://www.worldbank.org/en/webarchives/archive?url=http%3A%2F%2Fwww.worldbank.org%2Fen%2Fto pic%2Fforests%2Fbrief%2Fenhance-livelihoods-of-forest-communities&mdk=22806722 [Last visit: 30.04.2023]

The World Bank, Enhance Livelihoods of Forest Communities, The World Bank, 16 March 2016. Internet resource – – http://www.worldbank.org/en/webarchives/archive?url=http%3A%2F%2Fwww.worldbank.org%2Fen%2Ftopic%2Fforests%2Fbrief%2Fenhance-livelihoods-of-forest-communities&mdk=22806722 [Last visit: 30.04.2023]

The World Bank, Why Forests Are Key to Climate, Water, Health, and Livelihoods, the World Bank, 18 March 2016. Internet resource – http://www.worldbank.org/en/news/feature/2016/03/18/why-forests-are-key-to-climate-water-health-and-livelihoods [Last visit: 30.04.2023]

Wouters, Jan/Chané, Anne-Louise, Multinational Corporations in International Law, KU Leuven Working Paper No. 129, 2013.

Wright, David K., Humans as Agents in the Termination of the African Humid Period, in: Frontiers in Earth Science, Volume 5, 2017. Internet resource – https://www.frontiersin.org/articles/10.3389/feart.2017.00004/full [Last visit: 30.04.2023]

Wright, Glenn D./Andersson, Krister P./Gibson, Clark C./Evans, Tom P., Decentralization Can Help Reduce Deforestation When User Groups Engage with Local Government, in: Proceedings of the National Academy of Sciences of the United States of America, Volume 113, Issue 52, 2016, pp. 14958–14963.

Wright, Jonathon S./Fu, Rong/Worden, John R./Chakraborty, Sudip/Clinton, Nicholas E./Risie, Camille/Sun, Ying/Yin, Lei, Rainforest-initiated Wet Season Onset over the Southern Amazon, in: Proceedings of the National Academy of Sciences of the United States of America, Volume 114, Issue 32, 2017, pp. 8481–8486.

Wunder, Sven, The Efficiency of Payments for Environmental Services in Tropical Conservation, in: Conservation Biology, Volume 21, Issue 1, 2007, pp. 48–58.

Wunder, Sven, When Payments for the Environmental Services Will Work for Conservation, in: Conservation Letters, Volume 6, Issue 4, 2013, pp. 230–237.

Yeater, Marceil/Vasquez, Juan, Demystifying the Relationship between CITES and the WTO, in: RECIEL, Volume 10, Issue 3, 2001, pp. 271–276.

Yotova, Rumiana, The Principles of Due Diligence and Prevention, in: International Environmental Law, Volume 75, Issue 3, 2016, pp. 445–448.

Young, Robert N., Importance of Biodiversity to the Modern Pharmaceutical Industry, in: Pure and Applied Chemistry, Volume 71, Issue 9, 1999, pp. 1655–1661.

Young, Tom, Malaysian Palm Oil Destroying Forests, Report Warns, in: The Guardian, 2 February 2011. Internet resource – https://www.theguardian.com/environment/2011/feb/02/malaysian-palm-oil-forests/ [Last visit: 30.04.2023]

Zaelke, Durwood/Cameroon, James, Global Warming and Climate Change – An Overview of the International Legal Process, in: American University International Law Review, Volume 5, Issue 2, 1990, pp. 249–290.

Zahar, Alexander, Climate Change Finance and International Law, Earthscan, London, 2016.

Zimmerman, Barbara L./Kormos, Cyril F., Prospects for Sustainable Logging in Tropical Forests, in: BioScience, Volume 62, Issue 5, 2012, pp. 479–487.

www.ingramcontent.com/pod-product-compliance
Ingram Content Group UK Ltd.
Pitfield, Milton Keynes, MK11 3LW, UK
UKHW020154290926
15471UKWH00005B/865

9 783631 898109